T0215354

Communications in Computer and Information Science 1071

Commenced Publication in 2007
Founding and Former Series Editors:
Phoebe Chen, Alfredo Cuzzocrea, Xiaoyong Du, Orhun Kara, Ting Liu,
Krishna M. Sivalingam, Dominik Ślęzak, Takashi Washio, and Xiaokang Yang

More information about this series at http://www.springer.com/series/7899

Ying Tan · Yuhui Shi (Eds.)

Data Mining and Big Data

4th International Conference, DMBD 2019
Chiang Mai, Thailand, July 26–30, 2019
Proceedings

 Springer

Editors
Ying Tan
Department of Machine Intelligence
Peking University
Beijing, China

Yuhui Shi
Department of Computer Science
and Engineering
Southern University of Science
and Technology
Shenzhen, China

ISSN 1865-0929   ISSN 1865-0937 (electronic)
Communications in Computer and Information Science
ISBN 978-981-32-9562-9   ISBN 978-981-32-9563-6 (eBook)
https://doi.org/10.1007/978-981-32-9563-6

This Springer imprint is published by the registered company Springer Nature Singapore Pte Ltd.
The registered company address is: 152 Beach Road, #21-01/04 Gateway East, Singapore 189721, Singapore

Preface

This volume (CCIS vol. 1071) constitutes the proceedings of the 4th International Conference on Data Mining and Big Data (DMBD 2019), which was held at the Duangtawan Hotel in Chiang Mai, Thailand, during July 26–30, 2019.

The 4th International Conference on Data Mining and Big Data (DMBD 2019) serves as an international forum for researchers and practitioners to exchange the latest advantages in theories, technologies, and applications of data mining and big data. DMBD 2019 was "Serving Life with Data Science." DMBD 2019 was the fourth event after the successful first event (DMBD 2016) in Bali, Indonesia, the second event (DMBD 2017) in Fukuoka, Japan, and the third event (DMBD 2018) in Shanghai, China.

Data mining refers to the activity of going through big data sets to look for relevant or pertinent information. This type of activity is a good example of the axiom "looking for a needle in a haystack." The idea is that businesses collect massive sets of data that may be homogeneous or automatically collected. Decision-makers need access to smaller, more specific pieces of data from these large sets. They use data mining to uncover the pieces of information that will inform leadership and help chart the course for a business. Big data contains a huge amount of data and information and is worth researching in depth. Big data, also known as massive data or mass data, refers to the amount of data involved that are too large to be interpreted by a human. Currently, the suitable technologies include data mining, crowdsourcing, data fusion and integration, machine learning, natural language processing, simulation, time series analysis, and visualization. It is important to find new methods to enhance the effectiveness of big data. With the advent of big data analysis and intelligent computing techniques we are facing new challenges to make the information transparent and understandable efficiently.

DMBD 2019 provided an excellent opportunity and an academic forum for academia and practitioners to present and discuss the latest scientific results, and innovative ideas and advantages in theories, technologies, and applications in data mining and big data. The technical program covered many aspects of data mining and big data as well as intelligent computing methods applied to all fields of computer science, machine learning, data mining and knowledge discovery, data science, etc.

DMBD 2019 took place in Chiang Mai, Thailand, which was founded in 1296 as the capital of the ancient Lanna Kingdom, located 700 km north of Bangkok in a verdant valley on the banks of the Ping River. Chiang Mai is a land of misty mountains and colorful hill tribes, a playground for seasoned travelers, a paradise for shoppers, and a delight for adventurers. Chiang Mai can expand visitors' horizons with Thai massage, cooking courses, a variety of handicrafts, and antiques. Despite its relatively small size, Chiang Mai truly has it all. Today it is a place where the past and present seamlessly merge with modern buildings standing side by side with venerable temples.

DMBD 2019 was the Duangtawan Hotel in Chiang Mai, Thailand, which is located in the center of Night Bazaar, one of a famous shopping area in downtown Chiang Mai. Surrounded by a night market where is an ideal district for shopping, sightseeing, meeting, and commercial business. The hotel is only 15 minutes away from Chiang Mai International Airport, the main railway station, and the Chiang Mai bus station.

Guests can easily access the weekend walking streets, historical attractions, and traditional temples while indulging in fascinating northern eateries, original handicrafts, souvenirs, and local entertainment. The hotel offers comfortable and convenient guestrooms overlooking Chiang Mai's vibrant city view and a plentiful service of TAI-style restaurants and bars, and a complete service of MICE events towards a selection of our function rooms. Guests can enjoy the wide-panoramic view of an outdoor swimming pool, fully-equipped fitness center, and a well-being Varee Spa.

DMBD 2019 received 79 submissions and invited manuscripts from about 249 authors in 14 countries and regions (Brunei Darussalam, China, Colombia, Cuba, India, Indonesia, Japan, Malaysia, Nepal, Russia, South Korea, Chinese Taipei, Thailand, Venezuela). Each submission was reviewed by at least two reviewers, and on average 3.2 reviewers. Based on rigorous reviews by the Program Committee members and reviewers, 34 high-quality papers were selected for publication in this proceedings volume with an acceptance rate of 43.04%. The papers are organized in six cohesive sections covering major topics of data mining and big data.

On behalf of the Organizing Committee of DMBD 2019, we would like to express our sincere thanks to Peking University, Southern University of Science and Technology, and Mae Fah Luang University for their sponsorship, and to Computational Intelligence Laboratory of Peking University, School of Information Technology of Mae Fah Luang University, and IEEE Beijing Chapter for its technical cosponsorship, as well as to our supporters: International Neural Network Society, World Federation on Soft Computing, Beijing Xinghui Hi-Tech Co., and Springer Nature.

We would also like to thank the members of the Advisory Committee for their guidance, the members of the international Program Committee and additional reviewers for reviewing the papers, and the members of the Publications Committee for checking the accepted papers in a short period of time. We are particularly grateful to Springer for publishing these proceedings in the prestigious series of *Communications in Computer and Information Sciences*. Moreover, we wish to express our heartfelt appreciation to the plenary speakers, session chairs, and student helpers. In addition, there are still many more colleagues, associates, friends, and supporters who helped us in immeasurable ways; we express our sincere gratitude to them all. Last but not the least, we would like to thank all the speakers, authors, and participants for their great contributions that made DMBD 2019 successful and all the hard work worthwhile.

July 2019   Ying Tan
Yuhui Shi

Organization

General Co-chairs

Ying Tan   Peking University, China
Russell C. Eberhart   IUPUI, USA

Program Committee Chair

Yuhui Shi   Southern University of Science and Technology, China

Advisory Committee Chairs

Xingui He   Peking University, China
Gary G. Yen   Oklahoma State University, USA
Benjamin W. Wah   Chinese University of Hong Kong, SAR China

Technical Committee Co-chairs

Haibo He   University of Rhode Island Kingston, USA
Kay Chen Tan   City University of Hong Kong, SAR China
Nikola Kasabov   Aukland University of Technology, New Zealand
Ponnuthurai Nagaratnam   Nanyang Technological University, Singapore
   Suganthan
Xiaodong Li   RMIT University, Australia
Hideyuki Takagi   Kyushu University, Japan
M. Middendorf   University of Leipzig, Germany
Mengjie Zhang   Victoria University of Wellington, New Zealand
Qirong Tang   Tongji University, China

Plenary Session Co-chairs

Andreas Engelbrecht   University of Pretoria, South Africa
Chaoming Luo   University of Mississippi, USA

Invited Session Co-chairs

Andres Iglesias   University of Cantabria, Spain
Haibin Duan   Beihang University, China
Junfeng Chen   Hohai University, China

Special Sessions Chairs

Ben Niu	Shenzhen University, China
Yan Pei	University of Aizu, Japan
Yinan Guo	China University of Mining and Technology, China

Tutorial Co-chairs

Milan Tuba	Singidunum University, Serbia
Junqi Zhang	Tongji University, China
Shi Cheng	Shanxi Normal University, China

Publications Co-chairs

Swagatam Das	Indian Statistical Institute, India
Radu-Emil Precup	Politehnica University of Timisoara, Romania

Publicity Co-chairs

Yew-Soon Ong	Nanyang Technological University, Singapore
Carlos Coello	CINVESTAV-IPN, Mexico
Yaochu Jin	University of Surrey, UK
Rossi Kamal	GERIOT, Bangladesh
Dongbin Zhao	Institute of Automation, CAS, China

Finance and Registration Chairs

Andreas Janecek	University of Vienna, Austria
Suicheng Gu	Google Corporation, USA

Local Arrangement Chair

Tossapon Boongoen	Mae Fah Luang University, Thailand

Conference Secretariat

Xiangyu Liu	Peking University, China
Renlong Chen	Peking University, China

Program Committee

Mohd Helmy Abd Wahab	Universiti Tun Hussein Onn Malaysia, Malaysia
Miltos Alamaniotis	University of Texas at San Antonio, USA
Duong Tuan Anh	HoChiMinh City University of Technology, Vietnam
Carmelo J. A. Bastos Filho	University of Pernambuco, Brazil
Tossapon Boongoen	Mae Fah Luang University, Thailand

Manik Sharma	Indian Institute of Science, Bengaluru, India
Pramod Kumar Singh	ABV-IIITM Gwalior, India
Joao Soares	GECAD, Germany
Chaoli Sun	Taiyuan University of Science and Technology, China
Hung-Min Sun	National Tsing Hua University, Chinese Taipei
Yifei Sun	Shaanxi Normal University, China
Ying Tan	Peking University, China
Andrysiak Tomasz	University of Technology and Life Sciences, Poland
Paulo Trigo	ISEL, Germany
Milan Tuba	Singidunum University, Serbia
Agnieszka Turek	Warsaw University of Technology, Poland
Gai-Ge Wang	Chinese Ocean University, China
Guoyin Wang	Chongqing University of Posts and Telecommunications, China
Handing Wang	Xidian University, China
Rui Wang	National University of Defense Technology, China
Yan Wang	The Ohio State University, USA
Zhenzhen Wang	Jinling Institute of Technology, China
Ka-Chun Wong	City University of Hong Kong, SAR China
Michal Wozniak	Wroclaw University of Technology, Poland
Zhou Wu	Chongqing University, China
Ning Xiong	Mälardalen University, Sweden
Rui Xu	Hohai University, China
Yingjie Yang	De Montfort University, UK
Zhile Yang	Shenzhen Institute of Advanced Technology, Chinese Academy of Sciences, China
Wei-Chang Yeh	National Tsinghua University, Taiwan
Guo Yi-Nan	China University of Mining and Technology, China
Peng-Yeng Yin	National Chi Nan University, Chinese Taipei
Jie Zhang	Newcastle University, UK
Junqi Zhang	Tongji University, China
Qieshi Zhang	Shenzhen Institutes of Advanced Technology, Chinese Academy of Sciences, China
Xinchao Zhao	Beijing University of Posts and Telecommunications, China
Yujun Zheng	Zhejiang University, China
Zexuan Zhu	Shenzhen University, China

Additional Reviewers

Abujar, Sheikh	Rodríguez Henríquez, Francisco
Chai, Zhengyi	Song, Yulin
Kang, Kunpeng	Tam, Hiu-Hin
Lin, Jiecong	Tian, Yanlling
Mahmud, S. M. Hasan	Zhang, Peng
Márquez Grajales, Aldo	Zhang, Shixiong
Pérez Castro, Nancy	

Contents

Data Analysis

Exploring Music21 and Gensim for Music Data Analysis
and Visualization 3
Somnuk Phon-Amnuaisuk

A Visual Analytics on Mortality of Malignant Neoplasm
and Organic Food. 13
Chien-wen Shen, Irfandi Djailani, and Cheng-Wei Tang

What Are You Reading: A Big Data Analysis
of Online Literary Content. 23
Xiping Liu and Changxuan Wan

Factors Affecting the Big Data Adoption as a Marketing Tool in SMEs. 34
Jesús Silva, Lissette Hernández-Fernández,
Esperanza Torres Cuadrado, Nohora Mercado-Caruso,
Carlos Rengifo Espinosa, Felipe Acosta Ortega, Hugo Hernández P,
and Genett Jiménez Delgado

RETRACTED CHAPTER: Data Mining to Identify Risk Factors
Associated with University Students Dropout 44
Jesús Silva, Alex Castro Sarmiento, Nicolás María Santodomingo,
Norka Márquez Blanco, Wilmer Cadavid Basto, Hugo Hernández P,
Jorge Navarro Beltrán, Juan de la Hoz Hernández, and Ligia Romero

Citescore of Publications Indexed in Scopus: An Implementation
of Panel Data 53
Carolina Henao-Rodríguez, Jenny-Paola Lis-Gutiérrez, Carlos Bouza,
Mercedes Gaitán-Angulo, and Amelec Viloria

IBA-Buffer: Interactive Buffer Analysis Method for Big Geospatial Data. . . . 61
Ye Wu, Mengyu Ma, Luo Chen, and Zhinong Zhong

Semi-automated Augmentation of Pandas DataFrames 70
Steven Lynden and Waran Taveekarn

The Temporal Characteristics of a Wandering Along Parallel
Semi-Markov Chains. 80
Tatiana A. Akimenko and Eugene V. Larkin

Prediction

Analysis and Prediction of Heart Diseases Using Inductive Logic
and Image Segmentation   93
   S. Anand Hareendran, S. S. Vinod Chandra, and Sreedevi Prasad

Customer Retention Prediction with CNN   104
   Yen Huei Ko, Ping Yu Hsu, Ming Shien Cheng, Yang Ruei Jheng,
   and Zhi Chao Luo

Analysis of Deep Neural Networks for Automobile Insurance
Claim Prediction   114
   Aditya Rizki Saputro, Hendri Murfi, and Siti Nurrohmah

The Performance of One Dimensional Naïve Bayes Classifier
for Feature Selection in Predicting Prospective Car Insurance Buyers   124
   Dilla Fadlillah Salma, Hendri Murfi, and Devvi Sarwinda

Clustering

Clustering Study of Crowdsourced Test Report with Multi-source
Heterogeneous Information.   135
   Yan Yang, Xiangjuan Yao, and Dunwei Gong

Strategy for the Selection of Reactive Power in an Industrial Installation
Using K-Means Clustering   146
   Fredy Martínez, Fernando Marínez, and Edwar Jacinto

A Collaborative Filtering System Using Clustering
and Genetic Algorithms   154
   Soojung Lee

Using K-Means Algorithm for Description Analysis of Text
in RSS News Format.   162
   Paola Ariza-Colpas, Ana Isabel Oviedo-Carrascal,
   and Emiro De-la-hoz-Franco

RETRACTED CHAPTER: Differential Evolution Clustering and Data
Mining for Determining Learning Routes in Moodle   170
   Amelec Viloria, Tito Crissien Borrero, Jesús Vargas Villa,
   Maritza Torres, Jesús García Guiliany, Carlos Vargas Mercado,
   Nataly Orellano Llinas, and Karina Batista Zea

Student Performance Assessment Using Clustering Techniques   179
   Noel Varela, Edgardo Sánchez Montero, Carmen Vásquez,
   Jesús García Guiliany, Carlos Vargas Mercado, Nataly Orellano Llinas,
   Karina Batista Zea, and Pablo Palencia

Classification

Classification of Radio Galaxy Images with Semi-supervised Learning   191
Zhixian Ma, Jie Zhu, Yongkai Zhu, and Haiguang Xu

Multi-label Text Classification Based on Sequence Model   201
Wenshi Chen, Xinhui Liu, Dongyu Guo, and Mingyu Lu

A Survey of State-of-the-Art Short Text Matching Algorithms   211
Weiwei Hu, Anhong Dang, and Ying Tan

Neuronal Environmental Pattern Recognizer: Optical-by-Distance LSTM
Model for Recognition of Navigation Patterns in Unknown Environments . . .   220
Fredy Martínez, Edwar Jacinto, and Holman Montiel

Mining Patterns

Discovering Strategy in Navigation Problem.   231
*Nurulhidayati Haji Mohd Sani, Somnuk Phon-Amnuaisuk,
and Thien Wan Au*

Exploring Frequent Itemsets in Sweltering Climates   240
*Ping Yu Hsu, Chen Wan Huang, Ming Shien Cheng, Yen Huei Ko,
Cheng-Han Tsai, and Ni Xu*

Topic Mining of Chinese Scientific Literature Research About
"The Belt and Road Initiative" Based on LDA Model from
the Sub Disciplinary Perspective   248
Jie Wang, Yan Peng, Ziqi Wang, Chengyan Yang, and Jing Xu

Early Warning System Based on Data Mining to Identify Crime Patterns. . . .   259
*Jesús Silva, Stefany Palacio de la Cruz, Jannys Hernández Ureche,
Diana García Tamayo, Harold Neira-Molina, Hugo Hernandez-P,
Jairo Martínez Ventura, and Ligia Romero*

Fuzzy C-Means in Lower Dimensional Space for Topics Detection
on Indonesian Online News   269
Praditya Nugraha, Muhammad Rifky Yusdiansyah, and Hendri Murfi

Mining Tasks

A Novel Big Data Platform for City Power Grid Status Estimation   279
*Wuhan Lin, Yuanjun Guo, Zhile Yang, Juncheng Zhu, Ying Wang,
Yuquan Liu, and Yong Wang*

Taxi Service Pricing Based on Online Machine Learning   289
Tatiana Avdeenko and Oleg Khateev

PRESENT Cipher Implemented on an ARM-Based System on Chip   300
  Fernando Martínez Santa, Edwar Jacinto, and Holman Montiel

Enhancing Quality of Movies Recommendation Through Contextual
Ontological User Profiling   307
  Mohammad Wahiduzzaman Khan, Gaik-Yee Chan,
  and Fang-Fang Chua

Determinants of ResearchGate (RG) Score for the Top100 of Latin
American Universities at Webometrics.   320
  Carolina Henao-Rodríguez, Jenny-Paola Lis-Gutiérrez,
  Mercedes Gaitán-Angulo, Carmen Vásquez, Maritza Torres,
  and Amelec Viloria

BIOG: An Effective and Efficient Algorithm for Influence Blocking
Maximization in Social Networks   328
  Kuei-Sheng Lin and Bi-Ru Dai

Retraction Note to: Chapters. C1
  Ying Tan and Yuhui Shi

Author Index   339

Data Analysis

Exploring Music21 and Gensim for Music Data Analysis and Visualization

Somnuk Phon-Amnuaisuk[1,2](✉)

[1] Media Informatics Special Interest Group, Centre for Innovative Engineering, Universiti Teknologi Brunei, Gadong, Brunei
somnuk.phonamnuaisuk@utb.edu.bn
[2] School of Computing and Informatics, Universiti Teknologi Brunei, Gadong, Brunei

Abstract. Computational musicology has been garnering attention since the 1950s. Musicologists appreciate the utilisation of computing power to look for patterns in music. The bottlenecks in the early days were attributed to the lack of standardization of computer representation of music and the lack of computing techniques specialized for the music domain. However, due to the increase in computing power, advances in music technology and machine learning techniques in recent years; the field of computational musicology has been revitalized. In this paper, we explored *Music21* toolkit and Gensim, the recent open-source data analytical tool which includesd the *Word2Vec* model, for an analysis of Bach Chorales. The tools and techniques discussed in this paper have revealed many interesting exploratory fronts such as the semantic analogies of musical concepts which deserve a detailed investigation by computational musicologists.

Keywords: Word2Vec model · Bach's chorales · Music21 · Gensim · Music data analysis and visualization

1 Introduction

Computational musicology has a long history even before the electronic computing era. This approach gives an objective analytical description of music. A statistical study of the occurrences of pitches and chords in a piece of music reveals its statistical properties which can be understood as a model of the music being analysed. A comprehensive knowledge of a musical model may be employed to synthesize new pieces of music according to the parametrized model. Experiments in Musical Intelligence (EMI) [1], Explicitly Structured Control Model [2] and many recent works using deep learning technology [3] (see for examples deepbach [4] and deepjazz[1]) are examples of computational models built from music data.

---

[1] deepjazz: https://deepjazz.io/.

© Springer Nature Singapore Pte Ltd. 2019
Y. Tan and Y. Shi (Eds.): DMBD 2019, CCIS 1071, pp. 3–12, 2019.
https://doi.org/10.1007/978-981-32-9563-6_1

Recent advances in information technology, artificial intelligence and machine learning have introduced various tools and techniques to facilitate researchers in the extraction of knowledge from music data set. In this work, we explore *Music21* [5] as a symbolic music processing tool and explore *Gensim* [6] as a knowledge discovery tool. Both packages are freely available in *Python*. These tools are employed to (i) identify and count the frequency of pitches and chords; (ii) identify harmonic progression at the ending of chorales i.e. the cadence patterns, and (iii) explore knowledge gained from the application of the Word2Vec [7]. The results show that the *music analysis through machine learning process* discussed in this paper is effective and can reveal many interesting structural information.

The rest of the paper is organized into the following sections: Sect. 2 gives an overview of the symbolic music representation and the word embedding model; Sect. 3 provides the exploratory analysis and results; Sect. 4 provides critical discussion of the music data analysis and visualization; and finally, the conclusion and further research are presented in Sect. 5.

2   Music Analysis Through Machine Learning

In this paper, the domain of interest is the Bach's chorales which have been catalogued by music scholars, notably, the collections from *371 Harmonized Chorales and 69 Chorale melodies with Figured Bass* edited by Albert Riemenschneider [8] and *Bach-Werke-Verzeichnis* (BWV) edited by Wolfgang Schmieder. The chorale identification index in this paper is based on BWV index. Detailed information of the chorales' titles can be viewed at these urls[2].

2.1   Representing Music for Automated Reasoning Process

The representation of music can be at various grain sizes, from the representation of continuous sound wave to discrete symbols representing concepts such as pitch and chord. Musical knowledge representation has been extensively explored in many previous works [2,9–11]. The choice of representation often depends on the applications and it is important that the music must be abstracted to appropriate representation for reasoning process using computers. Here, music is abstracted as discrete symbolic representation of pitch, chord, and duration for analysis using computers.

Music21: Music21 is a symbolic music computational toolkit from MIT [5]. It provides tools for reasoning on musical structures such as pitch, duration, chord, harmony, etc. It also provides a built-in Bach's chorales corpus with over 400 Bcach's chorales in *xml* format. Hence, we have decided to exploit Music21 and its encoded chorales in this work.

---

[2] web.mit.edu/music21/doc/moduleReference/moduleCorpusChorales.html   and www.jsbchorales.net.

Music21 encodes each chorale as a score. The score contains metadata about the piece such as name, key signature, and the chorale in four parts: soprano, alto, tenor and bass. Each part describes a sequence of note (i.e. pitch and duration) at a similar expressive level as a MIDI file format. Music21 provides API to access note events and to *chordify* pitch combinations. That is, combinations of pitches at any time slice can be computed and described using traditional harmonic vocabulary.

What kind of musical concepts can be extracted using machine learning? It will be a challenge to come up with a definite list of answers for this question since (i) a large amount of compound concepts can be constructed from primitive concepts of pitch and duration and (ii) compound concepts are genre specific. Musicologists describe chorales using the following concepts: the occurrences of pitches, chords, movement of pitches (i.e., voice leading), movement of chords (i.e., harmonic progression), harmonic rhythm, harmonic modulation, patterns of chords at the beginning and ending of a musical phrase (i.e., opening and cadence), etc.

2.2   Word-Embedding Model

The *word-embedding model* is developed from the *natural language processing* research community where similarity among linguistic terms are computed based on their co-occurrence. The word-embedding model maps a set of vocabulary into a vector. Recently the implementation of Word2Vec model [7] offers a convenient tool for word embedding. Word2Vec is a shallow artificial neural network (ANN) model with word vectors input and output. Two model architectures: *Continuous Bag-of-Words (CBOW)* and *skip-grams* are publicly available in many open-source implementations. CBOW learns the model of the current word given its context. For example, with the context window of size one, the current word can be modelled based on the context of another word occurring before and after the current word. This is similar to the n-gram model $p(w_t|w_{t-1}, w_{t+1})$. However, the modelling using ANN greatly simplifies the model construction task. In the skip-gram model, the model learns to predict context words given an input word. Figure 1 shows the CBOW and skip grams architectures. In natural language processing (NLP) domain, the co-occurrence of words in the same context can be interpreted as they share similarity in meaning (synonym) or even a deeper degree of analogical reasoning e.g., Man is to Woman as Brother is to Sister, or as King is to Queen. Such relationships are fascinating and have been investigated in other syntactic relations e.g., the use of tenses.

3   Bach's Chorales Analysis and Visualization

In this paper, a total of 335 Bach chorales were employed. There were a total of 173 chorales in the major mode and 162 chorales in the minor modes. We leveraged on Music21 Bach's chorales corpus and extracted the following information from each chorale: *BWV number, pitch and duration of each note in soprano,*

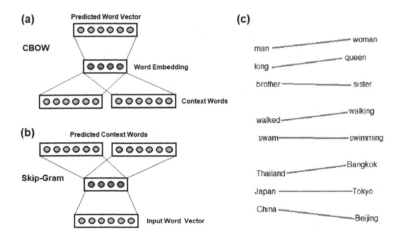

Fig. 1. Word2vec model: (a) Continuous Bag-Of-Words (CBOW) architecture; (b) Skip-Gram architecture; and (c) In natural language processing domain, a deeper relationship of words emerges from the model

alto, tennor and bass voices, and its key signature. Pitches and durations are primitive concepts and other compound concepts such as *chords* and *cadences* can be computed. Figure 2 summarizes the flow of music data analysis starting from symbolic representation of music from the corpus, to symbolic computation using *Music21*, to machine learning using *Gensim Word2Vec* model and finally provides informative visualization of music information.

3.1   Occurrences of Notes and Chords

Pitch and duration are the primitive concepts describing musical notes. A chord can be constructed from a combination of pitches. Figure 3 summarizes the frequency of occurrence of normalized scale degree in two common mode: major and minor modes. The pie chart graphically summarizes the appearances of different functional chords in the context of major and minor modes.

The left pane in Fig. 3 shows the histogram of the pitches observed in the chorales in the major mode and in the minor mode. The MIDI notes 48, 60, and 72 corresponded to the pitch class *c*. It was observed that the pitch class: *c, d, e, f, g, a,* and *b* were prominent in the major mode; and the pitch class from the melodic and harmonic minor scales were prominent in the minor mode i.e., *c, d, e♭, f, g, a♭,* and *b♭*.

The pie charts (right pane in Fig. 3) show clear patterns of the prominent functional chords *I, IV, V, i, ii* and *vi* in the major mode, and *i, iv, v, III* and *V* in the minor mode[3]. These observations about pitch, chord and functional

---

[3] The term prominent here is subjective. Here, we select chords that appear more than 75% from the major mode and the minor mode.

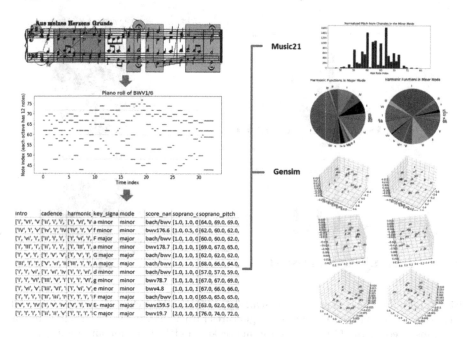

Fig. 2. Music Data Analysis and Visualization Process: Bach chorales are represented symbolically using standard western music notations. They are re-represented in textual format using pitch, duration, etc. Complex representations, such as chords and cadence patterns, can be computed from pitches and duration primitives. Music21 and Gensim provides various tools that support patterns extractions and visualization

harmony agree very well with theoretical music knowledge [12]. This illustrates the power of computational analysis applied to the music domain.

3.2   Markov Model for Cadence Chord Progressions

Pitches and chords describe music events at a specific local area. We can gain more insight by looking into the progression of music events. Here, the patterns of chord progression at the cadence (the last four chords) from both major and minor modes were analysed. Chord transition probabilities $P(s_t|s_{t-1})$ in the major mode and the minor mode were summarised in Table 1.

3.3   Similarity of Functional Harmony from Word2vec Model

In Table 1, the description of chord transition at the cadence was expressed using the first order Markov model. A higher order Markov model should give a better performance but it was computationally expensive to construct one.

The Word2Vec is a recent approach proposed by a team led by [7] at Google. The CBOW and the Skip-gram architectures can effectively learn relationships

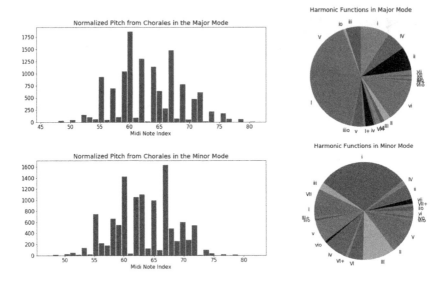

Fig. 3. Left: Summary of normalized scale degree from all the chorales in this study. All pitches from the chorales in other key signatures rather than the C major and C minor are transposed to the C major for the major key and to C minor for the minor key respectively. Right: Summary of functional chords that appeared in the major mode and the minor mode (the figure is best viewed in color)

between a word and its context from its surrounding words. In this work, functional chords found in the chorales were the words and we constructed both CBOW and Skip-gram from them. The CBOW and Skip-gram learned embedding representation of these chords and each chord could be thought of as occupying a point in a hyperspace. In order to view these points, we extracted only the first three principle components and plotted them.

All chord progressions of chorales in the major mode were grouped together and the Word2Vec model with 100 hidden nodes were constructed using one context chord (i.e., one chord before and one chord after the current chord). Figure 4 (top row) shows the positions of the chord as learned by CBOW architecture and the borrom row shows the positions of the chord as learned by Skip-gram architecture.

In the same fashion, all the chords from all the chorales in the minor mode were employed to train the Word2Vec model. Figure 5 (top tow) shows the positions of the chord as learned by CBOW architecture and the bottom row shows the positions of the chord as learned by Skip-gram architecture.

Figures 4 and 5 give us hints that some chords were similar to each other. Their spatial relationships in the hyperspace also revealed interesting semantic associations as those observed in Word2Vec model built using natural language. A much quoted example is that a trained Word2Vec model could associate *queen* to *king* to the query if *man–woman* then *king–?*.

Table 1. Chord transition probability at the cadence (read as a transition from row to column). Top: cadence progression in the minor mode; and Bottom: cadence progression in the major mode

	i	iv	v	III	V
i	0.51	0.15	0.05	0.05	0.24
iv	0.63	0.07	0.07	0.03	0.20
v	0.46	0.08	0.08	0.13	0.25
III	0.30	0.11	0.06	0.34	0.19
V	0.60	0.24	0.00	0.03	0.13

	I	IV	V	i	ii	vi
I	0.32	0.06	0.14	0.17	0.17	0.14
IV	0.29	0.05	0.13	0.26	0.03	0.24
V	0.50	0.11	0.11	0.07	0.04	0.17
i	0.65	0.06	0.06	0.13	0.00	0.10
ii	0.38	0.03	0.13	0.13	0.13	0.20
vi	0.40	0.03	0.16	0.12	0.11	0.18

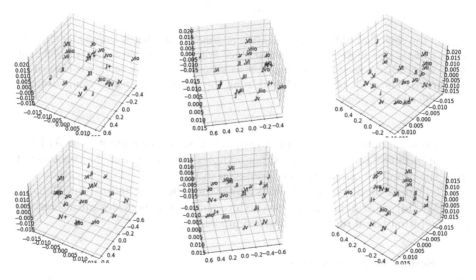

Fig. 4. Top row: Embedding representation of major mode chords learned by CBOW. These Chords were plotted using the first three principle components in three different viewpoints; Bottom row: Embedding representation of major mode chords learned by Skip-grams. These Chords were plotted using the first three principle components in three different viewpoints

4 Discussion

Music21 provides the API for counting the occurrences of pitch and duration. It also provide the API to calculate the chord from simultaneous pitches. This allows us to count the occurrences of pitches in each key signature and to transpose pitches from one key to another key. Although Music21 does not provide all the functions required by some operations e.g., one may wish to increase the

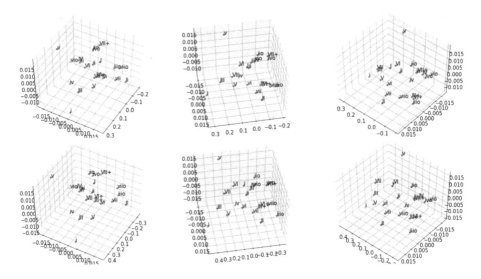

Fig. 5. Top row: Embedding representation of minor mode chords learned by CBOW. These Chords were plotted using the first three principle components in three different viewpoints; Bottom row: Embedding representation of minor mode chords learned by Skip-grams. These Chords were plotted using the first three principle components in three different viewpoints

duration of all notes by the factor of two, and many other requests, the API has enough primitive operations that other complex operations may be constructed from the primitive operation (see discussion in [13]).

The Word2Vec model in Gensim is lighter than the Long Short-Term Memory (LSTM) [14] since it has a shallow architecture but it provides an interesting interpretation based on its vector-space model.

4.1   Semantic Associations

We observed the following interesting outcomes in *semantic associations* of functional chord properties. For examples, the following semantic associations show a kind of 'similarity of musical quality' for the chords *I, IV, V, ii, vi* and *i*.

```
model.wv.most_similar( positive=['I'],topn=3) --> ['vi','IV', 'V']
model.wv.most_similar( positive=['IV'],topn=3) --> ['V', 'ii', 'vi']
model.wv.most_similar( positive=['V'],topn=3) --> ['IV','ii', 'iii']
model.wv.most_similar( positive=['ii'],topn=3) --> ['viio', 'iii', 'IV+']
model.wv.most_similar( positive=['vi'],topn=3) --> ['IV', 'I', 'V']
model.wv.most_similar( positive=['i'],topn=3) --> ['III', 'VII', 'v']
```

Similarity: We use the term 'similarity of musical quality' since the Word2Vec model captured co-occurrences of chords. The co-occurrence was from (i) the two chord performing the same function so they were interchanged and (ii) the two chords captured the antecedent-consequence of functional harmony e.g. V → I signifies a cadence pattern.

The results are interesting since they have valid musical interpretations, for example,

- I is close to vi since they share two common pitches so vi may replace I in certain contexts.
- I is close to IV and V since the I → IV progression and V → I progression are common. This fact is captured in the vector space model.

To summarize, the Word2Vec model successfully captures the fact that chords can be replaced by other chords if they share many common notes and chords are close if they occur together in the progression.

Dissimilarity: In the same fashion, the following semantic associations show a kind of 'dissimilarity' for the chord *I, IV, V, ii, vi* and *i*.

```
model.wv.most_similar( negative=['I'],topn=3) --> ['i', 'III','VII']
model.wv.most_similar( negative=['IV'],topn=3) --> ['i', 'III', 'VII']
model.wv.most_similar( negative=['V'],topn=3) --> ['i', 'III', 'VII']
model.wv.most_similar( negative=['ii'],topn=3) --> ['i', 'III', 'I']
model.wv.most_similar( negative=['vi'],topn=3) --> ['i','III', 'VII']
model.wv.most_similar( negative=['i'],topn=3) --> ['I', 'vi, 'IV']
```

Despite sharing two common pitches, I and i are far apart since they belong to two different modes and it is rare to have a progression like I → i or i → I. The Word2Vec model successfully captures this fact. Dissimilarity could be used to segment out a chord from a group of chords as well, for example:

```
model.wv.most_similar( negative=['I'],topn=3) --> ['i', 'III','VII']
```

- The ii^o is a diminished chord which is not in the context of other chords in the group which are in a major scale.

Semantic Analogies: One of the interesting outcomes from Word2Vec model trained with natural language is its ability to compute semantic analogies, for example, as *"man is to woman as king is to queen"*. In the same fashion, we can write the following from the outcomes below:

```
model.wv.most_similar( positive=['I','vi'],negative=['vi'],topn=1) --> ['IV']
model.wv.most_similar( positive=['vi','VI'],negative=['ii'],topn=1) --> ['II']
model.wv.most_similar( positive=['V','v'],negative=['I'],topn=1) --> ['i']
```

- "I is to vi as vi is to IV" This makes sense since chords I and vi are third apart and share two common notes, chords vi and IV are also third apart and share two common notes.
- "vi is to ii as VI is to II" This makes sense since it analogues the major and minor chords.
- "V is to I as v is to i" This makes sense since it shows a cycle of fifth pattern.

5   Conclusion and Future Directions

Symbolic music computing in musicology has been somewhat neglected even though this is a rich domain with many useful applications in music education. This work explored open source tools to analyse Bach Chorales. We showed that a combination of Music21, Word2vec, Python and visualization techniques provided effective tool sets for musicologists to perform music data analysis. The idea of music analogies was highlighted here and the concept could be further explored in greater details with longer chord sequences e.g., motifs and phrases.

Acknowledgments. We wish to thank anonymous reviewers for their comments that have helped improve this paper. We would like to thank the GSR office for their partial financial support given to this research.

References

1. Cope, D.: Computer and Musical Style. Oxford University Press, Oxford (1991)
2. Phon-Amnuaisuk, S.: Control language for harmonisation process. In: Anagnostopoulou, C., Ferrand, M., Smaill, A. (eds.) ICMAI 2002. LNCS (LNAI), vol. 2445, pp. 155–167. Springer, Heidelberg (2002). https://doi.org/10.1007/3-540-45722-4_15
3. LeCun, Y., Bengio, Y., Hinton, G.: Deep learning. Nature **521**, 436444 (2015)
4. Hadjeres, G., Pachet, F., Nielsen, F.: DeepBach: a steerable model for Bach chorales generation. arXiv:1612.01010 (2016)
5. Cuthbert, M., Ariza, C.: music21: a toolkit for computer-aided musicology and symbolic music data. In: Proceedings of the International Symposium on Music Information Retrieval, p. 63742 (2010)
6. Rehurek, R., Sojka, P.: Software framework for topic modelling with large corpora. In: Proceedings of the LREC 2010 Workshop on New Challenges for NLP Frameworks, pp. 45–50 (2010)
7. Mikolov, T., Chen, K., Corrado, G., Dean, J.: Efficient estimation of word representations in vector space. arXiv:1301.3781 (2013)
8. Riemenschneider, A.: 371 Harmonized Chorales and 69 Chorale Melodies with Figured Bass. G Schirmer Inc., New York (1941)
9. Wiggins, G., Harris, M., Smaill, A.: Representing music for analysis and composition. In: Proceedings of the International Joint Conferences on Artificial Intelligence, (IJCAI 1989), Detroit, Michigan, pp. 63–71 (1989)
10. West, R., Howell, P., Cross, I.: Musical structure and knowledge representation. In: Howell, P., West, R., Cross, I. (eds.) Representing Musical Structure, chap. 1, pp. 1–30. Academic Press (1991)
11. Huron, D.: Design principles in computer-based music representation. In: Marsden, A., Pople, A. (eds.) Computer Representations and Model in Music, pp. 5–40. Academic Press (1992)
12. Piston, W.: Harmony: Revised and expanded by Mark Devoto. Victor Gollancz Ltd. (1982)
13. Phon-Amnuaisuk, S., Ting, C.-Y.: Formalised symbolic reasonings for music applications. In: Abd Manaf, A., Zeki, A., Zamani, M., Chuprat, S., El-Qawasmeh, E. (eds.) ICIEIS 2011. CCIS, vol. 251, pp. 453–465. Springer, Heidelberg (2011). https://doi.org/10.1007/978-3-642-25327-0_39
14. Hochreiter, S., Schmidhuber, J.: Long short-term memory. Neural Comput. **9**(8), 1735–1780 (1997)

A Visual Analytics on Mortality of Malignant Neoplasm and Organic Food

Chien-wen Shen⑩, Irfandi Djailani⁽⊠⁾⑩, and Cheng-Wei Tang

Department of Business Administration, National Central University,
Taoyuan City, Taiwan
cwshen@ncu.edu.tw, irfandidjailani@gmail.com

Abstract. The rise of organic food consumption globally is the evidence of consumer expectation for better health. Despite the fact that many research favorable toward the idea that more organic food means more health there are research who reject this notion. The basis of rejection toward this idea are varied, from the perspective of efficiency in organic food cultivation process to the research methodology itself. Due to the preface, this research tapping into the issue with the approach of advanced visual analytics with cluster analysis that enables comparison and analyze over 37 million electronic invoice (e-invoice) data of organic food expenditure and 7 thousand data related to death caused by malignant neoplasm of trachea, bronchus and lung. Our results indicate the negative correlations between organic food expenditure and mortality rate of the case chosen. This finding can be a significant insight for government as a justification to accelerate the organic food consumption. This research also contributes to the academic dialectical narrative regarding this issue as well as for the practitioner.

Keywords: Visual analytics · Cluster analysis · Disease mortality ·
Organic food · Electronic invoice

1 Introduction

While the growing skeptical comment on the superiority of organic food versus conventional food is rising (Smith-Spangler et al. 2012), the organic food provider keeps making profit of this consumer believe which have become the phenomenon worldwide. It is recorder in US alone, by 2017, the market value of organic food related transaction reached $ 49.4 billion, or 6.4 percent higher than previous year (Organic Trade Association 2018). In the same year, the phenomenon is also found in Germany with market value reach 8.6 billion, and 5.5 billion in France (Willer and Lernoud 2017). In Asia, Taiwan also follows the trend, from 2002 to 2011 the organic farmland hectare increase from 1018 hectare to 5015 hectares (Portal Taiwan Organic Information 2013). Despite the fact of organic food rapid growth in market share, consumer spend and agricultural land conversion is still relatively low comparing to the total food spending. The reason behind it are many, but doubtfulness toward organic food is among many factors that held back the growth of organic food adoption globally. The doubts have been addressing around several issues such as the claim that organic food

© Springer Nature Singapore Pte Ltd. 2019
Y. Tan and Y. Shi (Eds.): DMBD 2019, CCIS 1071, pp. 13–22, 2019.
https://doi.org/10.1007/978-981-32-9563-6_2

is healthier, more environmental friendly, and other issues related to food safety. This is coherence with meta-analysis research of 71 papers by Tuomisto et al. (2012). The more academically correct claim is there are difficulties encountered by previous researches in connecting between organic food consumption and health benefit (Denve and Christensen 2015).

This research proposes the use of advanced visual analytics with cluster analysis in attempt to contribute in organic food discourse. The adopted methodology helps researcher to better understand the phenomenon because it opens door to large volume and heterogeneous data (Levin et al. 2016). Our data was acquired from 31 million electronic invoice data of organic food expenditure from Fiscal Information Agency, and then it was compared to the 7 thousand data of death toll caused by malignant neoplasm of trachea, bronchus, and lung from Taiwan Government Open Data Platform. The reason in choosing this disease because according to Taiwan Ministry of Health and Welfare (2017), malignant neoplasm is among top 10 death caused in Taiwan in which is 27.7% of total death for mortality rate of 203.1 per 100,000 populations. By utilizing this data, it allows us to measure the concrete behavior in purchasing organic food, rather than measuring the intention to purchase which usually adopt by many researches in this field. Our approach can also avoid green gap in which explained by Chekima et al. (2017) as the gap between intention behavior or attitude behavior that most organic food research use as variable with the purchasing behavior that actually happen and effecting the market. The main objective of this research is to find the correlation between organic food expenditure in Taiwan and its correlation with mortality rate caused by disease chosen using the visual analytics of big data. The contribution of this research is to encourage the adoption of organic food consumption either by industry or government. The organic food industry can refer to our results to promote the benefits of organic food consumption. This research can also provide a significant insight for government as a justification to accelerate the organic food consumption as well as the development of the related public welfare policy.

2 Literature Review

In Barański et al. (2014) study, it specified that the organic crops have higher antioxidants and lower pesticide residues compared with the non-organic foods. This research tried to explore the composition differences between the organic and conventional crops through the literature review of studies. Result demonstrates that the crops which cultivated through the organic farming system containing higher antioxidants but lower Cadmium and pesticide residues. A number of research around the world has shown that organic diet helps reduce the health risk that caused by unhealthy contaminant from food intake. Kummeling et al. (2008) research on 2,834 participants in Netherland regarding the developing of atopic manifestation for infants within two years after their birth find out that there are no correlations between the organic diets and the atopic sensitization. This research also find that the consuming more organic product could reduce the risk of getting eczema. Vogt et al. (2012) aimed to investigate the children from 21 counties in California who exposed to multiple contaminants in their food. Results concluded that having more organic dairy and selected fruits and vegetables can prevent you from pesticide intake while having lower animal food products included meat and fish could avoid the contact of the pollutants. To reduce the intake of acrylamide, having lower amount of the chips, cereal, crackers and other processed carbohydrate foods will have significant effect on it. In Zhang et al. (2014) study, organic food products are rich in antioxidant activity and bioactivity rather than conventional foods. Also, it suggested that the higher amount of organic food intake, it will protect you from getting the heart attack and cardiovascular diseases due to its natural antioxidants and vitamins in the contents.

Some researchers also try to find the in correlation between organic food and some disease, such as cancer. Bradbury et al. (2014) investigated the relationship between the organic food consumption and the cancer incidence regarding on the middle-aged women in the United Kingdom. Roughly 1.3 million middle-aged women were included in this research during 1996 to 2001 and they have been re-surveyed after 3, 8 and 12 years after joining the research in order to evaluate the relationship between the cancer incidence and the organic food consumption. 16 most common cancers were included in this research such as oral cavity, stomach, colorectal cancer and non-Hodgkin lymphoma. Result suggested that the association between the organic food consumption and the incidence of cancer is non-significant except the non-Hodgkin lymphoma. Previous study not only discussing from the perspective of health benefit of organic food, but also examining the harm that non-organic food can cause. It is Lerro et al. (2015) reported that the usage of organophosphates in non-organic food cultivation and the possibility of cancer for those people who consume it has a strong correlation. This research is conducted on around 30,003 females. It suggests that the women who often exposed under the organophosphates were easily having the hormonally-related cancer such as breast, thyroid and ovary.

With the rise of organic-food interest and usage worldwide, it is important to employ method which allows the understanding on dynamic nature of organic food development and impact. This need in-line with data mining and visual analytic method which according to Flood et al. (2016) is the analytical reasoning science empowered by large volume and heterogeneous data-driven software allowing the interactive-visualizations for improving the information-processing and analyzing capabilities in order to support micro-prudent decision-and policy making process, or in this case policy and decision related to organic food implementation.

3 Research Method

The E-invoice transaction data during 2014–2015 were utilized in this research, which are retrieved from the Fiscal Information Agency, Ministry of Finance in Taiwan. The Fiscal Information Agency provided the data of organic food expenditure from the whole Taiwan during year 2014–2015; there are 21,564,904,393 record which are the accumulation of 2014 data set (10,094,071,679) and data set of 2015 (11,470,832,714). This record includes 10 variables; PO Number, Product, Quantity, Price, Unit, Unit Price, Time, City and County. For a better result, data cleansing is conducted. These data are filtered by excluding several data set such as; data set without "organic" keyword, data with negative value in quantity or unit price which shows return after purchased, and redundancy data. This steep reduces the data set to 44,527,407. Next step is categorizing all these data into each brand, and the product that cannot be recognized will be categorized as "other". The last step is to classify the product into groups, which are; fruits and vegetables, drink, processed foods, condiment and sauces, cereal, staple merchandise, fertilizer, health supplement and others. Furthermore, the non-edible product is eliminated. Finally, we got 37,531,200 records which are 12,286,112 from 2014 and 25,245,088 from 2015.

This record is being analyzed and compared with the death tolls that acquired from open data platform in Taiwan within the same year. There are 183,090 data set record, which 91,452 records form 2014, and 91,638 records from 2015. From this data, there are 73 kinds of disease that lead to death, but only death that caused by diet will be included, which are; Endocrine, Nutritional, and Metabolic disease, diseases of circulatory system, neoplasm and the diseases of nervous system and sense organs. Five diseases are associated with the neoplasm, which are; malignant neoplasm of colon, rectum and anus, malignant neoplasm of liver and intrahepatic bile ducts, malignant neoplasm of mouth, malignant neoplasm of mouth and malignant neoplasm of trachea, bronchus and lung. After the process of data cleansing, there were 55,427 records containing inside the data set. From all these diseases we only look at one disease, malignant neoplasm of trachea, bronchus and lung which caused death for 3,916 people in 2014, and 3,910 people in 2015, which in total reach 7,826 death.

To analyze the death tolls for each disease and the organic food expenditure, the death tolls should be converted into the mortality rate in order to have more accurate analysis result. According to the Ministry of Health and Welfare in Taiwan, formula of mortality rate as presented below will be applied in this research. To calculate the mortality rate for a specific disease, the death tolls for that disease are required and from (1), the denominator, midyear population, means the combination of the population of this year and last year, and the outcome should be divided by two. Moreover, the final result should be multiplied by 100,000.

$$Rate = (\frac{number\ of\ death\ within\ a\ year}{midyear\ population}) \times 100,000 \qquad (1)$$

After data cleansing, the visual analysis method is utilized to illustrate and analyze the two datasets; per capita organic food expenditure of datasets from 2014–2015 and the disease mortality rate of data set from 2014 and 2015. There are two steps to be applied to get the data visualization result. First, mortality rate is divided into four groups. For each city and county belonged to the fourth cluster, it will be examined closely to understand the purchasing pattern of consumer toward organic foods. Second, the relationship between the per capita organic food expenditure and the disease mortality rate will be examined through the scatter plot and correlation analysis as well as the clustering analysis in order to gain a statistical result among those two variables.

4   Results of Visual Analytics

Figure 1 illustrates the mortality rates result caused by malignant neoplasm of trachea, bronchus, and lung from 2014 and 2015 at each cities and counties in Taiwan. This figure shows that Chiayi County has the highest mortality rate, which is 33.7 per 100,000 persons and the Hsinchu City has the lowest mortality rate caused by this disease which is at 14.2 per 100,000 persons. While in 2015 Chiayi County has the highest mortality rate, which is 31.6 per 100,000 persons and Lienchiang County has the lowest mortality rate which is 8 per 100,000 persons.

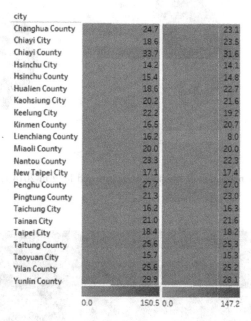

Fig. 1. Mortality rates by city and county for Malignant Neoplasm of tracheas, bronchus and lung in 2014 and 2015

For the purpose of understanding the correlation between mortality rate and organic food expenditure from available data, all 22 cities and counties in Taiwan was classified into 4 clusters based on the mortality rate of each disease. The clustering is represented by color from blue representing the highest mortality rate, follow by red for the second highest, green for the third one, and orange as fourth level which represent the lowest mortality rate. Figure 2 presents the results in 2014 and 2015, where there are 12 cities in the fourth cluster which includes Taipei City, New Taipei City, Taoyuan City, Hsinchu County, Hsinchu City, Miaoli County, Taichung City, Chiayi City, Kaohsiung City, Hualien County, Kinmen County and Lienchiang County. By comparing the figures in 2014 and 2015, there are some changes in the combination of each cluster over these two years. Chiayi City, Kaohsiung City, Hualien County, Miaoli County and Kinmen County were originally in the fourth cluster in 2014 while they were in second cluster in 2015. For the rest of areas belonged to the fourth cluster in

2014, they were grouped into third cluster in 2015. Keelung City, Nantou County, Changhua County, Tainan City, Pingtung County, Yilan County and Taitung County were classified into third cluster in 2014 while in 2015, the latter two counties were included in first cluster and the rest of the areas were included in second cluster. Yunlin County and Penghu County were contained in second cluster in 2014 but both of two counties were included in first cluster in 2015.

Fig. 2. The malignant neoplasm of trachea, bronchus and lung in 2014 (left) and 2015 (right) (Color figure online)

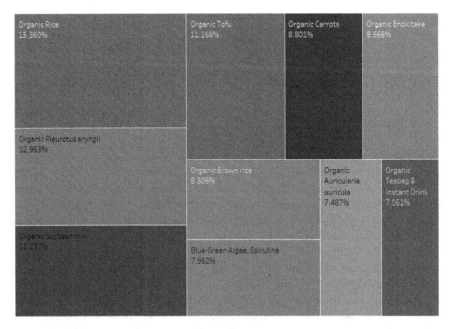

Fig. 3. The top 10 organic foods in New Taipei City, Taipei City, Taoyuan City, Hsinchu County, Hsinchu City, Taichung City, Kinmen County and Lienchiang County in 2014 and 2015 (Color figure online)

To specifically understand which organic food consumed and lead to reduction of malignant neoplasm of trachea, bronchus and lung, the analysis result shows top 10 organic food consumed in the cities and counties classified as the fourth cluster. As Fig. 3 presents above, it demonstrates the total amount of organic foods from the city and county such as the New Taipei City, Taipei City, Taoyuan City, Hsinchu County, Hsinchu City, Taichung City, Kinmen County and Lienchiang County from 2014 to 2015, and the top 10 organic foods are presenting below while organic rice is the best-selling organic products which accounts for 15.36%, followed by the organic pleurotus eryngii, 12.963%, organic soybean milk, 12.237%, organic tofu, organic carrots, organic enokitake, organic brown rice, blue-green algae spirulina, organic auricularia auricula, organic teabag and instant drink, which accounts for 11.168%, 8.801%, 8.666%, 8.306%, 7.962%, 7.487% and 7.051% respectively in the 2014 and 2015.

The next data analysis is set up for understanding the correlation between the clustered cities and counties based on mortality rate cities with the organic food expenditure. For this purpose, the cluster analysis, correlation analysis and scatter plot is utilized. It is important notes that the ratio of organic food expenditure is represented by the size of the circle. After the clustering analysis, the correlation analysis should be applied in order to find out the relationship between the mortality rate for each disease and the per capita organic expenditure. As shown in Fig. 4, it demonstrates the scatter plot between per capita organic food expenditure and the mortality rate of malignant neoplasm of trachea, bronchus and lung of 2014 and 2015. The clustering analysis result in 2014 demonstrates in Fig. 4(a). Cities and counties are including in first cluster can be expressed as follows: Changhua County, Chiayi County, Nantou County, Penghu County, Taitung County, Yilan County and Yunlin County. Second cluster comprises of the cities and counties such as Chiayi City, Hsinchu County, Hualien County, Kaohsiung City, Keelung City, Kinmen County, Miaoli County, New Taipei City, Pingtung County, Taichung City, Tainan City and Taoyuan City. Hsinchu City and Lienchiang County are included in the third cluster, which have relatively higher ratio of organic food expenditure. Taipei City lies in the fourth cluster which has the highest organic food expenditure. Figure 4(b) shows the clustering analysis result in 2015, which differs from Fig. 4(a) in regard to the components of each cluster. Following cities and counties such as Chiayi City, Tainan City, Kaohsiung City, Miaoli County, Pingtung County, Hualien County and Kinmen County were classified into the second cluster in 2014 while all of cities and counties grouped into first cluster in 2015. Hsinchu City originally was included in third cluster in 2014 but it was comprised in the second cluster in 2015.

Results of 2014 data on correlation analysis between the per capita organic food expenditure and the mortality rate caused by malignant neoplasm of trachea, bronchus, and lung reveals the correlation coefficient at −0.56216 and p-value 0.008. In the other hand, the correlation coefficient is −0.36755 and p-value is 0.1451 in 2015. The statistical results indicated that per capita organic food expenditure has been shown to correlate highly with the mortality rate of malignant neoplasm of trachea, bronchus and lung at 90% of confidence level. From the perspective of clustering analysis for the interaction between the organic food expenditure and the mortality rate, people in cities of southern Taiwan easily suffered from malignant neoplasm of trachea, bronchus and lug, which is not the case with the inhabitant lived in the central and southern Taiwan.

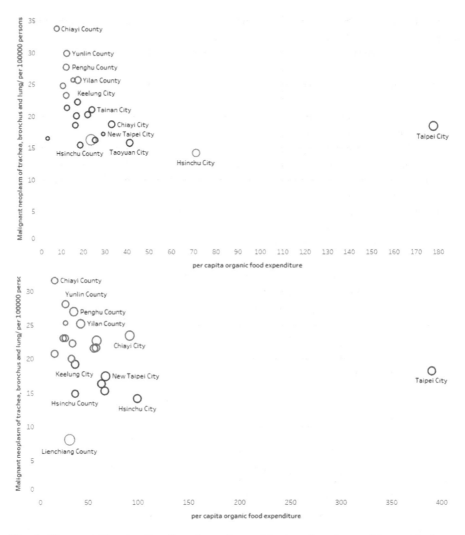

Fig. 4. The mortality rate of malignant neoplasm of trachea, bronchus and lung with the per capita organic food expenditure and the ratio of organic food expenditure in 2014 (upper) and 2015 (lower).

5   Conclusion

Much of the previous researches regarding the correlation between organic food expenditure and the health trend are conducted through the questionnaires, which by its nature can only gather small number of data that in some perspective cannot represent the population. This research conducts through the analyzing of big data gathered from the digital data of E-invoice transaction data and the raw data of death tolls retrieved from government agencies rather than the questionnaires. On these grounds, this research has arrived at the conclusion that from the perspective of correlation analysis of two factors, the organic food expenditure and the mortality rate caused by malignant neoplasm of trachea, bronchus and lung, it has negative relationship, which means spending much on the organic foods will have relatively lower mortality rate. Based on our analysis, the correlation coefficient is negative, in other words, people who spent more on the organic food the mortality rate would be lower.

From the visual analytics on the mortality rate of malignant neoplasm of trachea, bronchus and lung disease and the organic food expenditure, city and county which located at the northern part of Taiwan have the lowest mortality rate. Furthermore, the organic food products such as organic pleurotus eryngii, organic soybean milk, organic tofu, organic rice, organic carrots, and organic auricularia auricula were commonly purchased by the inhabitants who belonged to the fourth cluster of this disease. Results of this research was corresponded with Vogt et al. (2012) and Lerro et al. (2015) studies, consuming more organic foods could lower down the intake of pesticides, which related with the incidence of the cancer or in this research case specify to malignant neoplasm of trachea, bronchus and lung disease. This paper can exercise an influence on consumer, attracting them to get more organic foods. The objective of the national development program in Taiwan was to increase the hectare of the organic agriculture and promote the cultivation of the organic crops. With this research, it can provide consumer the benefit of organic food regarding on the health, enhancing the demand from consumer from the results of correlation analysis. Furthermore, the area of the organic agriculture would be increased and the idea of organic agriculture will be applied gradually instead of the conventional agriculture. As this research stated earlier, the development of organic is a way to protect our environment as well as the health of human beings, besides, it can achieve the food self-sufficiency. All in all, the result of this research recommends people to have more organic foods to avoid health risks.

There are some limitations of this research that need to be pointed out. Due to resource constraint, this research data is based on the E-invoice data from Fiscal Information Agency and overlook the data from the Computer Uniform Invoice in which still used by small number of store in Taiwan. In addition, this research only evaluated the organic food effect on human health from the perspectives of the organic food expenditure, it did not explore from the nutrition content of organic food. Future research could contain the index such as the healthcare quality, health-related quality of life, monitoring indicator, and exploring the interaction among those indexes with organic food and the mortality rate for the 22 cities and counties in Taiwan individually.

References

Barański, M., et al.: Higher antioxidant and lower cadmium concentrations and lower incidence of pesticide residues in organically grown crops: a systematic literature review and meta-analyses. Br. J. Nutr. 112(5), 794–811 (2014). https://doi.org/10.1017/s0007114514001366

Bradbury, K.E., et al.: Organic food consumption and the incidence of cancer in a large prospective study of women in the United Kingdom. Br. J. Cancer 110(9), 2321–2326 (2014). https://doi.org/10.1038/bjc.2014.148

Chekima, B., Oswald, A.I., Wafa, S.A.W.S.K., Chekima, K.: Narrowing the gap: factors driving organic food consumption. J. Clean. Prod. **166**, 1438–1447 (2017). https://doi.org/10.1016/j. jclepro.2017.08.086

Denve, S., Christensen, T.: Organic food and health concerns: a dietary approach using observed data. NJAS - Wageningen J. Life Sci. **74**, 9–15 (2015)

Flood, M.D., Lemieux, V.L., Varga, M., William Wong, B.L.: The application of visual analytics to financial stability monitoring. J. Financ. Stab. **27**, 180–197 (2016). https://doi.org/10.1016/j.jfs.2016.01.006

Kummeling, I., et al.: Consumption of organic foods and risk of atopic disease during the first 2 years of life in the Netherlands. Br. J. Nutr. **99**(3), 598–605 (2008). https://doi.org/10.1017/S0007114507815844

Levin, N., Salek, R.M., Steinbeck, C.: From database to big data. In: Nicholson, J., Darzi, A., Holmes, E. (eds.) Metabolic Phenotyping in Personalized and Public Healthcare. Elsevier Inc (2016)

Lerro, C.C., et al.: Organophosphate insecticide use and cancer incidence among spouses of pesticide applicators in the Agricultural Health Study. Occup. Environ. Med. **72**(10), 736–744 (2015). https://doi.org/10.1136/oemed-2014-102798

Organic Trade Association: Maturing U.S. organic sector sees steady growth of 6.4 percent in 2017 (2018). https://ota.com/news/press-releases/20236

Portal Taiwan Organic Information: Taiwan organic agricultural area. Retrieved from Taiwan Organic Information Portal (2013). http://www.organic.org.tw/supergood/front/bin/ptlist.phtml?Category=105937

Smith-Spangier, C., Brandeau, M.L., Hunter, G.E., Bavinger, J.C., Pearson, M.: Are organic foods safer of healthier than conventional alternatives? Ann. Internal Med. **157**, 348–370 (2012)

Taiwan Ministry of Health and Welfare: 2017 Taiwan Health and Welfare Report (2017). ISSN-24092630

Tuomisto, H., Hodge, I., Riordana, P., Macdonalda, D.: Does organic farming reduce environmental impacts? A meta-analysis of European research. J. Environ. Manag. **112**, 309–320 (2012)

Vogt, R., Bennett, D., Cassady, D., Frost, J., Ritz, B., Hertz-Picciotto, I.: Cancer and non-cancer health effects from food contaminant exposures for children and adults in California: a risk assessment, 9 November 2012. https://www.ncbi.nlm.nih.gov/pmc/articles/PMC3551655

Willer, H., Lernound, J.: The World of Organic Agriculture. FiBL & IFOAM - Organics International (2016)

Willer, H., Lernoud, J.: The World of Organic Agriculture, Statistics and Emerging Trends 2017 (2017). https://shop.fibl.org/CHen/mwdownloads/download/link/id/785/?ref=1

Zhang, P.-Y., Xu, X., Li, X.-C.: Cardiovascular diseases oxidative damage and antioxidant protection. European Review for Medical and Pharmacological Science (2014)

What Are You Reading: A Big Data Analysis of Online Literary Content

Xiping Liu[✉] and Changxuan Wan

Jiangxi University of Finance and Economics, Nanchang 330013,
People's Republic of China
lewislxp@gmail.com, wanchangxuan@263.net

Abstract. The combination of Internet and literature gives rise to the online literature, which is sharply distinct from traditional literature. Online literary content has become an important and active Internet resource and has now been formalized into a large and rapidly growing industry. Despite the great success of online literature platforms, little is known about the structure of literary content and patterns and characteristics in the creation and consumption of online literary content. In this paper, we conducted a deep analysis of online literary content on leading online literature platforms based on the real big data. From the crawled data, we obtained a set of interesting results, which describes the content-related and writing-related characteristics of online literature content. Our analysis results will shed light on the current status of online literary content and reveal some interesting research topics.

Keywords: Big data · Online literary content · Data analysis

1 Introduction

The Internet has become an indispensable part of our lives and plays an ever more important role. New technologies and media provide new opportunities for the development and enrichment of every aspect of our lives and society. A recent example is literature. The introduction of the Internet into the composition, publication, and consumption of literature directly nurtures and catalyzes the *online literature.*

Online literature consists of literary works composed, disseminated, and consumed online [1]. It is sharply distinct from traditional literature, and will have a profound influence on the development of literature and culture. China has experienced a boom in online literature. Reading online literary works has been one of the most popular Internet activities in China. According to the *39th China Statistical Report on Internet Development* [2] released by China Internet Network Information Center (CNNIC), as of December 2016, China's online literature users reached 333 million, an increase of 36.45 million over a year ago, accounting for 45.6% of overall Internet users. Among users of Internet entertainment applications, the fastest rate of growth has been registered among users of online literature. According to another more specialized report, *2016 China's Digital Reading Sector Report* [3], as of December 2015, the monthly users of China's online literature on PC and mobile platforms reached 141 and 148 million, respectively, and the daily users on the two platforms reached 11.97 million

© Springer Nature Singapore Pte Ltd. 2019
Y. Tan and Y. Shi (Eds.): DMBD 2019, CCIS 1071, pp. 23–33, 2019.
https://doi.org/10.1007/978-981-32-9563-6_3

and 32.97 million, respectively. Each month, PC users spent 162 million of hours reading online literary content; for the mobile platform, this figure was 803 million of hours. Indeed, for many young Chinese people, reading several chapters of their favorite novels online has become a routine of daily life.

Due to its huge user base, online literature has received a lot of attention and investments and has now formalized into an industry with a scale of 10 billion RMB by the end of 2016 [4]. A complete value chain has been shaped around online literature, which involves writers, online literature portals, readers, publishers, movie & TV makers, game developers, etc. Chinese online literature is also going global. The overseas market has mushroomed since 2015, primarily led by two translation sites, Wuxia World (wuxiaworld.com) and Gravity Tales (gravitytales.com). At present, the estimated traffic of Wuxia World is 23.7 million sessions per month [5]. Readers of Wuxia World come from more than 100 countries, with North Americans accounting for one-third of that number [6].

A lot of research has been conducted on online literature. Most of them fall into one of the two categories: theoretical research from the perspectives of literature [7–9], culture [1, 10], and industry [11–14], and case studies [15] of some online literary works and platforms. We find that research in this area lacks a data-driven perspective. Although there are some statistics about online literary content, most of them are high-level summary figures [2, 3] and are too general and rough to obtain deep insights. In this paper, we present a data-driven analysis of online literary content in China. We crawled data from the top-10 online literature portals and obtained hundreds of thousands of literary works written by hundreds of thousands of authors, containing millions of chapters. The literary works were created in 2003–2017, ranging from the foundation of these websites to the present, which witnessed the prosperity of Chinese online literature. Based on the data, we obtained some insightful results regarding several aspects of online literary content, e.g., the scale, topics and updating process of online literary content; the productivity and diversity of writers, etc. Our work can serve as an up-to-date overview of this new field, and we believe that the results we obtained will expand the knowledge about online literature and enhance the understanding of online literary content. So far as we know, we are the first to conduct a large-scale data-driven study on online literary content.

2   Data Preparation

There are many websites devoted to online literary content in China. In this work, we focus on the top-10 online literature websites, which are the main players in the field of online literature [16]. Almost all hot literary works originated from these websites. The websites are listed as follows: Qidian (qidian.com), JJWCX (jjwxc.net), Hongxiu (hongxiu.com), 17K (17k.com), Zongheng (zongheng.com), XS8 (xs8.cn), Chuangshi (chuangshi.com), QDMM (qdmm.com), XXSY (xxsy.net), and Faloo (b.faloo.com). Online literary works include online novels, cartoons, poetry, among others, but currently, the dominant form of online literature in China is the novel. Indeed, online novels account for more than 99% of the online literary content. We crawled descriptive data about the novels on the top-10 websites from May 2017 to July 2017.

For each novel, the following data were crawled: *title*, *author*, *size* (in terms of the number of chapters and characters), *category*, *latest update time*, and *status* (whether it is completed or not). In addition to these data, more information were obtained for a subset of the novels to perform a more detailed analysis. We chose Qidian, the largest website in terms of visits, as a typical example. For each novel, we went further to crawl its abstract and details about each chapter, including chapter title, chapter length, the time of uploading the chapter, whether the chapter is free to read or not, etc.

We conduct analysis at three different scales or granularities, corresponding to three subsets of the novels. The first set, denoted as ALL, includes all novels on Qidian. The second set, COMPLETE, includes all novels on Qidian that are completed. The last dataset, CHOICE, consists of the novels translated into English on Wuxia World, which represents the state of the art of online literary works in China. The cardinalities of the three subsets are 500,000, 30,000 and 33, respectively.

3 Analysis Results

In this section, we present the results of our analysis.

3.1 Data Volume

We have an overview of the data volume in this section. The top-10 websites include approximately 3.3 million online novels, where the duplicates are removed. We compare the figure with two other numbers. One is the number of the books on book.jd. com, a leading online bookstore in China. The other number is the number of print books published in 2016 in China [17]. The numbers are shown in Fig. 1. We can see that the number of online novels on these websites has exceeded the volume of the online bookstore and is approximately 6 times that of books published in 2016. Considering that most online novels have not been published yet, we can say that the online literature has become another source of readings parallel to print books.

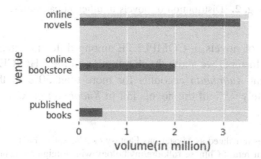

Fig. 1. Volume of novels.

3.2   Categories

In the following sections, we dive into Qidian for a deeper analysis of the online literary content. In this section, we explore their differences between categories on Qidian.

Qidian divides all novels into 15 categories. We first show the different popularity rates of these categories; more differences will be shown in the following sections. For each category, we plot the number of novels and writers under that category. The result is shown in Fig. 2. We can see that the number of authors and novels are very close in each category because each novel has only one author, and it is very common that an author writes only one novel. *Xuanhuan*[1] (Mysterious Fantasy) is the most popular category. Indeed, the number of novels in *Xuanhuan* almost doubles that of the second-largest category. The different shares of categories can be more clearly observed from Fig. 3(a). *Xuanhuan, Modern,* and *Xianxia*[2] are the top 3 categories in terms of number of novels and authors. Indeed, they account for 60% of the novels. The 15 categories can be divided into two groups: the major group and the minor group. The major group contains *Xuanhuan, Modern, Xianxia, Sci-fi, Fantasy, 2D World*[3], *Game, History, Wuxia* (martial heroes), and *Supernatural*, while the minor group contains *Short, Career, Competition, Military,* and *Others*. The major group has more than 96% of the novels.

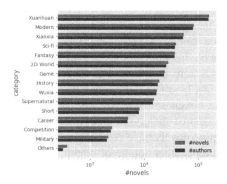

Fig. 2.   Distribution of novels in different categories.

The distribution of novels in COMPLETE among different categories are similar, so we omit it here. Instead, we show the distribution in the CHOICE dataset in Fig. 3 (b). We can see that *Xuanhuan* accounts for more than 50% of the novels in this dataset. Approximately 75% of the novels fall in *Xuanhuan* or *Xianxia*, which is very

---

[1] *Xunhuan* is sometimes translated as Mysterious Fantasy or Alternative Fantasy. It has a broad genre of fictional stories that remix Chinese folklore/mythology with foreign elements & settings.

[2] Xianxia literally means immortal heroes. It is about fictional stories featuring magic, demons, ghosts, immortal characters, and a lot of Chinese folklore/mythology. The novels are heavily inspired by Daoism.

[3] The 2D world category contains stories where the contents and writing style match up with various cartoons and comics.

interesting, as the novels under these two categories are related with Chinese folklore and thus not easy for foreign readers to understand. This is probably because those mysterious Chinese story backgrounds bring the readers different reading experiences.

3.3 Completion Ratio

Each novel on the website is marked either as *completed* or *ongoing*. Figure 4 shows the completion ratio, the ratio of completed novels to all novels, in each category. It is clear that, except for *Others*, all categories have completion ratios of approximately 0.1. That means completed novels only account for a small fraction of the novels in most categories. The reasons are three-fold: (1) it is very easy to start a novel, so there are a lot of new novels that are just created; (2) some novels received little attention and were thus abandoned and discontinued; (3) some novels received a great deal of attention, thus the authors were inclined to prolong the novels. Though *Others* has high completion ratio, it is of little significance due to its small cardinality.

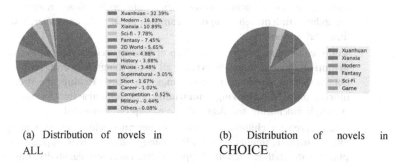

(a) Distribution of novels in ALL

(b) Distribution of novels in CHOICE

Fig. 3. Distribution of novels in ALL and CHOICE.

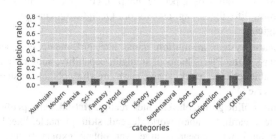

Fig. 4. Completion ratios of different categories.

3.4 Topics

In this section, we present the topic words of each category. We have crawled the summaries of all novels. We concatenated the summaries of all novels under each category, from which we then extracted the top-50 topic words using TextRank algorithm [18].

We found that some words appear frequently in many categories. If a word appears in more than 2/3 of the categories, we classify the word as a *common topic word*. We find that some common topic words are quite unique to online literature, such as *time-travel*, *rebirth*, *mankind*, etc. This is because novels with time traveling and rebirth plots are rather popular in online literary content. We then list the topic words other than the common topic words in each category in Table 1.

Table 1. Topic words of each category.

Category	Keywords
Xuanhuan	continent, the alien world, cultivation, the strong, magic, pinnacle, **reincarnation**, folklore, earth, strength, universe, soul, **genius**, all living things, set foot on, obtain, ancient times, legend, **Qi fighting, the Tao of heaven, firmament**, space, existence, **life and death**, appear, life, accomplishment, **in the distant past**, the human world, change, **inherit**
Modern	**metropolis**, love, system, beauty, obtain, adolescence, experience, happen, earth, change, supernatural power, society, university, obtain, reality, capability, **student**, return, appear, **campus**, faced with, encounter, **killer**, time, girl, **school**
Xianxia	world, cultivation, continent, **the Tao of heaven, reincarnation**, avenue, **primitive, ancient times, fairyland**, all living things, universe, set foot on, obtain, folklore, the human world, **longevity, Three Realms**, the strong, **mortal**, pinnacle, **achieving immortality, religious practice**, legend, accomplishment, life and death, disciple, step on, **immortal**
Sci-Fi	earth, universe, **eschatology**, zombie, **science and technology**, system, space, **evolve, survive**, infinity, game, life, **biology**, advent, **virus**, appear, war, **planet, interstellar**, strength, happen, enter, **variation**, capability, time and space, obtain, existence, change, obtain, **plane**, death
Fantasy	continent, magic, the alien world, strength, earth, folklore, war, empire, existence, soul, world, people, happen, **elf**, appear, fantasy, universe, legend, **demon**, life, the strong, cultivation, **Qi fighting**, change, hero, **race, magician**
2D World	**Naruto**, system, **animation**, game, like, girl, magic, continent, infinity, fantasy, **One Piece**, happen, strength, capability, author, appear, **plot**, possible, change, work, space, earth, **newbie**, obtain, existence, **figure**, universe
Game	game, **online game**, player, enter, reality, **virtual**, hero, **profession**, continent, system, earth, the alien world, legend, **skill**, pinnacle, like, king, brother, jianghu, magic, appear, change, war, obtain, **expert**, beauty, folklore, friend
History	history, **the Three Kingdoms, troubled times**, hero, **emperor**, China, return, change, empire, system, **century, period**, country, China, war, time and space, legend, nation, faced with, figure, **contend for hegemony**, happen, beauty, jianghu, **the Central Plains**, continent, experience
Wuxia	jianghu, hero, **gratitude and resentment**, troubled times, legend, continent, **love and hate**, folklore, history, the Central Plains, **conspiracy**, world, expert, love, **killer**, experience, **Imperial court**, world, dispute, happen, **unexpected situation, martial arts, groups of martial arts**, brother, obtain, the human world, disciple, like, faced with, **revenge**

(continued)

Table 1. (*continued*)

Category	Keywords
Supernatural	**yin and yang**, happen, **tomb-raiding**, incident, death, experience, secret, folklore, existence, **zombie**, appear, seek, the human world, enter, hell, soul, **ancient tomb**, **ghosts and monsters**, world, **female ghost**, **conceal**, protagonist, history, encounter, **mount Maoshan**, things, friend, strange, **curse**
Short	adolescence, love, experience, like, university, grow up, recall, happen, girl, **record**, **narrate**, dream, reality, author, time, **memory**, life, society, work, protagonist, encounter, student, friend, children, things, **choice**, school, people
Career	**entertainment**, **entertainment circle**, dream, **parallel**, **Korea**, experience, **company**, system, China, love, society, earth, career, job, movie, enter, like, work, university, reality, change, return, **music**, happen, grow up, **graduate**, legend, **development**
Competition	**basketball**, **football**, China, dream, player, team, legend, like, system, **coach**, **match**, **talent**, **league**, fan, adolescence, genius, court, profession, **sport**, grow up, warm blood, **champion**, pinnacle, **football world**, body, enter, brother, obtain, **passion**, **Europe**, **sport star**, obtain, **lead**
Military	war, China, history, empire, nation, soldier, hero, army, **battlefield**, **mission**, **special troops**, **military**, continent, **Japan**, system, **warm blood**, **King of soldiers**, **enemy**, grow up, legend, **Japanese army**, **battle**, experience, earth, metropolis, troops, change, faced with, **organization**, **devil**, brother, people, return
Others	cartoon, work, author, like, figure, **comment**, friend, **illustration**, **drawing**, love, **original**, girl, time, China, campus, existence, **reprint**, **kid**, **style**, **four-frame**, narrate, **capability**, belong to, process, **storyboard**, give up, sun, expect, enter, **classics**, beauty, **forgive**, **picture**

We can see that each category has some special, discriminative words, which are shown in bold. From these words, we can figure out the themes of different categories as well as their differences. For example, we find that Xuanhuan and Xianxia share many topic words. Indeed, the line between the two categories is blurry, both including a lot of stories with strong Chinese mythology backgrounds about reincarnation in an alien or fictional world.

3.5 Active Time

In this section, we investigate the time spent on writing novels. We define the *active time* of a novel as the time between its starting time and latest updating time. The data on the active time of the novels made it possible for us to analyze how much time it takes to write a novel, which is impossible with traditional novels.

We plot the active time in Fig. 5. For better comparison, we show the active time of novels in ALL, COMPLETE and CHOICE together. The active time is analyzed at the granularity of months, and we plot the cumulative percentage of #novels with different active times. Specifically, given an active time of n (months), the cumulative percentage $p(n)$ of #novels corresponding to n is computed as follows:

$$p(n) = \frac{|\{x|activetime(x) \leq n\}|}{N_{novels}} \tag{1}$$

where $activetime(x)$ is the active time of a novel x, and N_{novels} is the number of novels on the website.

We can see that more than 60% of the novels in ALL are active in just approximately 1 month. There are two reasons: (1) some of the novels were abandoned just 1 month after their start, and (2) others were started in the month just before we crawled the data. Obviously, as the active time gets longer, the number of novels decreases. Indeed, the active time of more than 96% of the novels is less than 30 months.

The line about novels in COMPLETE may be more interesting, as the active time of a completed novel is actually the time spent on completing the novel. From Fig. 5 we can see that approximately 30% of the novels were completed in just one month. Half of the novels were completed in 4 months, and 90% of the novels were completed in 25 months. There are also some extreme cases in which more than 170 months were used in writing a novel.

The line of CHOICE is also quite different. We can see that the median active time of CHOICE is approximately 32 months. Actually, under most percentage levels, the active time of CHOICE is much greater than that in ALL and COMPLETE. In simple words, writing novels in CHOICE generally costs more time. We think one of the reasons is that, as these novels received rather good feedback from the readers, the authors tended to continue the stories to maintain their popularity. The readers may also be glad to see that their favorite stories are updated all the time.

3.6   Sleep Time

This section investigates the *sleep time*, i.e. the time from the latest update time to the time we crawled the data, of the novels.

Figure 6 shows the #novels (number of novels) with different sleep time at the granularity of day. Obviously, there is no point in examining the sleep time for a novel that has already been completed. Therefore, completed novels are not taken into consideration in this figure. The sleep time of a great number of novels is very short, just one or several days; in other words, these novels were just updated a few days ago. This may be related to the fact that the platform encourages the authors to update their works actively, as it is believed that if an author is lazy in updating his novels, the readers may shift their interests. Indeed, the website created rules stipulating that authors updating a certain amount of characters every day will obtain certain rewards. However, there are also some novels that have been asleep for a long time, e.g., thousands of days. These novels probably are dead, i.e., they are abandoned.

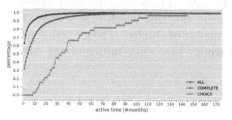

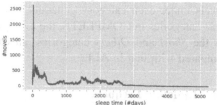

Fig. 5. Active time of novels.

Fig. 6. Sleep time of novels.

3.7 Length of Novels and Chapters

In this section, we focus on the lengths of the novels.

In Fig. 7, we plot the lengths of the novels in terms of #chapters (number of chapters) under different categories in ALL and COMPLETE. We employ violin plots to illustrate the distribution of #chapters. A violin plot shows an abstract representation of the probability distribution of a set of samples by computing an empirical distribution of the samples using kernel density estimation [19]. In Fig. 7, the shapes shown in red are violin plots. The widths of different parts in a violin plot show different probability densities. In each violin plot, there is also a box plot. Each box plot has a white circle indicating the median of the data, a box indicating the inter-quartile ranges, and a black line extending from the box indicating variability outside the upper and lower quartiles. For example, from the violin plot of *Xuanhuan* in ALL, we can see that the densities at small #chapters are higher than the densities at larger #chapters. In other words, there are more short novels than long novels. The median of #chapters of *Xuanhuan* is approximately 8, and the inter-quartile range is approximately 2 - 11. It can be observed from the figure that, in each category, there are some very long novels, e.g., with over thousands of chapters, but the majority of the novels have less than 100 chapters.

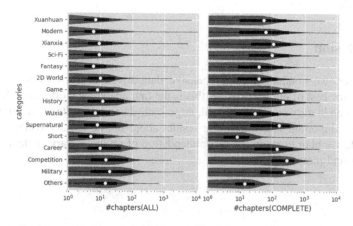

Fig. 7. #chapters of novels in ALL and COMPLETE. (Color figure online)

Comparing the violin plots of COMPLETE with ALL, we find that the completed novels under each category generally have more chapters, which makes sense. For most categories in COMPLETE, the average number of chapters is above 200. Except for the *Short* and *Others* categories, all categories have more medium-sized novels compared with that in ALL.

3.8   Updating Frequency

In this section, we explore how frequently the authors updated their novels by analyzing the gaps between successive updates. We compute the average gap between chapters for each novel, and then plot the #novels (number of novels) with different gaps. The three datasets are investigated. The result is shown in Fig. 8.

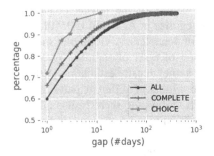

Fig. 8. Gap between successive updates.

We can see that most authors update their novels rather intensively. Approximately 60% of the novels in ALL, 65% of the novels in COMPLETE, and 70% of novels in CHOICE have gaps of one day. In addition, the gap for the majority of the novels in ALL and COMPLETE is less than 10 days. Generally, novels in COMPLETE have smaller gaps than that in ALL, and novels in CHOICE update more frequently than novels in other datasets.

4   Conclusion

In this paper, we conduct a comprehensive data-driven analysis of online literary content. To the best of our knowledge, this is the first analytical work about the current status of online literary content with real, large-scale data. As results of our analysis, we obtain some important and interesting results about online literary content. For the future work, we plan to study the recommendation of online literary content based on their topics and plots.

References

1. Zhang, Y.: Reflecting on online literature. Soc. Sci. China **32**(1), 182–190 (2011)
2. CNNIC. The 39th Statistical Report on Internet Development in China. http://cnnic.com.cn/IDR/ReportDownloads/201706/P020170608523740585924.pdf
3. iResearch. 2016 China's Digital Reading Sector Report. http://www.iresearchchina.com/content/details8_32861.html
4. Tian, X., Zhang, Q.: 500 billions, The Internet copyright industry is rising in China. http://media.people.com.cn/n1/2017/0928/c40606-29565262.html
5. wuxiaworld.com: Alexa ranking & traffic history for 8 years. http://www.rank2traffic.com/wuxiaworld.com
6. Hong, J.: China's Online Reading Craze Is So Big It's Challenging Amazon's Kindle. https://www.forbes.com/sites/jinshanhong/2017/07/17/chinas-online-reading-craze-is-so-big-its-challenging-amazons-kindle/#33ce1ebc4a8c
7. Ouyang, Y.: Chinese literature's transformation and digital existence in the new century. Soc. Sci. China **32**(1), 146–165 (2011)
8. Hockx, M.: Links with the past: mainland China's online literary communities and their antecedents. J. Contemp. China **13**(38), 105–127 (2004)
9. Chen, D.: Preservation and innovation in online literature: a brief discussion. Soc. Sci. China **32**(1), 129–145 (2011)
10. Lugg, A.: Chinese online fiction: taste publics, entertainment, and Candle in the Tomb. Chin. J. Commun. **4**(02), 121–136 (2011)
11. Zhao, E.J.: Writing on the assembly line: informal labour in the formalised online literature market in China. New Media Soc. **19**(8), 1236–1252 (2017)
12. Zhao, E.J.: Social network market: storytelling on a Web 2.0 original literature site. Convergence **17**(1), 85–99 (2011)
13. Ren, X., Montgomery, L.: Chinese online literature: creative consumers and evolving business models. Arts Mark. Int. J. **2**(2), 118–130 (2012)
14. Yan, W.: The reason, essence and influence of "IP" booming in online literature - a perspective from attention economy. China Publ. J. **24**, 37–41 (2016)
15. Tse, M.S.C., Gong, M.Z.: Online communities and commercialization of Chinese internet literature. J. Internet Commer. **11**(2), 100–116 (2012)
16. Analysys. 2016 Report on China's Online Literature Market. http://www.useit.com.cn/thread-12829-1-1.html
17. Big data report of books on jd.com. http://www.sohu.com/a/161255256_667892
18. Mihalcea, R., Tarau, P.: TextRank: bringing order into texts. In: Conference on Empirical Methods in Natural Language Processing (EMNLP) (2004)
19. Hintze, J.L., Nelson, R.D.: Violin plots: a box plot-density trace synergism. Am. Stat. **52**(2), 181–184 (1998)

Factors Affecting the Big Data Adoption as a Marketing Tool in SMEs

Jesús Silva[1](✉), Lissette Hernández-Fernández[2],
Esperanza Torres Cuadrado[2], Nohora Mercado-Caruso[2],
Carlos Rengifo Espinosa[2], Felipe Acosta Ortega[3],
Hugo Hernández P[4], and Genett Jiménez Delgado[5]

[1] Universidad Peruana de Ciencias Aplicadas, Lima, Peru
jesussilvaUPC@gmail.com
[2] Universidad de la Costa, St. 58 #66, Barranquilla, Atlántico, Colombia
{lhernand31, etorres27, nmercado1, creginfo}@cuc.edu.co
[3] Fundación Universitaria Popayán, Cauca, Colombia
felipe.acosta@fup.edu.co
[4] Corporación Universitaria Latinoamericana, Barranquilla, Colombia
{hhernandez, jdelahoz}@ul.edu.co
[5] Corporación Universitaria Reformada, Barranquilla, Colombia
g.jimenez@unireformada.edu.co

Abstract. The change brought by Big Data about the way to analyze the data is revolutionary. The technology related to Big Data supposes a before and after in the form of obtaining valuable information for the companies since it allows to manage a large volume of data, practically in real time and obtain a great volume of information that gives companies great competitive advantages. The objective of this work is evaluating the factors that affect the acceptance of this new technology by small and medium enterprises. To that end, the technology acceptance model called Unified Theory of Technology Adoption and Use of Technology (UTAUT) was adapted to the Big Data context to which an inhibitor was added: resistance to the use of new technologies. The structural model was assessed using Partial Least Squares (PLS) with an adequate global adjustment. Among the results, it stands out that a good infrastructure is more relevant for the use of Big Data than the difficulty of its use, accepting that it is necessary to make an effort in its implementation.

Keywords: Big data · Intention to use · UTAUT ·
Acceptance of technologies · Resistance to use · Partial least squares

1 Introduction

Talking about marketing and looking for a definition consistent with the digital era entails citing the American Marketing Association (AMA): "Marketing is the activity, set of institutions, and processes for creating, communicating, delivering, and exchanging offerings that have value for customers, clients, partners, and society at large" [1]. The previous statement is not only correct and indicative, but fully

© Springer Nature Singapore Pte Ltd. 2019
Y. Tan and Y. Shi (Eds.): DMBD 2019, CCIS 1071, pp. 34–43, 2019.
https://doi.org/10.1007/978-981-32-9563-6_4

applicable to the Big Data era in which the clients and partners and the community look for the generation of value in each and every process the companies execute [2].

The term Big Data began to be disseminated in the technological context by scientists and industry executives to the year 2008. At present, it not only represents a huge amount, variety, and volume of information, but the theme of "fashion" that daily appears in newspapers and magazines. Likewise, the economic sectors, the most important companies, and consultants try to show their possible applications and generate frequent reports in this regard [3].

In this digital age in a changing economic environment, Companies must investigate the tastes of customers, conduct market research, and know the actions of the competition with the main objective of launching products and services that generate higher revenues. In other words, the information is every day more relevant for companies to make decisions. Organizations not only need to collect data, but also look for the appropriate way to analyze them to conceive daily actions based on statistics and trends. However, companies currently lack the capacity to use Big Data and data analytics [4, 5].

With this study, the researchers intend to obtain data on the factors that affect the adoption and use of this new technology in small and medium enterprises (SMEs), as well as to understand the possible problems for its implementation in order to give pertinent recommendations to professionals that make decisions.

2   Theoretical Review

The adoption of a technology is decisive for its success. From the Theory of Planned Behavior-TPB to the widely used Technology Acceptance Model-TAM, many models of technology acceptance have been developed and tested. But the model proposed by the Unified Theory of Technology Adoption and Use of Technology, or UTAUT integrates different models and previous theories that have been proposed to analyze the acceptance of a technology [6].

The determinants of the model are [7]: (1) the Performance Expectancy (PE), defined as the degree to which using a technology offers benefits in the development of certain activities; (2) Effort Expectancy (EE), which measures the degree of ease associated with the use of technology; (3) Social Influence (SI) or how consumers perceive that friends and family believe that they should use a technology; and (4) Facilitating Conditions (FC), consumer's perceptions that resources and support are available to develop a behavior. The model proposes a direct influence of the first three determinants on the intention to use (Behavior Intention, BI). Facilitating conditions influence the use of new technology (Usage Behavior, UB).

As stated by [8], the value of this model is in its capacity to identify the main determinants of adoption, and allows to include and consider the effect of different moderators that affect in the influence of the key constructs of the model.

To the constructs of the UTAUT model, Resistance to Use (RE) is added since, in the adoption of new technologies and information systems, there is an adverse reaction or opposition to change or implementation of new technologies [6]. In this context, resistance is defined as the opposition to change associated with the implementation of a new technology or information system.

The hypotheses of this research emerge from the last premises based on the extension of the UTAUT model for the case of acceptance and use of Big Data by companies, which are explained below:

Performance Expectancy (PE) refers to the perception of the performance that the technology will have. Within the UTAUT, this is one of the most influential constructs in the intention to use. Several works, in addition to the original work itself, sustain this positive relationship [6].

Effort Expectancy (EE) refers to how easy it is to learn and use what this new technology will be. According to UTAUT, Big Data will be used more or less depending on how easy or difficult it is. Other studies reinforce the meaning and weight of this relationship that confirm [9] the effect of the expected effort on the intention to use.

The measure of Social Influence (SI) in the original proposal of [10], and extended in the UTAUT2 [11] has been used to measure the effect of the influence perceived by the users regarding what others -friends, family- think with respect to the use of a technology. In a business environment, it is also important what leaders and colleagues think.

Resistance to Use (RU) has been understood as opposition or negative reaction to the implementation of a new technology. As [12] points out, the use of many new technologies has failed because of the opposition of users to their implementation. And although the resistance to use is well studied in the literature [13–15], there are very few studies that use it, integrating it in the UTAUT model. However, there are precedents of resistance to use with the intention to use.

Facilitating Conditions (FC) highlight the ease of access to the resources needed to use a new technology, as well as the support and subsequent support. In a later work, the UTAUT2 [11], found that this construct has a significant effect on the intention to use a technology. Also, more recent studies have contrasted this positive effect on the intention to use [13].

In agreement with both the Theory of Planned Behavior TPB and so with the original UTAUT, it can be observed that facilitating conditions positively affect the use of a new technology. Subsequent studies [16] and [17] have contrasted this hypothesis.

From the widely used proposal of [18] of the Technology Acceptance Model (TAM) to predict and evaluate the acceptance and use of technologies, up to the proposed UATUT model [3] that predicted moderating effects on the antecedents of the intention to use the technologies, going through the Theory of Reasoned Action (TRA) [5], a direct relationship between the intention of behavior and the use of technologies is observed. In this case, it is unquestionable that the behavior intention of the Big Data user, positively and negatively influenced by the variables proposed in the model, favorably affects the final use of the service.

This influence has been contrasted in many contexts such as, for example, the adoption of Internet banking in Portugal [19], purchase of airline tickets in Spain [20], use of electronic document management systems [21], or adoption of ERPs in India [22].

Figure 1 presents the proposed model based on the hypotheses stated above.

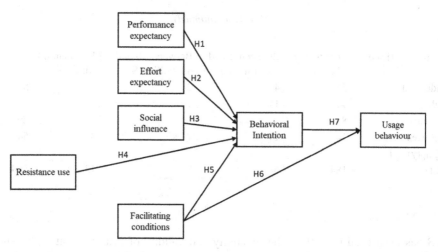

Fig. 1. Big data acceptance in companies

3 Materials and Methods

3.1 Database

The sample used in this work comes from managers responsible for an area such as HR, Financial, Marketing, and Sales of the different SMEs. The data was collected during the months of September and October 2018 through a survey carried out by Internet. To eliminate possible ambiguities in the questionnaire, it was previously reviewed, as a pre-test, with 5 volunteer managers and other researchers. The number of valid surveys was 564 and can be classified according to their turnover and sector of activity, as shown in Table 1a and b.

Table 1. Companies in the sample according to turnover and sector.

(a)

Sector	(Empty)	Less than 2 Million $	Between 2 and 10 Million $	Between 10 and 43 Million $	More than 43 Million $	Total
Farming	2	10	14	14	1	41
Commerce and distribution		20	8	12	18	58
Communications	1	35	4	45	51	136
Building		15		1	14	30
Education		12	3		20	35
Energy and Mining		10		1		11
Financial		10			42	52

(continued)

Table 1. (*continued*)

(b)

Sector	(Empty)	Less than 2 Million $	Between 2 and 10 Million $	Between 10 and 43 Million $	More than 43 Million $	Total
Industrial		36	4	2	15	57
Others	1	14	14	12	18	59
Sanitary		8	1		20	29
Services		14	19	7	14	54
(Empty)	1		1			2
Total	5	184	68	94	213	564

3.2 Methods

PLS has been used to analyze the reliability and validity of measurement scales and assess the structural model [23, 24]. Specifically, the SmartPLS 3 software package was used [25]. It was also previously checked that there were no errors due to measurement bias or Common Method Bias (CMB). For this, the indications of [26] and [27] were followed and a new latent variable called CMB variable was added as dependent of the previous ones of the model, measured with a previously unused indicator. All Variance Inflation Factors (VIF) obtained by this method should be less than 3 to confirm that the sample does not have a CMB. Compliance with the requirements is shown in Table 2.

Table 2. VIF extracted from the constructs to check the CMB

	Variable CMB
Behavioral intention	2,351
Effort expectancy	1,621
Facilitating conditions	2,027
Performance expectancy	1,971
Resistance to use	1,611
Social influence	1,710
Variable CMB	

4 Results

The measurement scales, mostly coming from the original model of [3] have been adapted to Big Data, according to different works, as shown in Table 3. The Resistance to Use variable was measured using the scale proposed by [28].

Table 3. Measurement scales for the sample SMEs

Construct	Scale
Performance expectancy	PE1: I believe that Big Data is useful to carry out the tasks of our company. PE2: I believe that with Big Data we could do the tasks of our company more quickly. PE3: I believe that with Big Data we could increase the productivity of our company. PE4: I believe that Big Data would improve the performance of our company. PE5: I believe that with Big Data you can obtain more information from our customers. PE6: I believe that with Big Data the quality of the information used in our company will be increased. PE7: I believe that with Big Data we will obtain new valuable information from our clients
Effort expectancy	EE1: Big Data would be clear and understandable to the people of our company. EE2: It would be easy for our company to become familiar with Big Data. EE3: For our company, it would be easy to use Big Data. EE4: I believe that with Big Data we could increase the productivity of our company. EE5: Generating valuable data using Big Data would be easy for our company
Social influence	SI1: The companies that influence ours use Big Data. SI2: Our reference companies use Big Data. SI3: The companies in our environment that use Big Data have more prestige than those that do not use it. SI4: The companies in our environment that use Big Data are innovative. SI5: Using Big Data is a status symbol in our environment
Facilitating conditions	FC1: Our company has the resources necessary to use Big Data. FC2: Our company has the knowledge necessary to use Big Data. FC3: Big Data is not compatible with other systems of our company. FC4: Our company has an available person (or group) for assistance with any difficulties that might arise
Resistance to use	RU1: We do not want to use Big Data to change the way we analyze our data. RU2: We do not want to use Big Data to change the way we make our decisions RU3: We do not want to use Big Data to change the way we interact with other people in our work. RU4: Above all, we do not want to use Big Data to change our current way of working
Behavioral intention	BI1: We intend to use Big Data in the coming months. BI2: We predict that we will use Big Data in the coming months. BI3: We plan to use Big Data in the coming months. BI4: We intend to obtain new and valuable data thanks to Big Data in the coming months
Usage behavior	UB: What is the current use of Big Data in your company? (i) We have never used. (ii) Once a year. (iii) Once in 6 months. (iii) Once in 3 months. (v) Once a month. (vi) Once a week. (vii) Once every 3–4 days. (vii) Once every 2–3 days. (ix) Daily

The next step was to analyze the reliability of the constructs using composed reliability and Cronbach's alpha indicators. In all cases, the indicators are greater than 0.7 as suggested by [22]. In addition, the convergent validity was ensured by analyzing the Average Variance Extracted (AVE). In this case, all the indicators offered levels above the proposed 0.5 [24]. These indicators are listed in Table 4 to verify that all constructs, including the Enabling Conditions construct, meet all the requirements.

Table 4. Composite reliability and convergent validity.

	Cronbach's alpha	Rho_A	Composite reliability	Average Variance Extracted (AVE)
Behavioral intention	0,979	0,978	0,998	0,948
Effort expectancy	0,873	0,925	0,914	0,659
Facilitating conditions	0,847	0,849	0,921	0,745
Performance expectancy	0,948	0,971	0,971	0,772
Resistance to use	0,951	0,978	0,987	0,861
Social influence	0,868	0,845	0,885	0,647
Usage behavior	1,000	1,000	1,000	1,000

The discriminant validity is assessed through the method of Heterotrait-Monotrait ratio (HTMT) [29] checking that, in all cases, levels below 0.9 were obtained, see Table 5.

Table 5. Discriminant validity (Heterotrait-Monotrait Ratio-HTMT)

	Behavioral intention	Effort expectancy	Facilitating conditions	Performance expectancy	Resistance to use	Social influence	Usage behavior
Behavioral intention							
Effort expectancy	0,322						
Facilitating conditions	0,678	0,556					
Performance expectancy	0,600	0,485	0,422				
Resistance to use	0,539	0,275	0,448	0,569			
Social influence	0,529	0,558	0,569	0,489	0,258		
Usage behavior	0,679	0,311	0,723	0,442	0,441	0,585	

Finally, Fig. 2 shows the values for each of the loads and the path of the model. Also, the R^2 of the constructs of second order is checked: Behavioral intention and use in Table 6.

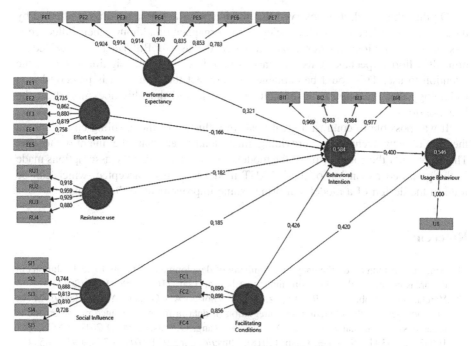

Fig. 2. Results of the model

Table 6. R^2 of the model

	R squared	R squared
Behavioral intention	0,574	0,555
Usage behavior	0,556	0,548

The results indicate that the proposed hypotheses are accepted with a high level of significance. So, in order of influence, it is observed how the Enabling Conditions is the construct that contributes the most to the Intention and Use, followed by the Expectations of Result. It was also found that the relationship between the Intention and Use were meaningful to the highest demands.

5 Conclusions

It was observed that the Intention to Use of Big Data on the part of SMEs is determined: (1) by the perception of getting good results with the use of this technology (Performance Expectancy); (2) by the positive effect posed in this technology that others consider important to use (Social Influence); and (3) mainly due to the fact that the company provides the support and resources to promote their use (Facilitating Conditions).

On the other hand, it is observed that the intention to use is negatively affected by the Resistance to Use new technologies in any organization although their influence is less than the previous relationships. Although the use of Big Data is perceived as difficult (Effort Expectancy), its influence is very low, with little significance over the intention to use. This could be explained by the fact that Big Data is perceived as a technology that presupposes a difficulty in its use and that this does not affect the intention to use.

It was possible to contrast a great positive influence of the enabling conditions on the use of new technology, providing more load even than the intention to use. Therefore, after the comparison of the model, it is concluded that the assumptions made in the proposed expansion of the UTAUT model have been accepted, which can be used to the design of a model that brings some improvement to the original.

References

1. Amelec, V.: Increased efficiency in a company of development of technological solutions in the areas commercial and of consultancy. Adv. Sci. Lett. **21**(5), 1406–1408 (2015)
2. Varela, I.N., Cabrera, H.R., Lopez, C.G., Viloria, A., Gaitán, A.M., Henry, M.A.: Methodology for the reduction and integration of data in the performance measurement of industries cement plants. In: Tan, Y., Shi, Y., Tang, Q. (eds.) DMBD 2018. LNCS, vol. 10943, pp. 33–42. Springer, Cham (2018). https://doi.org/10.1007/978-3-319-93803-5_4
3. Lis-Gutiérrez, M., Gaitán-Angulo, M., Balaguera, M.I., Viloria, A., Santander-Abril, J.E.: Use of the industrial property system for new creations in colombia: a departmental analysis (2000–2016). In: Tan, Y., Shi, Y., Tang, Q. (eds.) Data Mining and Big Data. Lecture Notes in Computer Science, vol. 10943, pp. 786–796. Springer, Cham (2018). https://doi.org/10.1007/978-3-319-93803-5_74
4. Anuradha, K., Kumar, K.A.: An E-commerce application for presuming missing items. Int. J. Comput. Trends Technol. **4**, 2636–2640 (2013)
5. Larose, D.T., Larose, C.D.: Discovering Knowledge in Data (2014). https://doi.org/10.1002/9781118874059
6. Pickrahn, I., et al.: Contamination incidents in the pre-analytical phase of forensic DNA analysis in Austria—Statistics of 17 years. Forensic Sci. Int. Genet. **31**, 12–18 (2017). https://doi.org/10.1016/j.fsigen.2017.07.012
7. Barrios-Hernández, K.D.C., Contreras-Salinas, J.A., Olivero-Vega, E.: La Gestión por Procesos en las Pymes de Barranquilla: Factor Diferenciador de la Competitividad Organizacional. Información tecnológica **30**(2), 103–114 (2019)
8. Prajapati, D.J., Garg, S., Chauhan, N.C.: Interesting association rule mining with consistent and inconsistent rule detection from big sales data in distributed environment. Futur. Comput. Inform. J. **2**, 19–30 (2017). https://doi.org/10.1016/j.fcij.2017.04.003
9. Abdullah, M., Al-Hagery, H.: Classifiers' accuracy based on breast cancer medical data and data mining techniques. Int. J. Adv. Biotechnol. Res. **7**, 976–2612 (2016)
10. Khanali, H.: A survey on improved algorithms for mining association rules. Int. J. Comput. Appl. **165**, 8887 (2017)
11. Ban, T., Eto, M., Guo, S., Inoue, D., Nakao, K., Huang, R.: A study on association rule mining of darknet big data. In: International Joint Conference on Neural Networks, pp. 1–7 (2015). https://doi.org/10.1109/IJCNN.2015.7280818

12. Vo, B., Le, B.: Fast algorithm for mining generalized association rules. Int. J. Database Theory Appl. **2**, 1–12 (2009)
13. Al-hagery, M.A.: Knowledge discovery in the data sets of hepatitis disease for diagnosis and prediction to support and serve community. Int. J. Comput. Electron. Res. **4**, 118–125 (2015)
14. Amelec, V., Alexander, P.: Improvements in the automatic distribution process of finished product for pet food category in multinational company. Adv. Sci. Lett. **21**(5), 1419–1421 (2015)
15. Cabarcas, J., Paternina, C.: Aplicación del análisis discriminante para identificar diferencias en el perfil productivo de las empresas exportadoras y no exportadoras del Departamento del Atlántico de Colombia. Revista Ingeniare **6**(10), 33–48 (2011)
16. Caridad, J.M., Ceular, N.: Un análisis del mercado de la vivienda a través de redes neuronales artificiales. Estudios de economía aplicada (18), 67–81 (2001)
17. Correia, A., Barandas, H., Pires, P.: Applying artificial neural networks to evaluate export performance: a relational approach. Rev. Onternational Comp. Manag. **10**(4), 713–734 (2009)
18. De La Hoz, E., González, Á., Santana, A.: Metodología de Medición del Potencial Exportador de las Organizaciones Empresariales. Información Tecnológica **27**(6), 11–18 (2016)
19. De La Hoz, E., López, P.: Aplicación de Técnicas de Análisis de Conglomerados y Redes Neuronales Artificiales en la Evaluación del Potencial Exportador de una Empresa. Información Tecnológica **28**(4), 67–74 (2017)
20. Escandón, D., Hurtado, A.: Los determinantes de la orientación exportadora y los resultados en las pymes exportadoras en Colombia. Estudios Gerenciales **30**(133), 430–440 (2014)
21. Kumar, G., Malik, H.: Generalized regression neural network based wind speed prediction model for western region of India. Procedia Comput. Sci. **93**(September), 26–32 (2016). https://doi.org/10.1016/j.procs.07.177
22. Obschatko, E., Blaio, M.: El perfil exportador del sector agroalimentario argentino. Las profucciones de alto valor. Estudio 1. EG.33.7. Ministerio de Economía de Argentina (2003)
23. Olmedo, E., Velasco, F., Valderas, J.M.: Caracterización no lineal y predicción no paramétrica en el IBEX35. Estudios de Economía Aplicada **25**(3), 1–3 (2007)
24. Paredes, D.: Elaboración del plan de negocios de exportación. Programa de Plan de Negocio, Exportador- PLANEX (2016). https://goo.gl/oTnARL
25. Qazi, N.: Effectof Feature Selection, Synthetic Minority Over-sampling (SMOTE) And Under-sampling on Class imbalance Classification (2012). https://doi.org/10.1109/UKSim. 116
26. Smith, D.: A neural network classification of export success in Japanese service firms. Serv. Mark. Q. **26**(4), 95–108 (2005)
27. Sharmila, S., Kumar, M.: An optimized farthest first clustering algorithm. In: Nirma University International Conference on Engineering, NUiCONE 2013, pp. 1–5 (2013). https://doi.org/10.1109/NUiCONE.2013.6780070
28. Sun, G., Hoff, S., Zelle, B., Nelson, M.: Development and Comparison of Backpropagation and Generalized Regression Neural Network Models to Predict Diurnal and Seasonal Gas and PM 10 Concentrations and Emissions from Swine Buildings, vol. 0300, no. 08 (2008)
29. Uberbacher, E.C., Mural, R.J.: Locating protein-coding regions in human DNA sequences by a multiple sensor-neural network approach. Proc. Natl. Acad. Sci. **88**(24), 11261–11265 (1991)

RETRACTED CHAPTER: Data Mining to Identify Risk Factors Associated with University Students Dropout

Jesús Silva[1]([⊠]), Alex Castro Sarmiento[2],
Nicolás María Santodomingo[2], Norka Márquez Blanco[3],
Wilmer Cadavid Basto[4], Hugo Hernández P[4], Jorge Navarro Beltrán[4],
Juan de la Hoz Hernández[4], and Ligia Romero[2]

[1] Universidad Peruana de Ciencias Aplicadas, Lima, Peru
jesussilvaUPC@gmail.com
[2] Universidad de la Costa, St. 58 #66, Barranquilla, Atlántico, Colombi
{acastro10,nmarial,lromeroll}@cuc.edu.co
[3] Universidad Libre Seccional Barranquilla, Atlántico, Colombia
norka.marquezb@unilibre.edu.co
[4] Corporación Universitaria Latinoamericana, Barranquilla, Colombia
{wcadavid,hhernandez,jjnavarro,jdelahoz}@.edu.co

Abstract. This paper presents the identification of university students dropout patterns by means of data mining techniques. The database consists of a series of questionnaires and interviews to students from several universities in Colombia. The information was processed by the Weka software following the Knowledge Extraction Process methodology with the purpose of facilitating the interpretation of results and finding useful knowledge about the students. The partial results of data mining processing on the information about the generations of students of Industrial Engineering from 2016 to 2018 are analyzed and discussed, finding relationships between family, economic, and academic issues that indicate a probable desertion risk in students with common behaviors. These relationships provide enough and appropriate information for the decision-making process in the treatment of university dropout.

Keywords: Knowledge extraction process · Tutoring · Decision making · Data mining

1 Introduction

A recent study by the World Bank reveals that Colombia is the second country in the region with the highest university dropout rate. The list is led by Bolivia, Ecuador is third, and Panama is in fourth place. The research shows that 42% of people in Colombia, between the ages of 18 and 24 who enter the higher education system, drop out [1, 2].

According to UNESCO [3], one of the reasons for this behavior may be the rising costs of higher education in the country, which averages US $ 5,000 per student, becoming one of the most expensive educations in the continent, after Mexico and

The original version of this chapter was retracted: The retraction note to this chapter is available at
https://doi.org/10.1007/978-981-32-9563-6_35

Chile. However, Colombia is also the country that offers the graduate the best return on investment in university studies [4].

"This conclusion was obtained through an analytical model called Mincer regression, which shows how incomplete higher education generates 40% return on investment to the graduate, compared to not studying, while completing university education generates the professional in Colombia up to 140% return on investment, compared to another educational level" [5].

According to several authors [6–10], the student dropout profiles obtained through data mining techniques involving classification, association, and grouping, indicate that these techniques can generate models consistent with the observed reality and the theoretical support based only on the databases of different universities. This work aims to study a database regarding students' dropout in a private university in Colombia, using data mining techniques in order to identify patterns that allow the development of retention strategies.

2 Theoretical Review

2.1 Techniques to Reduce Student's Dropout

SINNETIC developed a statistical analysis of the scientific literature with more than 1,726 research papers associated with the subject to find best practices that reduce university dropout [11].

2.1.1 Centralized Evaluation

It consists of generating standardized evaluation resources (scientifically developed exams) applied by the higher education institution, making 70% or more of the grades depend on this evaluation format, leaving the teacher only 30% of evaluation capacity with mechanisms such as homework, short tests, etc [12].

In universities that have applied this mechanism, student's dropout has been reduced by up to 34%, with institutions from India and Spain showing the best figures in this regard [13].

2.1.2 Recreate Proficiency Markets Among Teachers

This measure implies that a subject matter can be taught by two or more teachers, generating competitiveness and recreating an incentive scheme where the teachers receive an economical income that can be improved by higher level of enrolled to their courses. To avoid an indulgent position from the teacher, the evaluation is centralized in the university and are elements such as reputation, good comments from the previous cohorts which determine the flow of students enrolled by subject, semester to semester [14, 15].

In institutions where this measure has been applied, dropout reduction reaches levels of 32% throughout the training cycle, although this measure has a greater impact on early dropout, understood as desertion during the first two years, when the results of reduction is around 44% [16].

2.1.3   Reduction of Regulations

More than 13 studies that report the number of standards, the linguistic complexity of the policies, the number of statutes or articles within documents such as research regulations, teaching regulations, etc., tend to be positively correlated with student's dropout. This correlation oscillates around 0.43 and 0.67 on a scale ranging from −1 to 1 [17].

Universities that reduce the regulatory burden and simplify communication through strategies based on behavioral economics have managed to reduce student´s dropout by up to 23% [7].

2.1.4   Differential Registration Schemes with Fiduciary Incentives

This mechanism consists in raising the barriers of dropout by means of economic incentives that consist of charging during the first four semesters more than 7% of the cost of the whole program under the promise that 20% of extra charge in the first semesters will go to a savings program (fiducia, CDT, etc.) in order to generate interests for reducing the costs of finishing a career. The student who drops out before completing 50% of the program will lose the interests and over cost. Evidence of this mechanism can be found in only two studies showing reductions of 21% in students' dropout [18].

2.2   Data Mining and Classifiers

The process of extracting knowledge from large volumes of data has been recognized by many researchers as a key research topic in database systems, and by many industrial companies as an important area and opportunity for greater profits [15]. Fayyad et al. define it as "The non-trivial process of identifying valid, new, and potentially useful patterns from the data that are fundamentally understandable to the user".

The Knowledge Discovery in Databases (KDD) is basically an automatic process in which discovery and analysis are combined. The process consists of extracting patterns in the form of rules or functions from the data, for the user to analyze them. This task usually involves preprocessing data, doing data mining, and presenting results [17–19]. The KDD process is interactive and iterative involving several steps with the intervention of the user in making many decisions and is summarized in the following five stages.

Selection Stage. In the Selection Stage, once the relevant and priority knowledge is identified and the goals of the KDD process are defined from the point of view of the final user, an objective data set is created selecting the whole data set or a representative sample on which the discovery process will be carried out [20].

Preprocessing/Cleaning Stage. In the Preprocessing/Cleaning stage (Data Cleaning), the data quality is analyzed, and basic operations are applied such as the removal of noisy data. Strategies are selected for the management of unknown data (missing and empty), null data, duplicated data, and statistical techniques for its replacement [21].

Transformation/Reduction Stage. In the data transformation/reduction stage, useful features are searched to represent data depending on the goal of the process. Methods for dimensions reduction or transformation are used to decrease the effective number of variables under consideration or to find invariant representations of the data [16]. Dimensions reduction methods can simplify a table in a database in a horizontal or vertical way. The horizontal reduction involves the elimination of identical tuples as a result of the substitution of any attribute value for another of high level in a defined hierarchy of categorical values or by the discretization of continuous values. Vertical reduction involves the elimination of attributes that are insignificant or redundant with respect to the problem. Reduction techniques such as aggregations, data compression, histograms, segmentation, entropy-based discretization, sampling, etc. [19] are used.

Data Mining Stage. Data mining is the most important stage of the KDD process [20]. The objective of this stage is the search, extraction, and discovery of unsuspected and relevant patterns. Data mining consists of different tasks, each of which can be considered as a type of problem to be solved by a data mining algorithm as Adamo and Hernández et al. assert, where the main tasks are Classification, Association, and Clustering [21, 22].

Data Interpretation/Evaluation Stage. In the interpretation/evaluation stage, the discovered patterns are interpreted, and it is possible to return to the previous stages for subsequent iterations. This stage can include the visualization of the extracted patterns, the removal of redundant or irrelevant patterns, and the translation of useful patterns in terms that are understandable to the user. On the other hand, the discovered knowledge is consolidated to be integrated into another system for subsequent actions, or simply to document and report it to the interested parties, as well as to verify and solve potential conflicts with previously discovered knowledge [23].

3 Materials and Methods

3.1 Database

The data used in this study was obtained from a student tracking system aligned with the policies of the Colombian Institute for the Promotion of Higher Education. The sample is made up of 985 industrial engineering students from a private university in Colombia during the time period 2016–2018.

The questionnaires and tests included in the system are briefly described below [24].

1. Questions about family, socioeconomic, and academic topics of the students are included in order to know their background. The information is stored in the system database and can be modified in each application of the questionnaires if there is any change in the student's situation.
2. Questions are included to know specifically if the students work, the kind of work, who or how they pay their expenses, who they keep informed of their studies, and their general health conditions.
3. The tests of self-esteem, assertiveness, learning styles, and study skills are questionnaires of closed questions to recognize problems related to organization of activities, study techniques, and motivation to study.
4. Ratings. Optionally, there is a section of grades where students capture the grade obtained by each learning unit of the different educational programs they attend.

3.2 Methods

The data analysis was carried out using two important data mining tools: clustering and association rules. Particularly, the K-means clustering algorithm and the A priori algorithm were used for the association rules. To carry out the data analysis process, the KDD method was used, consisting of the following steps [14, 20, 21, 25]:

1. Selection of data and analysis of their properties. Once the questionnaires are applied to the students and the information is in the database, the relevant data are defined to analyze the available information such as academic, family, and socio-economic background. In this step for association rules, specific data from the database were used, consisting in sport practice, economic problems, whether they work or not, high school average, interruptions in their studies, and the career study field. In the case of grouping, the data used consisted of information about the reasons to choose the career, medical treatments, economic dependencies (if they are married or have children), and knowledge of programs such as scholarships.
2. Data pre-processing. In this stage, the database is cleaned when looking for inconsistent, out of range, missing data, or empty fields to later integrated and used in the analysis with data mining. The files created in this stage have the extension in .arff format for the analysis with Weka software, which is an application designed for the analysis of databases applying different algorithms for both supervised and unsupervised learning to obtain statistics and trend patterns according to the research objectives.
3. Application of data mining techniques. Once the files are obtained in .arff format, they are loaded to the Weka tool, and the option by which the information will be processed is selected. In the case of association rules, the *associate* option is chosen and then the a priori algorithm selected and parameters such as the number of rules, the output, the minimum coverage, among others, are adjusted to later start the execution and obtain the patterns.
   In the case of clustering, a similar procedure is followed, selecting the *cluster* option, then the algorithm to be used, k-means in this case, and the parameters are adjusted considering the number of clusters in which the information is to be grouped and the iterations that it will have.
4. Interpretation and evaluation based on the results obtained from the previous phase. The "patterns" obtained from Weka are analyzed and evaluated if they are useful. This task is performed by the analyst.

4 Results

This section presents the main results obtained from the analysis of the system data studied through data mining techniques. It is divided into two subsections, one for the discussion of results obtained with association rules and the other one for the results obtained from the analysis by means of grouping tools.

4.1   Association Rules

In order to generate strong rules that exceed the support and minimum trust, the minimum support was established at 3% and confidence at 80%. 1,957 rules were generated, from which, the rules were chosen with 100% confidence. The most representative association rules are the following:

Rule 1. 100% of students who drop out are single, their grade average is less than 2.4, they have failed courses in the first semesters (1 to 4) and they failed these courses only once.

Rule 2. 100% of the students who drop out have completed their secondary studies in a public school, they are single, their average grade is less than 2.4, they have failed courses in the first semesters (1 to 4) and all of them failed these courses only once.

Rule 3. 100% of students who drop out are single men, their average grade is less than 2.4, and they are from a private university.

According to the above results, among the factors associated with student's dropout are: being single, having a low average, having failed courses in the first semesters, and coming from a public school.

For the analysis by means of association rules, the information of the database that corresponds to the family, economic, and academic background sections was used. Based on this information, the following conclusions were reached.

1. Students who usually practice some type of sport like soccer have economic problems and those who do not practice any sport activity do not present economic problems.
2. Most of the students who drop out have economic problems, which is why they should look for a way to earn an income and decide to go to work and later they decide to abandon their studies to continue obtaining the income.
3. Students who drop out have got a favorable average in their previous studies, which indicates that the cause of their desertion is not directly related to academic situations.
4. Most of the students who drop out have never interrupted their studies at primary and secondary levels.
5. Most of the students entering the engineering career have a misconception of the approach to the moment they decide to enroll.

Based on the results obtained with the use of association rules, a general view was obtained about the situation of the students, the student's community, the industrial engineering career, and the probable causes for leaving studying. For example, the contrast of opinions between students (men and women) and the relationship that exists between students who practice sports and their economic situation, gives a reference of how students can act and what they think. These factors can influence the low academic performance of students and can have an impact on the decision of dropping out. However, these results are preliminary, and a more extensive study is necessary in order to obtain more information to enhance the decision making processes in the academic community.

4.2   Clustering or Grouping

Once the application of association rules was completed, the k-means clustering technique was used on a broader set of data since new information was included from the application of the questionnaire and self-esteem, assertiveness, and learning styles tests, besides the registration of new grades. From the application of this algorithm, the following results were obtained:

1. From a total of 940 students, 321 of them think they had liked to study another career. From these 321 students, 150 prefer not to continue studying and start working to obtain an income, while the rest prefer to change to another career they like.
2. From the 70% of students who take medication, their parents have higher educational studies, so they have medical attention, but it must be followed up to see if it is cause of desertion. In this case, the interest is because 100% of students who take medication only 70% have medical care, while the other 30% do not have it which, in some cases, is not enough or their parents do not know about the problem.
3. Students who have someone who depends economically on them -married or working- are strong candidates to drop out due to their economic and family situation.
4. All the students surveyed say they know the field of knowledge of the career, and 93% of them have not been enrolled in another engineering. However, only 3% of the students have no reason to finish their studies, so the career was their first choice and they do not have another engineering in mind. In this way, the causes of desertion are due to social, economic, and personal causes, or that they do not have enough commitment to complete their studies.
5. Students who have another career in mind have difficulty in the school situation since they are in classes only because they have no other option at the moment but when their opportunity arrives they will decide to leave the career and enroll in the career they really want to study.
6. The economic situation of the student causes desertion when they do not know what scholarships the student can access either by the institution or by a government agency. This is the job of tutors or the staff of academic services of the institution, however, this issue is sometimes left aside, and not enough attention is paid to the publication of calls among students.
7. From the sick students, it is observed that the institutional insurance helps to cover the cases, and not having this benefit can make difficult to cover their needs, which leads to some students being studying for keeping the insurance and not because they really want to do it. This is reflected in the qualifications at the end of each semester.

From the results obtained in this analysis, as more data is obtained from the application of the questionnaires, the analysis becomes deeper and more precise information is obtained about the needs of the students. It represents a benefit for the decision-making processes that allow the permanence of the students in the institution and can complete the career in a satisfactory way.

5 Conclusions

The research focuses on the capture of information on students of industrial engineering in a private university in Colombia. The data are stored in the database to be later analyzed by data mining techniques to obtain common behavior patterns among students. The resulting information can help identify problems in the student community that affect their performance and cause problems both for the students who drop out and for the institution in the search for terminal efficiency.

The main contribution of the work discussed in this paper is that, in addition to systematizing the tutoring process, it supports the analysis of the data through data mining tools and pattern recognition. With the analysis of the information that was carried out, preliminary results were obtained on the factors that can influence the dropout and underperformance of the students of industrial engineering in Colombia. Among these factors are the social, economic, and academic ones.

For future researches, it is necessary to try other techniques such as expert system based on artificial neural networks to improve the results obtained so far.

References

1. Caicedo, E.J.C., Guerrero, S., López, D.: Propuesta para la construcción de un índice socioeconómico para los estudiantes que presentan las pruebas Saber Pro. Comunicaciones en Estadística **9**(1), 93–106 (2016)
2. Torres-Samuel, M., Vásquez, C., Vitoria, A., Lis-Gutiérrez, J.P., Borrero, T.C., Varela, N.: Web visibility profiles of top100 latin american universities. In: Tan, Y., Shi, Y., Tang, Q. (eds.) DMBD 2018. LNCS, vol. 10943, pp. 254–262. Springer, Cham (2018). https://doi.org/10.1007/978-3-319-93803-5_24
3. Zhang, G.P.: Time series forecasting using a hybrid ARIMA and neural network model. Neurocomputing **50**(1), 159–175 (2003)
4. Duan, L., Xu, L., Liu, Y., Lee, J.: Cluster-based outlier detection. Ann. Oper. Res. **168**(1), 151–168 (2009)
5. Haykin, S.: Neural Networks a Comprehensive Foundation, 2nd edn. Macmillan College Publishing, Inc. USA (1999). ISBN 9780023527616
6. Haykin, S.: Neural Networks and Learning Machines. Prentice Hall International, New Jersey (2009)
7. Abbiw, K.A., Badal, N.A.: Novel approach for intelligent distribution of data warehouses. Egypt. Inf. J. **17**(1), 147–159 (2015)
8. Aguado-López, E., Rogel-Salazar, R., Becerril-García, A., Baca-Zapata, G.: Presencia de universidades en la Red: La brecha digital entre Estados Unidos y el resto del mundo. Revista de Universidad y Sociedad del Conocimiento **6**(1), 1–17 (2009)
9. Bontempi, G., Ben Taieb, S., Le Borgne, Y.-A.: Machine learning strategies for time series forecasting. In: Aufaure, M.-A., Zimányi, E. (eds.) eBISS 2012. LNBIP, vol. 138, pp. 62–77. Springer, Heidelberg (2013). https://doi.org/10.1007/978-3-642-36318-4_3
10. Isasi, P., Galván, I.: Redes de Neuronas Artificiales. Un enfoque Práctico. Pearson (2004). ISBN 8420540250
11. Kulkarni, S., Haidar, I.: Forecasting model for crude oil price using artificial neural networks and commodity future prices. Int. J. Comput. Sci. Inf. Secur. **2**(1), 81–89 (2009)

12. Mazón, J.N., Trujillo, J., Serrano, M., Piattini, M.: Designing data warehouses: from business requirement analysis to multidimensional modeling. In: Proceedings of the 1st International Workshop on Requirements Engineering for Business Need and IT Alignment, Paris, France (2005)

13. Jain, A.K., Mao, J., Mohiuddin, K.M.: Artificial neural networks: a tutorial. IEEE Comput. **29**(3), 1–32 (1996)

14. Kuan, C.M.: Artificial neural networks. In: Durlauf, S.N., Blume, L.E. (eds.) The New Palgrave Dictionary of Economics. Palgrave Macmillan, Basingstoke (2008)

15. Mombeini, H., Yazdani-Chamzini, A.: Modelling gold price via artificial neural network. J. Econ. Bus. Manag. **3**(7), 699–703 (2015)

16. Parthasarathy, S., Zaki, M.J., Ogihara, M.: Parallel data mining for association rule on shared-memory systems. Knowl. Inf. Syst. Int. J. **3**(1), 1–29 (2001)

17. Sekmen, F., Kurkcu, M.: An early warning system for Turkey: the forecasting of economic crisis by using the artificial neural networks. Asian Econ. Financ. Rev. **4**(1), 529–43 (2014)

18. Sevim, C., Oztekin, A., Bali, O., Gumus, S., Guresen, E.: Developing an early warning system to predict currency crises. Eur. J. Oper. Res. **237**(1), 1095–11 (201)

19. Singhal, D., Swarup, K.S.: Electricity price forecasting using artificial neural networks. IJEPE **33**(1), 550–555 (2011)

20. Vasquez, C., Torres, M., Viloria, A.: Public policies in science and technology in Latin American countries with universities in the top 100 of web ranking. J. Eng. Appl. Sci. **12**(11), 2963–2965 (2017)

21. Vásquez, C., et al.: Cluster of the latin american universities top100 according to webometrics 2017. In: Tan, Y., Shi, Y., Tang, Q. (eds.) DMBD 2018. LNCS, vol. 10943, pp. 276–283. Springer, Cham (2018). https://doi.org/10.1007/978-3-319-93803-5_26

22. Viloria, A., Lis-Gutiérrez, J.P., Gaitán-Angulo, M., Godoy, A.R.M., Moreno, G.C., Kamatkar, S.J.: Methodology for the design of a student pattern recognition tool to facilitate the teaching – learning process through knowledge data discovery (big data). In: Tan, Y., Shi, Y., Tang, Q. (eds.) DMBD 2018. LNCS, vol. 10943, pp. 670–679. Springer, Cham (2018). https://doi.org/10.1007/978-3-319-93803-5_63

23. Prodromidis, A., Chan, P.K., Stolfo, S.J.: Meta learning in distributed data mining systems: Issues and approaches. In: Kargupta, H., Chan, P. (eds.) Book on Advances in Distributed and Parallel Knowledge Discovery. AAAI/MIT Press (2000)

24. Savasere, A., Omiecinski, E., Navathe, S.: An efficient algorithm for data mining association rules in large databases. In: Proceedings of 21st Very Large Data Base Conference, vol. 5, no. 1, pp. 432–44 (1995)

25. Stolfo, S., Prodromidis, A.L., Tselepis, S., Lee, W., Fan, D.W.: Java agents for metalearning over distributed databases. In: Proceedings of 3rd International Conference on Knowledge Discovery and Data Mining, vol. 5, no. 2, pp. 74–81 (1997)

Citescore of Publications Indexed in Scopus: An Implementation of Panel Data

Carolina Henao-Rodríguez[1]([⊠]), Jenny-Paola Lis-Gutiérrez[2],
Carlos Bouza[3], Mercedes Gaitán-Angulo[2], and Amelec Viloria[4]

[1] Corporación Universitaria Minuto de Dios, Bogotá, Colombia
linda.henao@uniminuto.edu
[2] Fundación Universitaria Konrad Lorenz, Bogotá, Colombia
{jenny.lis,mercedes.gaitana}@konradlorenz.edu.co
[3] Universidad de La Habana, La Habana, Cuba
bouza@matcom.uh.cu
[4] Universidad de la Costa, Barranquilla, Colombia
aviloria7@cuc.edu.co

Abstract. This article is intended to establish the variables that explain the behavior of the CiteScore metrics from 2014 to 2016, for journals indexed in Scopus in 2017. With this purpose, journals with a CiteScore value greater than 11 were selected in any of the periods, that is to say, 133 journals. For the data analysis, a model of standard corrected errors for panel was used, from which a coefficient of determination of 77% was obtained. From the results, it was possible to state that journals of arts and humanities; business; administration and accounting; economics, econometrics, and finance; immunology and microbiology; medicine and social sciences, have the greatest impact.

Keywords: CiteScore · Publications · Journals · Indexing · Scopus

1 Introduction

When considering the impact of publications, two approaches can be identified. The first one associated with the analysis of impact indicators of journals, and the second one related to the importance given by the institutions and researchers regarding the impact assessments. Among the empirical studies included in the first group, the one carried out by [1] shows the results of the implementation of a quantile regression for predicting a probability distribution to set the future number of quotations from a publication. In the same way, [2] reports that the publications of the first quarter of the year had a number of citations higher than the papers published in the last quarter. A similar finding associated with the month of publication was found by [3], for the citations during the two years following the publication.

For their part, [4] reported that self-citations of journals greatly affect the impact factors of publications at meso and micro levels, since it is a way of artificially increasing the impact assessment indicators. This fact severely undermines the authenticity of the indicators. [5] discerned on five types of citations: application, affirmation, negation, review and perfunctory mention. Their results encourage us to go

© Springer Nature Singapore Pte Ltd. 2019
Y. Tan and Y. Shi (Eds.): DMBD 2019, CCIS 1071, pp. 53–60, 2019.
https://doi.org/10.1007/978-981-32-9563-6_6

beyond the citation counts to assess the scholar scientific contribution of a paper, through the implementation of a panel data model with fixed effects. At the same time, [6], propose other impact assessments, based on the journal editor's data for evaluating the scholarly impact of an academic institution.

Within the second approach stands out the study of [7] who identified that researchers do not interpret the number of citations and the perceived impact to the same extent. Also, they stated that, in contexts of expertise in a specific area, there is a bias to prefer their own publications. [8], based on a survey applied to researchers in the United Kingdom and the development of two logit models, concluded that institutional factors, the contexts of intensive research, and non-academic work experience, explain the preference of researchers for high impact of their publications.

It is worth mentioning some other studies of multivariate techniques for measuring the impact of journals, such as [5, 6, 8, 9], among others.

This paper seeks to establish the variables explaining the behavior of the CiteScore from 2014 to 2016, for journals indexed in Scopus with a higher value in this indicator at 11. For this purpose, a panel data model was applied with CiteScore as dependent variable, and with the following explanatory variables: coverage years of the journal, languages in which the journal publishes, type of access to the journal, H-index of the journal origin country, and the dichotomous variables associated with the journal study area.

2   Method

2.1   Data

The data compiled to build the model were obtained from the Scopus Web Page in the following link: https://www.scopus.com/sources.uri?DGCID=Scopus_blog_post_check2015, and the Scopus Source List, available in October 2017. Firstly, a filter was applied to get a list of active publications and leaving the inactive ones out of the search. Secondly, a filter was applied regarding the type of source, choosing the "Journal" and "Trade Journal" options for this study.

2.2   Variables

An econometric exercise was conducted to analyze the CiteScore determinants. To avoid bias due to the heterogeneity of the individuals analyzed, the panel data econometric technique was applied for avoiding issues related to the identification of the models. The study period is between the years 2014 and 2016, and the technique was applied to 133 journals indexed in Scopus during 2017, with a CiteScore value greater than eleven, in any of the years of the mentioned period.

As mentioned above, the dependent variable used was the CiteScore, and the explanatory variables are: (i) The journal coverage years. A positive relationship with the CiteScore is expected since the journal has more years of coverage, so there will be a greater likelihood of citation. (ii) The journal coverage years. A positive relationship with the CiteScore is expected since the journal has more years of coverage, so there

will be a greater likelihood of citation. (iii) The languages in which the journal publishes. It was intended to confirm if there is any empirical evidence if the greater the
number of languages of publication, the more citations it has. (iv) A dichotomous
variable that specifies whether it is an open-access journal or not. In general, it is
expected that open-access journals are most often cited. (v) The H-index of the journal
origin country as a control variable. It is expected that the citation shows a positive
relationship with the quality and the impact of research in its environment.
(vi) Dichotomous variables associated with the study area of the journal: biological and
agricultural sciences; arts and humanities; biochemistry; genetics and molecular biology; business administration, and accounting; chemical engineering; chemistry; decision sciences; earth and planetary sciences; economics, econometrics and finance;
engineering, environmental sciences; immunology and microbiology; sciences of
materials; mathematics; medicine, neuroscience, nursing, pharmacology, toxicology
and pharmacy; psychology, social sciences, and health professions. The degree of
citation is expected to be affected by the science of study[1].

2.3 Model

The model is specified as follows:

$$
\begin{aligned}
\text{logciteScore}_{jt} = {} & \beta_0 + \beta_1 \text{logcobertura}_{jt} + \beta_2 \text{logidiomas}_{jt} + \beta_3 \text{open}_{jt} + \beta_4 \text{loghindex}_{it} + \beta_5 x_{it} \\
& + \beta_6 c_{it} + \beta_7 v_{it} + \beta_8 b_{jt} + \beta_9 n_{jt} + \beta_{10} m_{jt} + \beta_{11} h_{jt} + \beta_{12} f_{jt} + \beta_{13} g_{jt} + \beta_{14} k_{jt} + \beta_{15} d_{jt} \\
& + \beta_{16} s_{jt} + \beta_{17} a_{jt} + \beta_{18} q_{jt} + \beta_{19} w_{jt} + \beta_{20} e_{jt} + \beta_{21} r_{jt} + \beta_{22} t_{jt} + \beta_{23} u_{jt} + \beta_{24} i_{jt} + \beta_{25} qa_{jt} \\
& + \eta j + \delta t + \varepsilon_{jt}.
\end{aligned}
\tag{1}
$$

j corresponds to the journal; t represents the year; i is the journal origin country;
logciteScore is the logarithm of CiteScore; logcobertura is the logarithm of the journal
coverage years; logidiomas is the logarithm of the number of languages in which the
journal publishes; open is a dichotomous variable which specifies if it is open-access;
loghindex is the H-index logarithm; while the dichotomous variables that specify the
study area are: agricultural and biological sciences (x); arts and humanities (c); biochemistry; genetics, and molecular biology (v); business; administration and
accounting (b); chemical engineering(n); chemistry (m); decision sciences (h); earth
and planetary sciences (f); economics, econometrics and finance (g); engineering (k);
environmental sciences (d); immunology and microbiology (s); sciences of materials
(a); mathematics (q); medicine (w); neuroscience (e); nursing (r); pharmacology, toxicology and pharmacy (t); psychology (u); social sciences (i); health professions (qa);
δt represents the effects that vary with time not observed, ηj captures a common
deterministic trend and ε_{jt} is a random disturbance that is supposed $\varepsilon_{jt} \sim N(0, \sigma^2)$.

---

[1] The dichotomous variables specifying the following study areas were omitted: General; Computer
Science; Energy; Physics and Astronomy; Veterinary Medicine, and Odontology, since they caused
multicollinearity in the proposed model.

3   Results

To establish the best model, some tests were carried out, with the results shown below. When completing the Hausman test, the null hypothesis of the difference between the coefficients of random and fixed effects is rejected. Therefore, the estimation of fixed effects is performed.

Test: Ho: Difference in coefficients not systematic.
Chi2(1) = (b-B)'[(V_b-V_B)^(-1)](b-B) = 5.92
Prob > chi2 = 0.0149

When completing the Wooldridge test to detect autocorrelation, the null hypothesis that there is no first order autocorrelation at a significance level of 1% is rejected. Wooldridge test for autocorrelation in panel data.

H0: no first-order autocorrelation
F (1, 122) = 17.592
Prob > F = 0.0001

When completing the Wald test to detect if there is heteroskedasticity in the model, the null hypothesis of homoskedasticity at the significance level of 1% is rejected. Modified Wald test for groupwise heteroskedasticity in fixed effect regression model.

H0: sigma(i)^2 = sigma^2 for all i
Chi2 (127) = 7.4E + 07
Prob > chi2 = 0.0000

When completing the Pesaran test (2015) of cross-sectional dependence for not balanced panel data set; the null hypothesis cannot be rejected. Therefore, the errors of the cross-section are weakly dependent. Pesaran (2015) test for weak cross-sectional dependence unbalanced panel detected, test adjusted.

H0: errors are weakly cross-sectional dependent.
CD = 0,431
P-value = 0,667

The problems of contemporary correlation, heteroskedasticity and autocorrelation detected, can be solved along with estimates of standard corrected errors for panel. For this reason, the estimation was performed, and time dichotomous variables were included, due to the inability to directly capture the fixed effects (see Algorithm 1).

Algorithm 1. The Prais-Winsten regression; heteroskedastic panels corrected standard errors.

	logcitescore
logcobertura	-0.010
	(0.16)
logidiomas	1.042
	(2.06)*
open	-0.294
	(2.00)*
loghindex	0.299
	(3.69)**
x	0.127
	(1.88)
c	0.208
	(3.56)**
v	0.222
	(6.07)**
b	0.027
	(0.55)
n	0.090
	(0.24)
	(0.04)
u	-0.442
	(2.90)**
i	0.215
	(5.48)**
qa	-0.262
	(2.52)*
_Iyear_2015	0.026
	(2.24)*
_Iyear_2016	-0.039
	(0.22)
_cons	0.326
	(0.56)
R2	0.78
N	375

$* p<0.05; ** p<0.01$

The model of standard corrected errors for panel, has a determination coefficient of 78%. The significant variables in the model were the dichotomous variables that specify if the journal is open-access, and the following study areas: arts and humanities (c); biochemistry; genetics and molecular biology (v); engineering (k); economics, econometrics and finance (g); environmental sciences (d); science of materials (a);

medicine (w); neuroscience (e); nursing (r); psychology (u); social sciences (i); health professions (qa), and the dichotomous variable of time associated with the year 2015.

Given the above, empirical evidence shows that journals about arts and humanities; business; biochemistry; genetics and molecular biology, engineering, econometrics, and finance; medicine and social sciences, have the greatest impact, considering the assessment by CiteScore. In addition, the results show a negative relationship between the citation and the journals of environmental sciences; sciences of materials; neuroscience; nursing; psychology and health professions. This is consistent with the results shown by other authors. [10] presented the characteristics of the citations in the text in more than five million full-text articles from two databases (PMC, open-access subset and Elsevier journals) and found that the fields of biomedical and health sciences; life and earth sciences; and physics, science and engineering have similar reference distributions, although they vary in their specific details, while the fields of mathematics and informatics; and social sciences and humanities, have distributions of reference different from the other three.

In this context, [11] conclude that the average values of reference density in some categories of Social Sciences and Arts and Humanities, were equal to or higher than the "hard sciences" since the citations to the references occur at least with the same frequency in these two areas of knowledge, despite the potentially less impact of the journal.

Contrary to what was expected, if a journal is open-access, the number of citations does not increase, since the coefficient of the dichotomous variable "open" presents a negative sign that was not expected. Despite this, the finding is supported by other studies. For instance, [11], and [12] found no significant differences in the average values or the growth rates between Gold open-access and Non-Gold open-access journals, getting bibliometric and bibliographical data collections from 27.141 journals (indexed between 2001 and 2015 in the SCImago Journal & Country Rank (SJR)).

For their part, [13] conducted a study where they documented the growth in number of journals and articles, along with the increased rates of standardized citations of open-access journals listed in the Scopus bibliographic database, from 1999 to 2010. They concluded that the open-access journals and articles had grown faster than the subscription journals, but even so, they represented a very low percentage among the Scopus journals. [14] found that, by using a model of standard corrected errors for panel and a model of feasible generalized least squares, the open-access was not significant when considering the journals with an SJR greater than 8.

The variable "languages" is significant, meaning that the greater the number of languages in which the journal publishes, the greater the likelihood of citation. Finally, the H-index of the journal origin country is significant as a control variable and makes possible to confirm that the citation features a positive relationship with the quality and the impact of the research in its environment. According to the results found by [15], they argued that the number of citations is used to assess the impact of academic research or the quality of an academic department and reported that there are other important factors different from the journal, including the length of the article, the number of references, and the status of the first author's institution. Therefore, it is expected that the relevance of the research in the journal origin country is greater, due to a higher quality of the educational institutions.

4 Conclusions

The model of standard corrected errors for panel has a determination coefficient of 77%, and the significant variables in the model at 5% of the level of significance are the dichotomous variables that specify if the journal is open-access, and the following study areas: arts and humanities, business administration and accounting; economics, econometrics and finance; immunology and microbiology; sciences of materials; medicine; neuroscience; nursing; psychology; social sciences; health professions, and the dichotomous variable of time associated with the year 2015.

The model shows that open-access does not present the expected sign; therefore, it cannot be said that an open-access journal presents a greater impact indicator, confirming this way the findings of Sánchez, Gorraiz, and Melero (2018). In addition, empirical evidence shows that journals of arts and humanities; business; administration and accounting; economics, econometrics, and finance; immunology and microbiology; medicine and social sciences, have the greatest impact.

A similar analysis is suggested to be applied in future researches on journals indexed in WoS; as well as the use of other indicators as independent variable (like in the case of CiteScore, SNIP, among others); comparative analysis between different areas of knowledge is finally suggested in order to identify whether the findings are maintained.

References

1. Stegehuis, C., Litvak, N., Waltman, L.: Predicting the long-term citation impact of recent publications. J. Inform. **9**(3), 642–657 (2015)
2. Donner, P.: Effect of publication month on citation impact. J. Inform. **12**(1), 330–343 (2018)
3. Kosteas, V.D.: Journal impact factors and month of publication. Econ. Lett. **135**, 77–79 (2015). https://doi.org/10.1016/j.econlet.2015.08.010
4. Yu, T., Yu, G., Wang, M.Y.: Classification method for detecting coercive self-citation in journals. J. Inform. **8**(1), 123–135 (2014)
5. Stremersch, S., Camacho, N., Vanneste, S., Verniers, I.: Unraveling scientific impact: citation types in marketing journals. Int. J. Res. Mark. **32**(1), 64–77 (2015)
6. Wu, D., Li, J., Lu, X., Li, J.: Journal editorship index for assessing the scholarly impact of academic institutions: an empirical analysis in the field of economics. J. Inform. **12**(2), 448–460 (2018). https://doi.org/10.1016/j.joi.2018.03.008
7. Radicchi, F., Weissman, A., Bollen, J.: Quantifying perceived impact of scientific publications. J. Inform. **11**(3), 704–712 (2017)
8. Salter, A., Salandra, R., Walker, J.: Exploring preferences for impact versus publications among UK business and management academics. Res. Policy **46**(10), 1769–1782 (2017)
9. Lando, T., Bertoli-Barsotti, L.: Measuring the citation impact of journals with generalized Lorenz curves. J. Inform. **11**(3), 689–703 (2017). https://doi.org/10.1016/j.joi.2017.05.005
10. Boyack, K.W., Van Eck, N.J., Colavizza, G., Waltman, L.: Characterizing in-text citations in scientific articles: a large-scale analysis. J. Inform. **12**, 59–73 (2018). https://ac.els-cdn.com/S17511557717303516/1-s2.0-S17511557717303516-main.pdf?_tid=a31a573c-aa04-43b7-abe8-d38f230004be&acdnat=1523576882_830f710bace5ba4add6778c9ac486eba

11. Sánchez, S., Gorraiz, J., Melero, D.: Reference density trends in the major disciplines. J. Inform. **12**, 42–58 (2018). https://ac.els-cdn.com/S1751157717301487/1-s2.0-S17511577 17301487-main.pdf?_tid=919a6ae3-2f8c-45f6-b2a2-7d61f4927ab5&acdnat=1523577929_71 c9094e2952f12c35958d1d3b02495e

12. Zhang, L., Watson, E.M.: Measuring the impact of gold and green open-access. J. Acad. Librariansh. **43**(4), 337–345 (2017). https://doi.org/10.1016/j.acalib.2017.06.004

13. Solomon, D., Laakso, M., Björk, B.: A longitudinal comparison of citation rates and growth among open-access and subscription journals. J. Inform. **7**(3), 642–650 (2013)

14. Henao-Rodríguez, C., Lis-Gutiérrez, J.-P., Gaitán-Angulo, M., Malagón, L.E., Viloria, A.: Econometric analysis of the industrial growth determinants in Colombia. In: Wang, J., Cong, G., Chen, J., Qi, J. (eds.) ADC 2018. LNCS, vol. 10837, pp. 316–321. Springer, Cham (2018). https://doi.org/10.1007/978-3-319-92013-9_26. https://www.springerprofessional. de/en/data-mining-and-big-data/15834080

15. Mingers, J., Xu, F.: The drivers of citations in management science journals. Eur. J. Oper. Res. **205**(2), 422–430 (2010)

IBA-Buffer: Interactive Buffer Analysis Method for Big Geospatial Data

Ye Wu(✉)🆔, Mengyu Ma, Luo Chen, and Zhinong Zhong

College of Electronic Science, National University of Defense Technology,
Changsha 410073, China
yewugfkd@nudt.edu.cn

Abstract. Efficient buffer analysis over large geospatial datasets is of
great importance for Geographic Information System (GIS) applications.
Although modern GISs have introduced many optimizing techniques like
parallel processing, their performance deteriorates when the volume of
data increases greatly. Quick interactive analysis of such data is significant in order to make the most of them. This paper presents IBA-Buffer,
an interactive buffer analysis method for huge geospatial data that provides faster buffer analysis answers. The key features include (1) prune-
based strategies for buffer generation, and (2) parallel pipeline for a quick
response. Extensive experiments using real geospatial datasets show that
our algorithm outperforms traditional approaches by an order of magnitude and is superior to the current real-time algorithm for interactive
buffer analysis.

Keywords: Interactive analysis · Buffer analysis ·
Prune-base strategy · Geographic Information System

1 Introduction

Given the growing flood of location acquisition devices, a large amount of spatial
data is available from many different sources. However, the high complexity of
geospatial data and its operations make it difficult to perform online or interactive analyses. Fast response time queries are quite important for such tasks.

Buffer analysis is crucial for business marketing, impact analysis of public
facilities, site selection and many more. It is one of the fundamental operations because many spatial analyses are concerned with distance constraints.
For example, one wants to find a suitable location for housing. Given several
datasets about roads, rivers, parks and so on, he can "find the houses 500 m
from any park", "find the houses that are 1 km away from railway" and "find
a community with large supermarkets within 200 m", with the buffer analysis
method.

Supported by National Natural Science Foundation of China (Grant No. 41871284) and
National Natural Science Foundation of Hunan Province (Grant No. 2019JJ50718).

Given any set of objects, which may include points, lines, or areas, a buffer operation builds a new object or objects by identifying all areas that are within or without a certain specified distance of the original objects [8]. Euclidean buffers are the most common type of buffer, and work well when analyzing distances around features in a projected coordinate system. Unless otherwise stated, buffers in the paper are based on Euclidean distance, which can be easily extended to other kinds of types.

GIS systems are often used to visualize results for end-users to help decision-making processes. With the rapid development and popularity of World Wide Web, we are apt to use online maps to interact with and view the analysis process. Unfortunately, the large dataset size makes interactive buffer analysis for online maps impossible. Most of the existing techniques focus on using and returning geometry features [1,3,13], and propose several optimizations to improve the efficiency [11,14]. They mainly try to accelerate buffer zone for each geometric feature and self-intersection tackled among simple buffer zones. The calculation speed has increased by tens of times. However, when the scale of data increases sharply, all the aforementioned stand-alone algorithms fall into disaster.

Although plenty of methods based on parallel computing [5,6,10] and distributed computing framework [12,15], such as Spark [16] and Hadoop [2], are introduced to boost the performance of buffer generation, they can not meet the interacting requirements. The other class of methods focuses on raster-based buffer analysis by rasterizing vector data [4], then the buffers are calculated by raster image. However, the resolution is too low to be acceptable while zooming in due to blurring pixels.

This paper proposes a completely different approach, called IBA-Buffer (Interactive **B**uffer **A**nalysis **B**uffer), to the online buffer generation problem through in-memory indexing, hierarchical grid partition, and parallelization. The algorithm absorbs the advantages of both raster-based and vector-based techniques. The basic idea is to generate buffers from the aspect of visualization, rather than getting buffers from geometry features as usual. It also provides buffer results visualization supporting drilling down and rolling up with infinite precision.

The remaining of the paper is organized as follows. Section 2 contains a detailed description of the proposed method. We provide experimental results for large real GIS datasets that demonstrate the efficiency of our methods in Sect. 3. Finally, Sect. 4 concludes the paper.

2 Methodology

In this section, we present a procedural framework for interactive buffer analysis, followed by an explanation of the concept of our method, i.e., the prune-based buffer generation approach.

2.1   Parallel Processing Infrastructure

Figure 1 depicts the typical flow of interactive buffer analysis using R-tree indexes, which can be pre-built using fast bulk loading methods [7]. For online maps exploration, after parsing a received HTTP request from a front web browser, several map tile tasks are generated with square size (*e.g.* 256×256 pixels). It is desirable to cache all tasks in a pool. The parallel fetching unit fetches one task at a time from the task pool, decides what tiles to split (a tile is also divided into $n_p \times n_p$ grids for parallel acceleration purpose), and sends them to the R-tree index for pruning (detailed in Sect. 2.2). Finally, pixels at corresponding positions with *true* or *false* marking make up the resulting tile. It's also worth mentioning that, there are n tiles being processed simultaneously, although tasks are taken in an FIFO mode. Pruning pixels in a tile by distance is substantiated to be very efficient by our experiments, especially when the buffer size r is rather small or large compared with the real distance represented by the tile grid.

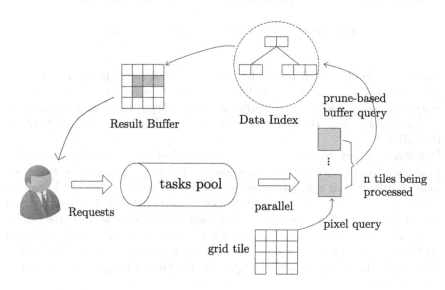

Fig. 1. Framework of interactive buffer analysis

2.2   Prune-Based Buffer Generation

The kernel idea of IBA-Buffer is the prune-based buffer generation, as illustrated in Fig. 2. Given a geometry dataset D and a visual window B, the basic steps of prune-based buffer generation include: (1) building a spatial index for D; (2) hierarchically partitioning B into equal size rectangles b_i; and (3) extending every pixel p in b_i to a circle with radius r and determining if it is in the buffer with approximation.

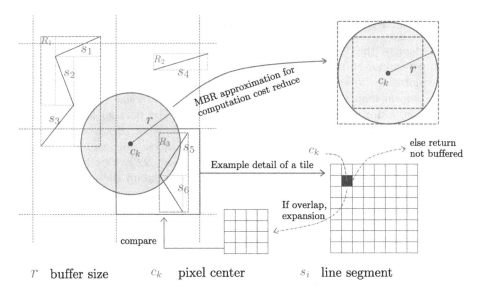

r buffer size c_k pixel center s_i line segment

Fig. 2. Illustration of prune-based buffer generation

- **Index**: Generally, the R-tree index is built on the input point dataset. For lines and polygons, in order to reduce search dead space, each line or polygon feature is broken into several segments. Then a segment is approximated by an MBR (**M**inimum **B**ounding **R**ectangle). and the R-tree index is built on the MBRs.
- **Partition**: Every tile contains $k = n \times n$ pixels. We need k intersection judgments using brute-force nested loops. To avoid redundant accesses, we make a hierarchical grid partition for the input tile. In detail, a tile can be decomposed into $m \times m$ smaller partitions, with size $(n/m) \times (n/m)$. Each time a partition is used to make intersection comparisons with R-tree nodes. If a grid element is not intersected with any node of the R-tree or falls into an R-tree node completely, it is returned immediately as part of the result. Otherwise, the partition will be expanded for further comparisons until the origin pixels.
- **Extending pixel**: As queries in R-tree are performed by rectangles. The inner and outer boxes are used to deal with different situations. In the situation where there are lots of spatial objects within the distance r from p, we use the inner box for intersection query, as a high density of spatial objects in the neighbor is very likely to intersect with the inner box. For the situation where there are few spatial objects in the neighbor of p, we use the outer box to filter the spatial objects which are far from p.

The parallel prune-based buffer generation algorithm is described in Algorithm 1.

Algorithm 1. Parallel prune-based buffer generation algorithm

---

Input   : buffer size r, bounding box of request tile B, geometry dataset D
Output: T: requested tile indicates the buffer area
1 **Algorithm** *BufferfGeneration(r,B,D)*
2   split bounding box B into $m \times m$ grids (satisfying $k = m^2 \times n^2$)
3   **forall** $b_i \in B$ *(i = 1, 2, $\cdots$, m^2)* **do**
4          $R = r + r(b_i)$ // $r(b_i)$ is the lenght of grid b_i
5   **if** $CenterBuffer((b_i).center.x, (b_i).center.y, R)$ **then**
6   **for** *pixel* $p \in b_i$ **do**
7                  $p = CenterBuffer(p.x, p.y, r)$
8   **else**
9              $p = false \ \forall p \in b_i$
10   combine every calculated p into tile T
    Result: T
11 **Procedure** *CenterBuffer(x, y, r, Rtree)*:
12     $r' = \frac{\sqrt{2}}{2}r$
13     $InnerBox = BOX(x - r', y - r', x + r', y + r')$
14   **if** $InnerBox.Intersects(Rtree)$ **then**
15   **return** *true*
16   **else**
17         $OuterBox = BOX(x - r, y - r, x + r, y + r)$
        // *rt* is any geometry in *Rtree*
18   **if** $OuterBox.Interserct(Rtree)$ and $Circle(x, y, r).Intersect(rt)$ **then**
19   **return** *true*
20   **return** *false*

---

3   Evaluation

In this section, we present experimental results of the *pruned-based interactive buffer analysis (IBA-Buffer)* algorithm. As in our previous work, *HiBuffer* [9] outperforms most of the state-of-the-art works, we only compare the performance of *IBA-Buffer* with that of *HiBuffer*. We present only the most illustrative subset of our experimental results due to space consideration. All the experiments are conducted on a symmetric multi-processor server with four 8-core Intel(R) Xeon(R) E5-4620@2.60GHz CPUs and totally 256GB main memory. The algorithm was implemented in C++ language and was compiled with GNU C++ compiler set for maximum optimization ($-O3$).

In our experiments, we consider real geometry datasets from Open Street Map[1] (**OSM**) detailed in Table 1.

---

[1] www.openstreetmap.org.

Table 1. Experimental dataset description

Dataset name	Abbreviation	# Records	# Size
Beijing roads	L_1	40,927	203,413 segments
Spain roads	L_2	3,132,496	42,497,196 segments
China roads	L_3	21,898,508	163,171,928 segments
Spain points	P_1	355,105	355,105 points
China points	P_2	20,258,450	20,258,450 points

Performance Comparison. To evaluate the performance of *IBA-Buffer*, we first compare it with that of *HiBuffer* on all the five datasets. We generate 10,000 tiles randomly, with different regions and zooming levels. The buffer size is 200 m and the grid size of *IBA-Buffer* is set to 16×16. In Fig. 3, we report the average time spent by each algorithm. Form the result, we can draw the following observation that *IBA-Buffer* takes about only 20%–80% time of *HiBuffer*, especially for point and sparse datasets. Note that the intersection detection by grid partitions of tile clips most of the non-buffered regions. Hence many of the unnecessary nearest neighbor or intersection operations are eliminated, resulting in much better performance.

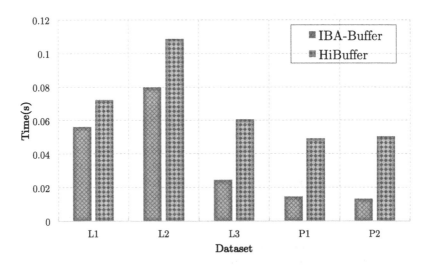

Fig. 3. Performance comparison of IBA-Buffer and HiBuffer

Varying Grid Size. We evaluated the performance of *IBA-Buffer* with different grid size: 8×8, 16×16, 32, 64×64 and the original 256×256. As depicted in Fig. 4, when the grid partition becomes finer, the computing time decreases gradually. However, when the partition reaches a certain extent, finer partitioning does not lead to performance improvements. The cause of this phenomenon is that finer partition itself increase additional computational cost.

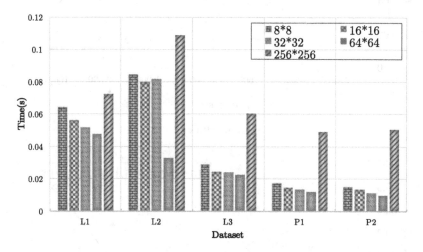

Fig. 4. Average time to generate a tile with different grid size

Varying Buffer Size. We now exhibit the execution time for *IBA-Buffer* (Fig. 5a) and *HiBuffer* (Fig. 5b) varying the buffer size from 100 m to 10 km. We calculated the average time of 10,000 tiles by each buffer size. For Fig. 5, there exhibit two completely opposite trends. With the increase of buffer, more pixels in tiles belong to buffers of spatial objects. As for *HiBuffer*, the average time of each tile shows general uptrends as more computation are required to generate the buffer tiles. Thanks to our pruning strategy, *IBA-Buffer* diminishes a lot of unnecessary comparisons when the majority of pixels fall into or outside the buffer. Therefore, the performance of *IBA-Buffer* shows a tendency to rise first and then down. Overall, compared with *HiBuffer*, the *IBA-Buffer* algorithm costs less time.

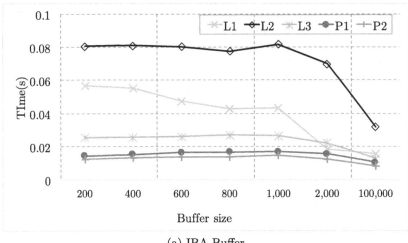

(a) IBA-Buffer

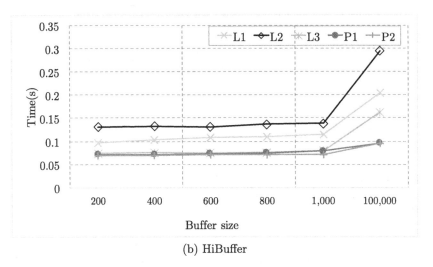

(b) HiBuffer

Fig. 5. Average time of *IBA-Buffer* and *HiBuffer* to generate a tile with different buffer size

4   Conclusion

Due to the large dataset size, it may take an extremely long time to compute the buffer of GIS data. The traditional buffer method does not provide any idea of how the final result will look like until the buffer is completed. Hence, we present an interactive, visualization-oriented buffer method to get fast result. To achieve better performance, we partition the tile into hierarchical grids to eliminate unnecessary intersection operations. Experimental evaluations on real datasets showed that our proposed *IBA-Buffer* method resulted in buffer analysis results with near zero response time.

References

1. Bhatia, S., Vira, V., Choksi, D., Venkatachalam, P.: An algorithm for generating geometric buffers for vector feature layers. Geo-Spat. Inf. Sci. **16**, 130–138 (2013)
2. Eldawy, A., Mokbel, M.F.: Spatialhadoop: a mapreduce framework for spatial data. In: 2015 IEEE 31st International Conference on Data Engineering, pp. 1352–1363, April 2015
3. Er, E., Kilinç, I., Gezici, G., Baykal, B.: A buffer zone computation algorithm for corridor rendering in GIS, pp. 432–435, September 2009. https://doi.org/10.1109/ISCIS.2009.5291855
4. Fan, J., He, H., Hu, T., Li, G., Qin, L., Zhou, Y.: Rasterization computing-based parallel vector polygon overlay analysis algorithms using OpenMP and MPI. IEEE Access **8**, 21427–21441 (2018). https://doi.org/10.1109/ACCESS.2018.2825452
5. Fan, J., Ma, T., Zhou, C., Ji, M., Zhou, Y., Xu, T.: Research on a buffer algorithm based on region-merging of buffered geometry components and its parallel optimization. Acta Geodaetica Cartogr. Sin. **43**(9), 969 (2014). https://doi.org/10.13485/j.cnki.11-2089.2014.0122
6. Huang, X., Pan, T., Ruan, H., Fu, H., Yang, G.: Parallel buffer generation algorithm for GIS. J. Geol. Geosci. **02**, 115 (2013). https://doi.org/10.4172/2329-6755.1000115
7. Leutenegger, S.T., Lopez, M.A., Edgington, J.: STR: a simple and efficient algorithm for r-tree packing. In: Proceedings 13th International Conference on Data Engineering, pp. 497–506, April 1997. https://doi.org/10.1109/ICDE.1997.582015
8. Longley, P.A., Goodchild, M.F., Maguire, D.J., Rhind, D.W.: Geographic Information Science and Systems. Wiley, Hoboken (2015)
9. Ma, M., Wu, Y., Luo, W., Chen, L., Li, J., Jing, N.: HiBuffer: buffer analysis of 10-million-scale spatial data in real time. ISPRS Int. J. Geo-Inf. **7**(12), 467 (2018). https://doi.org/10.3390/ijgi7120467
10. Pang, L., Li, G., Yan, Y., Ma, Y.: Research on parallel buffer analysis with grided based HPC technology. In: 2009 IEEE International Geoscience and Remote Sensing Symposium, vol. 4, pp. IV-200–IV-203, July 2009
11. Puri, S., Prasad, S.K.: A parallel algorithm for clipping polygons with improved bounds and a distributed overlay processing system using MPI. In: 2015 15th IEEE/ACM International Symposium on Cluster, Cloud and Grid Computing, pp. 576–585, May 2015. https://doi.org/10.1109/CCGrid.2015.43
12. Shen, J., Chen, L., Wu, Y., Jing, N.: Approach to accelerating dissolved vector buffer generation in distributed in-memory cluster architecture. ISPRS Int. J. Geo-Inf. **7**(1), 26 (2018). https://doi.org/10.3390/ijgi7010026
13. Žalik, B., Zadravec, M., Clapworthy, G.J.: Construction of a non-symmetric geometric buffer from a set of line segments. Comput. Geosci. **29**(1), 53–63 (2003)
14. Yao, Y., Gao, J., Meng, L., Deng, S.: Parallel computing of buffer analysis based on grid computing. Geospatial Inf. **1**, 35 (2007)
15. Yu, J., Zhang, Z., Sarwat, M.: Spatial data management in apache spark: the geospark perspective and beyond. GeoInformatica **23**(1), 37–78 (2019)
16. Zaharia, M., Chowdhury, M., Franklin, M.J., Shenker, S., Stoica, I.: Spark: cluster computing with working sets (2010). http://dl.acm.org/citation.cfm?id=1863103.1863113

Semi-automated Augmentation
of Pandas DataFrames

Steven Lynden[1][✉] and Waran Taveekarn[2]

[1] Artificial Intelligence Research Center, AIST Tokyo Waterfront, Tokyo, Japan
steven.lynden@aist.go.jp
[2] Faculty of Information and Communication Technology, Mahidol University,
Salaya, Thailand
waran.tav@gmail.com

Abstract. Creative feature engineering is an important aspect within machine learning prediction tasks which can be facilitated by augmenting datasets with additional data to improve predictions. This paper presents an approach towards augmenting existing datasets represented as pandas dataframes with data from open data sources, semi-automatically, with the aims of (1) automatically suggesting data augmentation options given an existing set of features, and (2) automatically augmenting the data when a suggestion is selected by the user. This paper demonstrates the performance of the approach in terms of aligning typical machine learning datasets with open data sources, suggesting useful augmentation options, and the design and implementation of a software tool implementing the approach, available as open-source software.

Keywords: Open data · Semantic Web · Data enrichment

1 Introduction

Creative feature engineering has the potential to benefit data scientists involved in various machine learning tasks, where complementing existing datasets with additional attributes or properties is beneficial. A large amount of open data sources exist with the potential to achieve this, for example, large cross-domain knowledge bases such as Wikidata [13], in addition to many domain specific databases such as OpenStreetMap [7]. Interest in performing data science and machine learning tasks using Python tools such as pandas [6] and Scikit learn [9] has exploded in recent years due to the simple, time-saving, intuitive way in which such tools can be utilized to achieve impressive results. In this paper we describe a tool developed in order to be seamlessly integrated with such tools to semi-automate the task of performing creative feature engineering using open data sources to potentially improve the performance of machine learning algorithms. Selecting useful data with which to augment datasets is challenging and time consuming, so we introduce an approach towards automatically suggesting augmentations that are of use in improving typical machine learning

© Springer Nature Singapore Pte Ltd. 2019
Y. Tan and Y. Shi (Eds.): DMBD 2019, CCIS 1071, pp. 70–79, 2019.
https://doi.org/10.1007/978-981-32-9563-6_8

performance. The motivation for our approach is to provide a fast, efficient way of augmenting typical pandas dataframes to provide an alternative to existing data wrangling tools, with a focus on:

– Suggesting augmentations. Rather than making it necessary to browse and select potential data with which to augment, suggest augmentations likely to be of use.
– Speed. Make the suggestion and augmentation process as fast and easy to perform as possible.
– Seamless integration. Focus on integration with popular Python tools which are currently widely used.

The remainder of this paper is structured as follows. Section 2 discusses the background and motivation for this work. Subsequently, Sect. 3 presents the design and implementation of the proposed approach. Section 4 presents example usage scenarios and an evaluation of the benefits of the approach, followed by a discussion of related work in Sect. 5 and concluding remarks in Sect. 6.

2   Background

Recent years have seen a rapid increase in the volume of open data available via various government initiatives and collaboratively edited community-based efforts. Such data sources include domain-specific sources such as meteorological (e.g. the Japan Meteorological Agency), geographical (e.g. OpenStreetMap) and general-purpose knowledge bases such as DBpedia and Wikidata. Furthermore, access to such data sources has been facilitated by interoperable web service APIs via the wide-scale adoption of the REST protocol and the output of the Semantic Web effort to represent data using standards such as the Research Description Framework (RDF).

Recent interest in artificial intelligence and machine learning is exemplified by data science competition communities such as Kaggle[1]. Such communities encourage the sharing of notebooks, which provide an insight into the tools and methods frequently employed in such tasks. Python is widely used along with the data manipulation and analysis tool pandas. This is reinforced by the 2018 figure eight Data Scientist Report [1] which surveyed data scientists and concluded pandas to be the most widely adopted software framework in use by data scientists.

Machine learning and data mining tasks involve, among other tasks, making predictions based on data. Such prediction tasks include regression and classification, where the process of feature engineering involves modeling domain-specific data available in a way that it can be utilised by machine learning algorithms effectively. Sometimes it may be desirable to have move features available, for example, in a recent Kaggle Competition[2] on Web traffic forecasting, some entries combined the competition dataset with social media and open data sources to achieve better results.

---

[1] https://www.kaggle.com/.
[2] https://www.kaggle.com/c/web-traffic-time-series-forecasting.

While manually enhancing feature sets with alternative data can be time consuming, the possibility exists of achieving it in an automated, or semi-automated way by using open data sources. Assuming a prediction task relates to entities for which data exists in knowledge bases such as Wikidata, linking column values to the entities and adding additional features is possible, for example as illustrated in Fig. 1 where it is assumed that the ubiquity of mobile phones is being predicted given a country name, region and population using the data in the top-left portion of the figure. Data on each country's Human Development Index (HDI) is a feature with the potential to improve the performance of typical machine learning algorithms to make predictions in this scenario, as illustrated by the additional column added to the data in the bottom-left part of the figure from Wikidata (shown on the right).

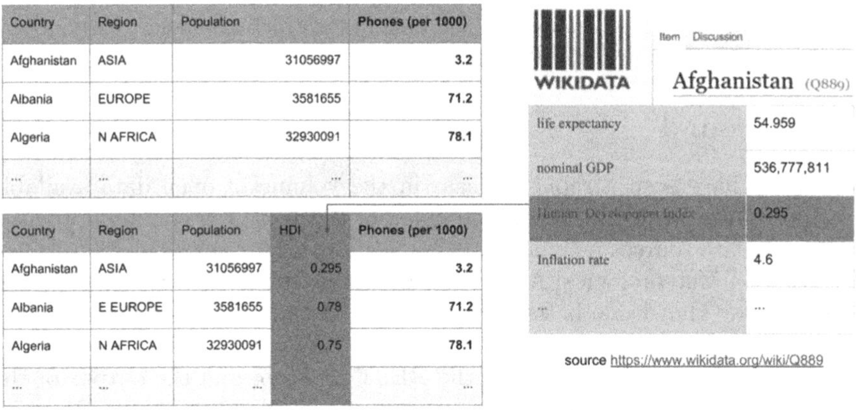

Fig. 1. Augmentation using Wikidata.

3 Design and Implementation

The software provides the following two operations:

- suggest: Takes a pandas dataframe and column number as an argument and returns a dataframe containing ordered suggested augmentations.
- augment: Allows the augmentation of the user's original dataframe by specifying one of the aforementioned suggestions and the column to augment.

Both operations are designed to be compatible with multiple open data sources, however currently only support for Wikidata has been developed so far. The design and implementation of these two operations are now discussed. Design decisions were motivated by the following factors:

1. Aim of making both operations fast.
2. Aim of making the storage space required for the tool small.
3. Seamless integration with pandas.

3.1   The suggest Function

The **suggest** operation requires examining values in a column of a dataframe, resolving column values to determine if they match entities in a knowledge base, followed by generating possible properties with which the column can be augmented and returning them in order of likely usefulness within a feature engineering task. The **suggest** operation requires linking column values in a dataframe with data from open data sources to construct augmentation options, and subsequently ranking the augmentation options to present ones that are likely to be of use as features prominently.

Linking to Open Data Sources. Joining records between data sources, referred to as record linkage, data matching, entity recognition, reconciliation etc. depending on the context, is a well-studied problem and various web services exist suitable for the purpose of linking entities to open data sources. However, calls to remote web services can be time-consuming or rate-limited so the **suggest** operation is designed with a hybrid approach, attempting first to link columns locally using a space-efficient probabilistic data structure consisting of multiple Bloom Filters [2] which are capable of recognising frequently occurring labels in data sources, e.g. country names, people names, and various other identifiers. For many use-cases, this suffices however it is less robust than some reconciliation services such as the OpenRefine reconciliation API[3]. Therefore if linking locally fails the **suggest** operation will attempt to link columns using such APIs.

Ranking Augmentation Options. For a given dataframe column, numerous augmentation options may be available meaning that it is necessary to filter out those that are not fit-for-purpose with respect to feature engineering and to rank augmentations that are likely to be of benefit higher than those less likely to be useful. The problem here is that it is difficult to know which features are likely to be of use prior to augmenting the dataset, and it can be prohibitively time-consuming to perform all augmentations. The ranking of suggested augmentations is therefore based on characteristics that can be known in advance or effectively sampled for the data source in question. In the case of Wikidata, the following characteristics of each property are utilized to construct a feature vector:

1. Facet: Wikidata properties are organised into a heirarchy[4], allowing the construction of an element in the feature vector representing what the property pertains to, for example science, art, etc.
2. Frequency: how frequently the property occurs in the Wikidata knowledge base.

---

[3] https://github.com/OpenRefine/OpenRefine/wiki/Reconciliation-Service-API.
[4] https://www.wikidata.org/wiki/Wikidata:List_of_properties.

3. Cardinality: the median number of occurrences of a property if it exists for a given type of entity. For example the for entities of type country, the currency property is equal to 1 as almost every country has one currency.
4. Variance: the average variance of property values.
5. Sub-properties: the number of sub-properties that a property has.
6. Parent-properties: a list of the parent properties that the property has.
7. Datatype: the datatype of the property.

The number of distinct properties in Wikidata numbers in the thousands, meaning that the above information about each property can be obtained from Wikidata and stored locally without requiring prohibitively large storage space for installation of the software tool. The **suggest** operation utilizes a predictive model constructed by obtaining a set of machine learning datasets from Kaggle and DataHub, performing the possible augmentations, and training a model, utilizing k-nearest neighbors (KNN) regression, to estimate the ratio of information gain [10], as typically used in decision tree algorithms, of a given augmentation based on the aforementioned characteristics. Categorical values, such as the datatype and facet are converted into numerical values using a label encoder. The predictive model trained using 540 augmentations based on sample datasets from Kaggle and Datahub results in an R^2 [11] (coefficient of determination) score of 60%(0.6) when predicting information gain (using a training dataset of 80% of the augmentations and a 20% test set to evaluate and compute the R^2 score). This indicates that reasonable predictions can be made that are of use in predicting information gain from the constructed feature vector, as R^2 values range between 0 and 1, with 1 being perfectly accurate predictions. The utility of this approach towards ranking suggestions is presented in the subsequent Sect. 4. The suggest operation can be utilized as follows:

```
df = pd.read_csv("dataset.csv")
suggest_df = suggest(df,1)
```

where the second argument taken by the **suggest** function is a column number from the input dataframe and the output is a dataframe with the ranked augmentation options, as illustrated in Fig. 2 allowing the user to quickly select interesting features.

Property	Label	Datatype	Frequency	Cardinality	Variance
P345	IMDb ID	wikibase-item	1	1	1
P495	country of origin	wikibase-item	1	1.05	0.19
P577	publication	time	1	1	0.85
P161	cast member	wikibase-item	0.9	12.7	0.85
P57	director	wikibase-item	0.8	1.05	0.83

Fig. 2. Example suggestion dataframe

3.2   The augment Function

Performing the **augment** operation on a column involves querying the open data source to obtain the additional properties using a SPARQL endpoint or other web service API. For example, Wikidata provides a SPARQL endpoint which can be queried directly or alternative options, such as the OpenRefine Data Extension API[5] may be utilized. Where multiple options exist this may be specified in a configuration file utilized by the package. Example usage of the augment operation is as follows:

```
df1 = augment(df,"P495",suggest_df)
```

The result of the operation is a new dataframe with an additional column labeled with the property with which the dataframe has been augmented.

4   Evaluation

In order to evaluate the benefit of the approach, three datasets are utilized:

1. https://datahub.io/world-bank/sh.tbs.incd
   (Countries-TB) from DataHub, prediction of the rates of tuberculosis per country.
2. https://www.kaggle.com/mauryashubham/linear-regression-to-predict-market-value/data
   (EPL-market-value) from Kaggle, prediction of the market values of soccer players in the English Premier League.
3. https://archive.ics.uci.edu/ml/datasets/Movie
   (movie-ratings) from the UCI Machine Learning Repository, prediction of Internet Movie Database (IMDb) ratings of various movies.

A machine learning process is employed as follows, representing a typical workflow that often used in Kaggle workbooks:

1. Perform each possible augmentation suggested by the **suggest** operation.
2. Split the dataset into train (80%) and test (20%) sets.
3. Train using the KNN regression module from Scikit Learn, with all configuration parameters set to their default, meaning the number of neighbors (K) = 5.
4. Evaluate the difference in predicting the target variable in each dataset by comparing the R^2 values of predictions made with and without the augmentation made in step 1, labeled as "improvement" in Tables 1, 2 and 3 for each of the datasets utilized. Augmentations that improve prediction performance are highlighted in these tables.

Results show that the ranking approach suggested top three augmentation options that are able to improve machine learning performance using KNN in all three datasets. The potential of the approach is clearly demonstrated, with at least some improvement in 45 out of the 60 suggested augmentations in the explored datasets.

---

[5] https://github.com/OpenRefine/OpenRefine/wiki/Data-Extension-API.

Table 1. Countries TB dataset

	Label	Property	Improvement
1	currency	P38	0.0442137
2	replaces	P1365	0.0464215
3	continent	P30	0.0155412
4	central bank	P1304	-0.0265574
5	part of	P361	0.0252332
6	described by source	P1343	0.0155412
7	award received	P166	-0.0155342
8	category for films shot at this location	P1740	0.0087858
9	number of out of school children	P2573	0.0425534
10	follows	P155	-0.0025082
11	located in or next to body of water	P206	-0.0093862
12	mains voltage	P2884	-0.0231408
13	head of government	P6	-0.0606893
14	official language	P37	-0.0681875
15	executive body	P208	-0.0352012
16	basic form of government	P122	-0.022289
17	ethnic group	P172	-0.035625
18	official website	P856	-0.1055928
19	inception	P571	-0.1067783
20	head of state	P35	-0.1786623

5   Related Work

Various data wrangling tools exist such as Trifacta[6] which emerged from the Stanford Data Wrangler [4], generally such tools offer graphical user interface-based approaches towards fixing errors and in some cases annotating or augmenting data sets with additional information. OpenRefine [12] supports various data wrangling functionality including the utilization of open data sources to augment data, which can be used, for example, to semi-automate data profiling as demonstrated in [5]. In terms of usage, OpenRefine offers a web browser-based interface or the option to invoke REST API calls directly, and a third-party Python library[7] is available. FeGeLod [8], eventually incorporated into RapidMiner [3], also proposes feature generation using Linked Open Data. The approach used involves a label-based linker that looks for entities in the knowledge base that have labels identical to fields in a given dataset and allows the user to select additional attributes with which the existing dataset is augmented via the use of SPARQL queries issued to the knowledge base endpoint. The principle difference between the approaches described above and the proposal presented in this paper is that the proposed tool supports ranking of the proposed augmentations and easy-to-use integration with pandas.

---

[6] https://www.trifacta.com/.
[7] https://github.com/maxogden/refine-python.

Table 2. EPL player market value dataset

	Label	Property	Improvement
1	child	P40	0.1805309
2	inception	P571	0.1794941
3	coordinate location	P625	0.1805309
4	feast day	P841	0.1805309
5	different from	P1889	-0.8589553
6	work location	P937	0.1805309
7	sport number	P1618	0.1146113
8	official website	P856	0.1682713
9	medical condition	P1050	0.1805309
10	sexual orientation	P91	0.1805309
11	sport	P641	0.1805309
12	family name	P734	0.3403272
13	date of birth	P569	0.204892
14	languages spoken written or signed	P1412	0.1871929
15	topic's main category	P910	0.1805309
16	employer	P108	0.1805309
17	religion	P140	0.1805309
18	member of	P463	0.1805309
19	member of political party	P102	0.1805309
20	instrument	P1303	0.1805309

Table 3. Movies ratings dataset

	Label	Property	Improvement
1	duration	P2047	0.3263041
2	publication date	P577	0.228311
3	country of origin	P495	0.1292513
4	original language of work	P364	0.1219773
5	cast member	P161	0.1132523
6	color	P462	0.1233706
7	filming location	P915	0.1563969
8	genre	P136	0.0830894
9	narrative location	P840	0.0405694
10	RTC film rating	P3834	0.0819171
11	producer	P162	0.2061141
12	screenwriter	P58	0.2487206
13	director of photography	P344	0.0355535
14	production designer	P2554	0.13002
15	composer	P86	0.2937582
16	sport	P641	0.1251002
17	director	P57	0.0055829
18	production company	P272	0.0885954
19	distributor	P750	-0.0413285
20	Prisma ID	P4515	-0.8116681

6   Conclusions

In this paper an approach has been described for augmenting pandas dataframes for efficient support of creative feature engineering in typical machine learning prediction tasks. The design and implementation of a tool implementing the approach is presented, showing that augmentations with the potential to improve the performance of such prediction tasks can be efficiently recommended to users using a predictive model, so far only demonstrated in the case of Wikidata, however we believe the approach can be generalized to other data sources.

We believe the approach can offer an alternative to data wrangling tools in scenarios where it is necessary to quickly and efficiently recommend augmentation options directly to data scientists and allowing them to select and then perform the augmentations. The advantages of the proposed approach include:

- Only semi-reliant on invoking remote APIs for matching data with Wikidata entities, which enables faster performance for many datasets.
- Ranking of suggestions to avoid looking for a needle in a haystack when it comes to what data to augment with.
- Integration with widely-used tools to quickly augment datasets.

In future work, use of more open data sources, and further evaluation of the approach is planned. The tool will serve as the basis for investigating the potential of more sophisticated augmentation options, for example using neural network autoencoders to perform augmentations. The described tool is available at https://gitlab.com/slynden/aug_df in the form of an open source software tool.

Acknowledgment. This work is supported by the New Energy and Industrial Technology Development Organization (NEDO), Japan.

References

1. Data Scientist Report (2018). https://www.datasciencetech.institute/wp-content/uploads/2018/08/Data-Scientist-Report.pdf
2. Bloom, B.H.: Space/time trade-offs in hash coding with allowable errors. Commun. ACM **13**(7), 422–426 (1970)
3. Hofmann, M., Klinkenberg, R.: RapidMiner: Data Mining Use Cases and Business Analytics Applications. Chapman & Hall/CRC, Boca Raton (2013)
4. Kandel, S., Paepcke, A., Hellerstein, J., Heer, J.: Wrangler: interactive visual specification of data transformation scripts. In: Proceedings of the SIGCHI Conference on Human Factors in Computing Systems, CHI 2011, pp. 3363–3372. ACM, New York (2011). https://doi.org/10.1145/1978942.1979444
5. Kusumasari, T.F., Fitria: Data profiling for data quality improvement with openrefine. In: 2016 International Conference on Information Technology Systems and Innovation (ICITSI), pp. 1–6, October 2016. https://doi.org/10.1109/ICITSI.2016.7858197
6. McKinney, W.: Data structures for statistical computing in python. In: van der Walt, S., Millman, J. (eds.) Proceedings of the 9th Python in Science Conference, pp. 51–56 (2010)

7. OpenStreetMap contributors: Planet dump (2018). https://planet.osm.org. https://www.openstreetmap.org
8. Paulheim, H.: Exploiting linked open data as background knowledge in data mining. In: Proceedings of the 2013 International Conference on Data Mining on Linked Data, DMoLD 2013, pp. 1–10, vol. 1082. CEUR-WS.org, Aachen (2013). http://dl.acm.org/citation.cfm?id=3053776.3053778
9. Pedregosa, F., et al.: Scikit-learn: machine learning in Python. J. Mach. Learn. Res. **12**, 2825–2830 (2011)
10. Quinlan, J.R.: Induction of decision trees. Mach. Learn. **1**(1), 81–106 (1986). https://doi.org/10.1023/A:1022643204877
11. Skovgaard, L.: Applied regression analysis. Stat. Med. **19**(22), 3136–3139 (1998)
12. Verborgh, R., Wilde, M.D.: Using OpenRefine, 1st edn. Packt Publishing, Birmingham (2013)
13. Vrandečić, D., Krötzsch, M.: Wikidata: a free collaborative knowledgebase. Commun. ACM **57**(10), 78–85 (2014). https://doi.org/10.1145/2629489

The Temporal Characteristics of a Wandering Along Parallel Semi-Markov Chains

Tatiana A. Akimenko$^{(\boxtimes)}$ and Eugene V. Larkin$^{(\boxtimes)}$

Tula State University, 92 "Lenina" prospect, Tula 300012, Russia
{tantan72, elarkin}@mail.ru

Abstract. A general method is proposed for calculating the probabilistic and temporal characteristics of a walk through a complex semi-Markov process, formed by combining M ordinary semi-Markov processes, based on the fact that a semi-Markov matrix is transformed into a characteristic matrix, which is raised to a power, then the degrees of the characteristic matrix are summed.

Keywords: Semi-Markov matrix ·
Temporal and probabilistic characteristics of wanderings · Matrices ·
Semi-Markov process

1 Introduction

Evaluation of the temporal and probabilistic characteristics of walks along parallel semi-Markov chains is the main task solved in the study and development of optimal cyclograms of digital control of mobile robots groups. The initial information for the calculation is the semi-Markov matrix $\boldsymbol{h}_\Sigma(t)$, formed as a result of the Cartesian product of semi-Markov matrices of the components of ordinary semi-Markov processes [1–7].

One of the determining factors for ensuring the sustainable functioning of the MR in solving target problems is the time factor, which must be taken into account and planned when developing a digital control system.

The proposed method will allow scheduling the operation of digital control systems, defining the requirements for temporary inter-shafts when coordinating the operation of digital control loops and developing programs for the operation of a mobile robot laid down at a strategic-level computer.

2 General Method

The main operation performed on the semi-Markov matrix [7]

$$\boldsymbol{h}_\Sigma(t) = \left\lfloor \mathrm{h}_{\mathrm{m}(\Sigma),\mathrm{n}(\Sigma)}(t) \right\rfloor \tag{1}$$

where $\mathrm{h}_{\mathrm{m}(\Sigma),\mathrm{n}(\Sigma)}(t)$ - element of semi-Markov matrix; $1(\Sigma) \leq \mathrm{m}(\Sigma)$, $\mathrm{n}(\Sigma) \leq \mathrm{N}(\Sigma)$.

Y. Tan and Y. Shi (Eds.): DMBD 2019, CCIS 1071, pp. 80–89, 2019.
https://doi.org/10.1007/978-981-32-9563-6_9

Elements $n(\Sigma) \times n(\Sigma)$ semi-Markov matrix are defined as follows:

$$h_{m(\Sigma),n(\Sigma)}(t) = p_{m(\Sigma),n(\Sigma)} \cdot f_{m(\Sigma),n(\Sigma)}(t) \qquad (2)$$

where $p_{m(\Sigma),n(\Sigma)}$ - element of the stochastic matrix of a complex semi-Markov process; $f_{m(\Sigma),n(\Sigma)}(t)$ - pure density matrix element.

Matrices, stochastic and pure distribution densities, are defined as

$$\boldsymbol{p_\Sigma} = \left\lfloor p_{m(\Sigma),n(\Sigma)} \right\rfloor; \qquad (3)$$

$$\boldsymbol{f_\Sigma}(t) = \left\lfloor f_{m(\Sigma),n(\Sigma)}(t) \right\rfloor. \qquad (4)$$

We assume that the states of the combined semi-Markov process can only be absorbing and non-absorbing (there are no semi-absorbing states).

For non-absorbing states $m(\Sigma)$, the following properties are valid:

$$\sum_{n(\Sigma)=1(\Sigma)}^{N(\Sigma)} p_{m(\Sigma),n(\Sigma)} = 1$$
$$0 \leq \arg\left\lfloor f_{m(\Sigma),n(\Sigma)}(t) \right\rfloor < \infty, 1(\Sigma) \leq n(\Sigma) \leq N(\Sigma) \qquad (5)$$

For absorbing states $m(\Sigma)$, the following properties are valid:

$$\sum_{n(\Sigma)=1(\Sigma)}^{N(\Sigma)} p_{m(\Sigma),n(\Sigma)} = 0;$$
$$f_{m(\Sigma),n(\Sigma)}(t) = \lim_{T_{m(\Sigma),n(\Sigma)} \to \infty} \left\lfloor \delta\left(t - T_{m(\Sigma),n(\Sigma)}\right) \right\rfloor, \ 1(\Sigma) \leq n(\Sigma) \leq N(\Sigma), \qquad (6)$$

where $\delta\left(t - T_{m(\Sigma),n(\Sigma)}\right)$ - δ- Dirac function.

The main task of determining the parameters of a walk along a semi-Markov chain is the problem of determining the weighted density of the time distribution of a walk from the state $m(\Sigma)$ in state $n(\Sigma)$. This task can be solved only for the case $m(\Sigma)$ - non-absorbing state. The only limitation on the trajectory when determining the wander time is that $m(\Sigma)$, nor to the state $n(\Sigma)$ the process should not fall twice. In order to satisfy this constraint, state $m(\Sigma)$ must receive the status of the start state, and state $n(\Sigma)$ must receive the status of absorbing one. To do this, the semi-Markov matrix $\boldsymbol{h_\Sigma}(t)$ must be deleted references to state $m(\Sigma)$ in all rows, the matrix, and from the line defining state $n(\Sigma)$ all references to all states of matrix $\boldsymbol{h_\Sigma}(t)$ should be removed, i.e. the $m(\Sigma)$-th column and the $n(\Sigma)$-th row should be cleared. Denote the specified transformation as follows.

$$\boldsymbol{h_\Sigma}(t) \to \boldsymbol{h'_\Sigma}(t) \qquad (7)$$

In order to obtain the required distribution density, it is necessary to stochastically add up the distribution densities, which are learned from wandering along all possible paths.

$$^{\Sigma}h'_{m(\Sigma),n(\Sigma)}(t) = {}^{r}\boldsymbol{I}_{m(\Sigma)} \cdot L^{-1}\left[\sum_{k=1}^{\infty} \{L[\boldsymbol{h}'_{\Sigma}(t)]\}^{k}\right] \cdot {}^{c}\boldsymbol{I}_{n(\Sigma)}, \tag{8}$$

where $1\ {}^{r}\boldsymbol{I}_{m(\Sigma)}$ - the row vector, the $m(\Sigma)$-th element of which is equal to one, and the remaining elements are equal to zero; ${}^{c}\boldsymbol{I}_{n(\Sigma)}$ - column vector, the $n(\Sigma)$-th element of which is equal to one, and the remaining elements are equal to zero; $L[\ldots]$ and $L^{-1}[\ldots]$- respectively, the direct and inverse Laplace transform, introduced to replace the matrix convolution operation with the multiplication operation of the characteristic matrices formed from $\boldsymbol{h}'_{\Sigma}(t)$.

The investigated semi-Markov process is described by the matrix $\boldsymbol{h}_{\Sigma}(t)$ of the most general form, therefore in the matrix $\boldsymbol{h}'_{\Sigma}(t)$ formed as a result of performing operation (7) there can be other absorbing states, besides the artificially created state $n(\Sigma)$. It follows from this that not all trajectories of a walk through a complex semi-Markov process, which begin in state $m(\Sigma)$, end in state $n(\Sigma)$. [8] This, in turn, means that the group of events attaining state $n(\Sigma)$ of state $m(\Sigma)$ is not understood, and therefore, in the general case, $h'_{m(\Sigma),n(\Sigma)}(t)$ has the character of a weighted rather than pure distribution density. The probability and net density of the time to reach state $n(\Sigma)$ from state $m(\Sigma)$ is defined as

$$^{\Sigma}p'_{m(\Sigma),n(\Sigma)} = \int_{0}^{\infty} {}^{\Sigma}h'_{m(\Sigma),n(\Sigma)}(t)dt; \tag{9}$$

$$^{\Sigma}f'_{m(\Sigma),n(\Sigma)}(t) = \frac{{}^{\Sigma}h'_{m(\Sigma),n(\Sigma)}(t)}{{}^{\Sigma}p'_{m(\Sigma),n(\Sigma)}}. \tag{10}$$

Another characteristic of a walk along an integrated semi-Markov chain, which requires a slightly different approach than the above, is the time and probability of the process returning to a state, for example, to $m(\Sigma)$ [8–16]. To determine the specified parameters $m(\Sigma)$ is divided into two: $m(\Sigma, b)$ and $m(\Sigma, e)$. State $m(\Sigma, b)$ is given the status of the starter, and state $m(\Sigma, e)$ is given the status of absorbing. Taking into account the fact that during wanderings the process should not fall into either state $m(\Sigma, b)$ or state $m(\Sigma, e)$ twice, matrix $\boldsymbol{h}_{\Sigma}(t)$ should be transformed as follows:

one row and one column should be added to the matrix, and so it increases in size with $N(\Sigma) \times N(\Sigma)$ to $[N(\Sigma) + 1] \times [N(\Sigma) + 1]$;

state $m(\Sigma)$ should be assigned the status of the start state, and state $N(\Sigma) + 1$ should be assigned the status of absorbing one, i.e.

$$m(\Sigma) = m(\Sigma, b); \tag{11}$$

$$N(\Sigma) + 1 = m(\Sigma, e); \tag{12}$$

due to (11), the column of matrix $\boldsymbol{h}_{\Sigma}(t)$ with the number $m(\Sigma)$ must be transferred to the column with the number $N(\Sigma) + 1$, and the column with the number $m(\Sigma)$ must be filled with zeros;

due to (12), the added line with the number $N(\Sigma)+1$ must be filled with zeros. Thus, the conversion is performed

$$h_\Sigma(t) \rightarrow h''_\Sigma(t) \tag{13}$$

The time interval for the return of the semi-Markov process to state $m(\Sigma)$ is determined by the dependence

$$^\Sigma h''_{m(\Sigma),m(\Sigma)}(t) = {}^r I_{m(\Sigma)} \cdot L^{-1} \left[\sum_{k=1}^{\infty} \{ L[h''_\Sigma(t)] \}^k \right] \cdot {}^c I_{N(\Sigma)+1}, \tag{14}$$

where $^r I_{m(\Sigma)}$ the row vector, the $m(\Sigma)$- th element of which is equal to one, and the remaining elements are equal to zero; $^c I_{n(\Sigma)}$ - column vector, $(N(\Sigma)+1)$ -th element of which is equal to one, and the rest of the matrices formed from $h'_\Sigma(t)$.

The investigated semi-Markov process is described by the matrix $h_\Sigma(t)$ of the most general form, therefore, in the matrix $h''_\Sigma(t)$ formed as a result of performing operation (13) there can be other absorbing states besides the artificially created state $N(\Sigma)+1$. It follows from this that not all trajectories of a walk through a complex semi-Markov process, which begin in state $m(\Sigma)$, end in state $N(\Sigma)+1$. This, in turn, means that the group of events attaining state $N(\Sigma)+1$ from state $m(\Sigma)$ is not understood, and therefore, in the general case, $h''_{m(\Sigma),m(\Sigma)}(t)$ has the character of a weighted rather than pure distribution density. The probability and net density of the time it takes to reach state $N(\Sigma)$ from state $m(\Sigma)$ is defined as

$$^\Sigma p''_{m(\Sigma),m(\Sigma)} = \int_0^\infty {}^\Sigma h''_{m(\Sigma),m(\Sigma)}(t)dt; \tag{15}$$

$$^\Sigma f''_{m(\Sigma),m(\Sigma)}(t) = \frac{^\Sigma h''_{m(\Sigma),m(\Sigma)}(t)}{^\Sigma p''_{m(\Sigma),m(\Sigma)}}. \tag{16}$$

From (10) and (16) the numerical characteristics of the corresponding distribution densities can be obtained, namely, the mathematical expectation and variance:

$$^\Sigma T'_{m(\Sigma),n(\Sigma)} = \int_0^\infty t \cdot {}^\Sigma f'_{m(\Sigma),n(\Sigma)}(t)dt; \tag{17}$$

$$^\Sigma D''_{m(\Sigma),n(\Sigma)} = \int_0^\infty \left(t - {}^\Sigma T'_{m(\Sigma),n(\Sigma)} \right)^2 \cdot {}^\Sigma f'_{m(\Sigma),n(\Sigma)}(t)dt; \tag{18}$$

$$\Sigma T''_{m(\Sigma),m(\Sigma)} = \int\limits_0^\infty t \cdot {}^\Sigma f''_{m(\Sigma),m(\Sigma)}(t)dt; \tag{19}$$

$$\Sigma D''_{m(\Sigma),m(\Sigma)} = \int\limits_0^\infty \left(t - {}^\Sigma T''_{m(\Sigma)m(\Sigma)}\right)^2 \cdot {}^\Sigma f''_{m(\Sigma)m(\Sigma)}(t)dt. \tag{20}$$

The dependences obtained are difficult to implement on a computer, since they first involve analytical transformations of mathematical expressions, and then the actual calculation of the basic numerical characteristics of distribution densities, expectation and dispersion .. Therefore, to solve the problems of analyzing the temporal and probabilistic characteristics of mobile robot control systems obtaining mathematical expressions that allow you to directly calculate the main numerical characteristics of the corresponding n lot numbers according to the initial numerical characteristics of the densities included in the matrix $h_\Sigma(t)$.

3   Direct Calculation of Numerical Characteristics

The following matrices can be obtained from the semi-Markov matrix (1) [9]:
  probability matrix (3)

$$p_\Sigma = \int\limits_0^\infty h_\Sigma(t)dt = \left[p_{m(\Sigma),n(\Sigma)}(t)\right]; \tag{21}$$

net density distribution matrix (4)

$$f_\Sigma(t) = h_\Sigma(t)/p_\Sigma = \left[f_{m(\Sigma),n(\Sigma)}(t)\right], \tag{22}$$

where /- the inverse operation of the direct multiplication of matrices, such that if $h_\Sigma(t) = f_\Sigma(t) \otimes p_\Sigma$, that $h_\Sigma(t) = [h_\Sigma(t)/p_\Sigma] \otimes p_\Sigma$.
  From (22) the following matrices can be obtained:
  pure expectation matrix of distribution densities

$$T_\Sigma = \int\limits_0^\infty t \cdot f_\Sigma(t)dt = \left[T_{m(\Sigma),n(\Sigma)}\right]; \tag{23}$$

matrix of pure dispersions of distribution densities

$$D_\Sigma = \int\limits_0^\infty t^2 \cdot f_\Sigma(t)dt - T_\Sigma \otimes T_\Sigma = \left[D_{m(\Sigma),n(\Sigma)}\right]; \tag{24}$$

weighted expectation matrix

$$\tilde{T}_\Sigma = T_\Sigma \otimes p_\Sigma = \left[p_{m(\Sigma),n(\Sigma)} T_{m(\Sigma),n(\Sigma)} \right] = \left[\tilde{T}_{m(\Sigma),n(\Sigma)} \right]; \qquad (25)$$

weighted dispersion matrix

$$\tilde{D}_\Sigma = D_\Sigma \otimes p_\Sigma = \left[p_{m(\Sigma),n(\Sigma)} B_{m(\Sigma),n(\Sigma)} \right] == \left[\tilde{D}_{m(\Sigma),n(\Sigma)} \right]. \qquad (26)$$

The matrices (21), (23), (24), (25) and (26) create a basic data set for the numerical analysis of the complex semi-Markov process.

We introduce the convolution operation of semi-Markov matrices $h_\alpha(t)$ and $h_\beta(t)$, having the same size

$$h_\alpha(t) * h_\beta(t) = L^{-1}[\{L[h_\alpha(t)] \cdot L[h_\alpha(t)]\}]. \qquad (27)$$

If $h_\alpha(t) = h_\beta(t) = h_\Sigma(t)$, then $h_\alpha(t) * h_\beta(t) = h_\Sigma^{*2}(t)$, and physically the element of the matrix $h_\Sigma^{*2}(t)$, standing at the intersection of the $m(\Sigma)$-th row and the $n(\Sigma)$-th column, determines the weighted density of the time distribution for which the element with the column number is reached from the element with the row number in two switchings. If $h_\alpha(t) = h_\Sigma^{*k-1}(t)$, $h_\beta(t) = h_\Sigma(t)$, then the convolution of these matrices means that the element of matrix $h_\alpha(t) * h_\beta(t)$, standing at the intersection of the $m(\Sigma)$-th row and the $n(\Sigma)$-th column, determines the weighted density of time distribution for which the element with the column number is reached from the element with the row number for k switching.

Denote

$$h_\alpha(t) = \lfloor p_{m,n,\alpha} \cdot f_{m,n,\alpha}(t) \rfloor; \quad h_\beta(t) = \lfloor p_{m,n,\beta} \cdot f_{m,n,\beta}(t) \rfloor \qquad (28)$$

Then

$$h_\alpha(t) * h_\beta(t) = L^{-1}\left[\left\{L[h_\alpha(t)] \cdot L[h_\beta(t)]\right\}\right]$$
$$= \left[\sum_{l=1}^{N} (p_{m,l,\alpha} \cdot q_{l,n,\beta}) \cdot L^{-1}\{L[f_{m,l,\alpha}(t)] \cdot L[f_{l,n,\beta}(t)]\} \right], \qquad (29)$$

where the external square brackets represent the matrix element at the intersection of the m-th row and the n-th column.

Find the expectation and variance of (29).

$$\int_0^\infty t \cdot [h_\alpha(t) * h_\beta(t)]dt = \left[\sum_{l=1}^{N} (p_{m,l,\alpha} \cdot p_{l,n,\beta}) \cdot (T_{m,l,\alpha} + T_{l,n,\beta}) \right]$$
$$= [p_{m,n,\alpha\beta} T_{m,n,\alpha\beta}] = p_{\alpha\beta} \otimes T_{\alpha\beta}; \qquad (30)$$

$$\int\limits_{0}^{\infty} t^2 \cdot \left[\boldsymbol{h}_\alpha(t) * \boldsymbol{h}_\beta(t)\right] dt - \boldsymbol{T}_{\alpha\beta} \otimes \boldsymbol{T}_{\alpha\beta} =$$

$$\left[\sum\limits_{l=1}^{N} \left(p_{m,l,\alpha} \cdot p_{l,n,\beta}\right) \cdot \left(D_{m,l,\alpha} + D_{l,n,\beta} + T_{m,l,\alpha}^2 + T_{l,n,\beta}^2\right)\right] - \left[T_{m,n,\alpha\beta}^2\right] = \qquad (31)$$

$$= \left[p_{m,n,\alpha\beta}\tilde{D}_{m,n,\alpha\beta}\right] - \boldsymbol{T}_{\alpha\beta} \otimes \boldsymbol{T}_{\alpha\beta},$$

where $T_{m,l,\alpha}$, $D_{m,l,\alpha}$ $p_{m,l,\alpha}$ - expectation, variance and probability of a semi-Markov matrix element $\boldsymbol{h}_\alpha(t)$, находящегося на пересечении m-й строки и l-го столбца; $T_{l,n,\beta}$, $D_{l,n,\beta}$ $p_{l,n,\beta}$ - expectation, variance, and probability of an element of semi-Markov matrix $\boldsymbol{h}_\beta(t)$ located at the intersection of the l-th row and the n-th column; $T_{m,n,\alpha\beta}$, $p_{m,n,\alpha\beta}$ - mathematical expectation, and the probability of the elements of the semi-Markov matrix of the product located at the intersection of the m-th row and the n-th column; $D_{m,n,\alpha\beta}$ - the initial moment of the second order is an element of the semi-Markov matrix of the product located at the intersection of the m-th row and the n-th column.

Using (30) and (31), we introduce the operation of convolution of numerical characteristics of semi-Markov matrices

$$\boldsymbol{T}_\alpha * \boldsymbol{T}_\beta = \int\limits_{0}^{\infty} t \cdot \left[\boldsymbol{h}_\alpha(t) * \boldsymbol{h}_\beta(t)\right] dt \qquad (32)$$

$$\boldsymbol{D}_\alpha * \boldsymbol{D}_\beta = \int\limits_{0}^{\infty} t^2 \cdot \left[\boldsymbol{h}_\alpha(t) * \boldsymbol{h}_\beta(t)\right] dt - \boldsymbol{T}_{\alpha\beta} \otimes \boldsymbol{T}_{\alpha\beta} \qquad (33)$$

Using the operations entered from (8) and (14), expressions for direct calculation of the following numerical characteristics can be obtained:

probability of reaching state n(Σ) of state m(Σ) during wanderings -.

$$^{\Sigma}p'_{m(\Sigma),n(\Sigma)} = {}^r\boldsymbol{I}_{m(\Sigma)} \cdot \left[\sum\limits_{k=1}^{\infty} \left(\boldsymbol{p}'_\Sigma\right)^k\right] \cdot {}^c\boldsymbol{I}_{n(\Sigma)}; \qquad (34)$$

weighted expectation of achieving state n(Σ) of state m(Σ) when wandering -.

$$^{\Sigma}\widetilde{T}'_{m(\Sigma),n(\Sigma)} = {}^r\boldsymbol{I}_{m(\Sigma)} \cdot \left[\sum\limits_{k=1}^{\infty} \left(\widetilde{\boldsymbol{T}}'_\Sigma\right)^{*k}\right] \cdot {}^c\boldsymbol{I}_{n(\Sigma)}; \qquad (35)$$

weighted dispersion of achievement of state N(Σ) of state m(Σ) during wanderings -.

$$^{\Sigma}\widetilde{D}'_{m(\Sigma),n(\Sigma)} = {}^r\boldsymbol{I}_{m(\Sigma)} \cdot \left[\sum\limits_{k=1}^{\infty} \left(\widetilde{\boldsymbol{D}}'_\Sigma\right)^{*k}\right] \cdot {}^c\boldsymbol{I}_{n(\Sigma)} \qquad (36)$$

pure mathematical expectation of reaching state n(Σ) of state m(Σ) when wandering -

$$^{\Sigma}T'_{m(\Sigma),n(\Sigma)} = \frac{^{\Sigma}\tilde{T}'_{m(\Sigma),n(\Sigma)}}{^{\Sigma}p'_{m(\Sigma),n(\Sigma)}};$$

(37)

pure dispersion of reaching state $n(\Sigma)$ from state $m(\Sigma)$ during wanderings -.

$$^{\Sigma}D'_{m(\Sigma),n(\Sigma)} = \frac{^{\Sigma}\tilde{D}'_{m(\Sigma),n(\Sigma)}}{^{\Sigma}p'_{m(\Sigma),n(\Sigma)}};$$

(38)

probability of returning to state $m(\Sigma)$ when wandering -.

$$^{\Sigma}p''_{m(\Sigma),n(\Sigma)} = {}^{r}I_{m(\Sigma)} \cdot \left[\sum_{k=1}^{\infty} \left(p''_{\Sigma} \right)^{k} \right] \cdot {}^{c}I_{n(\Sigma)+1};$$

(39)

weighted expectation of returning to state $m(\Sigma)$ when wandering -.

$$^{\Sigma}\tilde{T}''_{m(\Sigma),m(\Sigma)} = {}^{r}I_{m(\Sigma)} \cdot \left[\sum_{k=1}^{\infty} \left(\tilde{T}''_{\Sigma} \right)^{*k} \right] \cdot {}^{c}I_{n(\Sigma)+1};$$

(40)

weighted variance return to state $m(\Sigma)$ when wandering - -.

$$^{\Sigma}\tilde{D}''_{m(\Sigma),n(\Sigma)} = {}^{r}I_{m(\Sigma)} \cdot \left[\sum_{k=1}^{\infty} \left(\tilde{D}''_{\Sigma} \right)^{*k} \right] \cdot {}^{c}I_{n(\Sigma)+1};$$

(41)

pure expectation of returning to state $m(\Sigma)$ on wandering -

$$^{\Sigma}T''_{m(\Sigma),m(\Sigma)} = \frac{^{\Sigma}\tilde{T}''_{m(\Sigma),m(\Sigma)}}{^{\Sigma}p''_{m(\Sigma),m(\Sigma)}};$$

(42)

pure dispersion of return to state $m(\Sigma)$ when wandering -.

$$^{\Sigma}D''_{m(\Sigma),m(\Sigma)} = \frac{^{\Sigma}\tilde{D}''_{m(\Sigma),m(\Sigma)}}{^{\Sigma}p''_{m(\Sigma),m(\Sigma)}},$$

(43)

where p'_{Σ} - stochastic matrix derived from matrix $h'_{\Sigma}(t)$ using the operation (21); 3 $\tilde{T}'_{\Sigma}$- weighted expectation matrix derived from matrix $h'_{\Sigma}(t)$ using the operation (25); $\tilde{D}'_{\Sigma}$ weighted variance matrix derived from the matrix 6 $h'_{\Sigma}(t)$ using operation (26); p''_{Σ} - stochastic matrix derived from the matrix 8 $h''_{\Sigma}(t)$ using operation (21); 9 $\tilde{T}''_{\Sigma}$- weighted expectation matrix derived from the matrix 10 $h''_{\Sigma}(t)$ using operation (25); 11 $\tilde{D}''_{\Sigma}$ weighted variance matrix derived from the matrix 12 $h''_{\Sigma}(t)$ using operation (26);

$$\left(\widetilde{T}'_\Sigma\right)^{*k} = \left(\widetilde{T}'_\Sigma\right)^{*k-1} * \widetilde{T}'_\Sigma;$$

(44)

$$\left(\widetilde{D}'_\Sigma\right)^{*k} = \left(\widetilde{D}'_\Sigma\right)^{*k} * \widetilde{D}'_\Sigma;$$

(45)

$$\left(\widetilde{T}''_\Sigma\right)^{*k} = \left(\widetilde{T}''_\Sigma\right)^{*k-1} * \widetilde{T}'_\Sigma;$$

(46)

$$\left(\widetilde{D}''_\Sigma\right)^{*k} = \left(\widetilde{D}''_\Sigma\right)^{*k} * \widetilde{D}'_\Sigma.$$

(47)

Thus, we obtain general matrix analytical expressions that allow, without calculating the integrals, calculate the probabilistic and temporal characteristics of walks in a semi-Markov process using the temporal and probabilistic characteristics of semi-Markov matrix $h_\Sigma(t)$ and its derivatives, $h'_\Sigma(t)$ and $h''_\Sigma(t)$.

4   Conclusion

A general method is proposed for calculating the probabilistic and temporal characteristics of a walk through a complex semi-Markov process, formed by combining M ordinary semi-Markov processes, based on the fact that a semi-Markov matrix is transformed into a characteristic matrix, which is raised to a power, then the powers of the characteristic matrix are summed.

Using the general method, the problems of determining the probabilistic and temporal characteristics of a walk through a semi-Markov process from one state to another, as well as a return to one of the states are solved.

A method has been developed for direct calculation of the time and probabilistic characteristics of a complex walk between states and return to the state using only operations with numerical matrices: stochastic, mathematical expectations and variances characterizing the elements of the original semi-Markov matrix.

References

1. Brambilla, M., Ferrante, E., Birattari, M., Dorigo, M.: Swarm robotics: a review from the swarm engineering perspective. Swarm Intell. **7**(1), 1–41 (2013)
2. Morin, P., Samson, C.: Motion Control of Wheeled Mobile Robots. Springer, Heidelberg (2008). https://doi.org/10.1007/978-3-540-30301-5_35
3. Larkin, E.V., Antonov, M.A.: Semi-Markov model of a swarm functioning. In: Tan, Y., Shi, Y., Tang, Q. (eds.) ICSI 2018. LNCS, vol. 10941, pp. 3–13. Springer, Cham (2018). https://doi.org/10.1007/978-3-319-93815-8_1

4. Kahar, S., Sulaiman, R., Prabuwono, A.S., Akma, N. Ahmad, S.A., Abu Hassan, M.A.: A review of wireless technology usage for mobile robot controller. In: 2012 International Conference on System Engineering and Modeling (ICSEM 2012), vol. 34, pp. 7–12. IACSIT Press, Singapore (2012)

5. Larkin, E., Kotov, V., Privalov, A., Bogomolov, A.: Multiple swarm relay-races with alternative routes. In: Tan, Y., Shi, Y., Tang, Q. (eds.) ICSI 2018. LNCS, vol. 10941, pp. 361–373. Springer, Cham (2018). https://doi.org/10.1007/978-3-319-93815-8_35

6. Akimenko, T.A.: Formation of the image on the receiver of thermal radiation. In: Proceedings of SPIE - The International Society for Optical Engineering, vol. 10696, p. 1069627 (2018)

7. Korolyuk, V., Swishchuk, A.: Semi-Markov Random Evolutions. Springer, Dordrecht (1995). https://doi.org/10.1007/978-94-011-1010-5

8. Limnios, N., Swishchuk, A.: Discrete-time semi-Markov random evolutions and their applications. Adv. Appl. Probab. 45(1), 214–240 (2013)

9. Markov, A.A.: Extension of the law of large numbers to dependent quantities. Izvestiia Fiz.-Matem. Obsch. Kazan Univ. (2-nd Ser.) 15, 135–156 (1906)

10. Bielecki, T.R., Jakubowski, J., Niewęgłowski, M.: Conditional Markov chains: properties, construction and structured dependence. Stochast. Processes Appl. 127(4), 1125–1170 (2017)

11. Janssen, J., Manca, R.: Applied Semi-Markov Processes. Springer, Heidelberg (2005). https://doi.org/10.1007/0-387-29548-8

12. Larkin, E.V., Lutskov, Y., Ivutin, A.N., Novikov, A.S.: Simulation of concurrent process with Petri-Markov nets. Life Sci. J. 11(11), 506–511 (2014)

13. Larkin, E., Bogomolov, A., Privalov, A., Antonov, M.: About one approach to robot control system simulation. In: Ronzhin, A., Rigoll, G., Meshcheryakov, R. (eds.) ICR 2018. LNCS (LNAI), vol. 11097, pp. 159–169. Springer, Cham (2018). https://doi.org/10.1007/978-3-319-99582-3_17

14. Larkin, E.V., Ivutin, A.N., Kotov, V.V., Privalov, A.N.: Simulation of relay-races. Bull. South Ural State Univ. Ser.: Math. Model. Program. Comput. Softw. 9(4), 117–128 (2016)

15. Bauer, H.: Probability Theory. Walter de Gruyter, Berlin (1996)

16. Shiryaev, A.N.: Probability. Springer, Heidelberg (1996). https://doi.org/10.1007/978-1-4757-2539-1

Prediction

Analysis and Prediction of Heart Diseases Using Inductive Logic and Image Segmentation

S. Anand Hareendran[1]([✉])[iD], S. S. Vinod Chandra[2], and Sreedevi Prasad[3]

[1] Muthoot Institute of Technology and Science, Kochi, India
anandhs06@gmail.com
[2] University of Kerala, Trivandrum, India
[3] RIET, Trivandrum, India

Abstract. The human heart's electrical activity produces currents that radiate through the surrounding tissue to the skin. When electrodes are attached to the skin, they sense these electrical currents and transmit them to an electrocardiograph monitor. The currents are then transformed into waveforms that represent the heart's depolarisation - repolarisation cycle. An electro cardiogram (ECG) shows the precise sequence of electrical events occurring in the cardiac cells throughout the process. It can be used to monitor phases of myocardial contraction and to identify rhythm and conduction disturbances. In this work an automated ECG analysis and prediction of the heart conditions are carried out using a continuous image segmentation phase using Inductive Logic Programming (ILP) system. ECG taken from the patients is fed to the system as input and it uses the background knowledge to predict the possible underlying heart disorders. A continuous image segmentation module runs through out the process which identifies the leads and graph distortions by eliminating the outliers. The extracted lead graphs are compared with the in store ILP system database and possible predictions are made. The system performance was automatically compared (by Brier score method) with the MIT-BIH ECG database (PhysioNet) and a prediction accuracy of 97.8% was obtained in lead-lead prediction.

Keywords: Electrocardiogram · Myocardial infarction · Medical prediction · Image segmentation · Logic programming

1 Introduction

An electrocardiogram (ECG) is an electrical recording of the heart that depicts the cardiac cycle. It is a practical approach to detect any abnormalities in functioning of the heart. It is basically used as a first course of choice in diagnosing many cardiovascular diseases. To get important information about cardiac rhythm, acute myocardial injury, etc. we use the standard 12-lead electrocardiogram (ECG), which is risk free, simple to perform, and inexpensive. Over years,

© Springer Nature Singapore Pte Ltd. 2019
Y. Tan and Y. Shi (Eds.): DMBD 2019, CCIS 1071, pp. 93–103, 2019.
https://doi.org/10.1007/978-981-32-9563-6_10

ECG readers have been trained in clinical electrocardiography and cardiology. However, guiding of cardiology trainees in clinical electrocardiography has been superseded by emerging diagnostic and treatment modalities such as imaging techniques, invasive procedures and cardiac device therapies.

Waveforms produced by the heart's electrical current are recorded on graphed ECG paper by a stylus. ECG paper consists of horizontal and vertical lines forming a grid. A piece of ECG paper is called an ECG strip or tracing (Fig. 1).

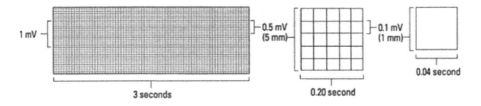

Fig. 1. ECG grid.

The horizontal axis of the ECG strip represents time. Each small block equals 0.04 s, and five small blocks form a large block, which equals 0.2 s. This time increment is determined by multiplying 0.04 s (for one small block) by 5, the number of small blocks that compose a large block. Five large blocks equal 1 s. When measuring or calculating a patient's heart rate, a 6-s strip consisting of 30 large blocks is usually used [5,6].

The ECG strip's vertical axis measures amplitude in millimetres (mm) or electrical voltage in millivolts (mV). Each small block represents 1 mm or 0.1 mV; each large block, 5 mm or 0.5 mV. To determine the amplitude of a wave, segment, or interval, count the number of small blocks from the baseline to the highest or lowest point of the wave, segment, or interval.

A normal ECG waveform has a characteristic shape as shown below (Fig. 2). The segment P represents the phase of atrial contraction/depolarisation when the deoxygenated blood enters the heart from right atrium and the oxygenated blood enters the heart from lungs into left atrium. The QRS segment (Fig. 2) depicts the phase of ventricular contraction/depolarisation when blood enters the right and left ventricles for discharging into the pulmonary artery and the Aorta respectively.

Finally, the T labeled segment represents ventricular repolarisation where ventricles relax to allow the cycle to begin again. Many disturbances in the functioning of heart shown as characteristic variations in the sinus rhythm waveform and it can be used as primary diagnosis for various heart problems.

For an automatic ECG analysis/prediction system, the raw input graph need to be converted to a processed outlier free representation for doing comparison. The system is equipped with an ILP database, which has been equipped by a trained model using first order logics. The segmented noise free graph is fed to the ILP system and it makes use of the background knowledge for doing accurate prediction [7].

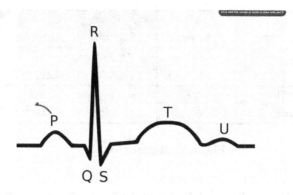

Fig. 2. PQRS waveform.

The analysis and prediction using the waveform is a tedious task even for experienced consultants because of the numerous conditions/facts that need to be considered before drawing a conclusion. There are 12 lead conditions that need to be studied before making a prediction regarding the area and type of heart condition. The lead details are shown in Table 1 and a detailed colour scheme is shown in Fig. 3.

Table 1. Lead details.

Lead names	Lead no	Activity
Inferior leads	Leads II, III, aVF	Look at electrical activity from the vantage point of the inferior surface
Lateral leads	I, aVL, V5, V6	Look at the electrical activity from the vantage point of the lateral wall of left ventricle
Septal leads	V1, V2	Look at electrical activity from the vantage point of the septal surface of the heart
Anterior leads	V3, V4	Look at electrical activity from the vantage point of the anterior wall of the right and left ventricles

Any two precordial leads next to one another are considered to be contiguous. For example, though V4 is an anterior lead and V5 is a lateral lead, they are contiguous because they are next to one another.

The major thumb rules while diagnosing the ECG are:

– PR interval should be 120 to 200 mS
– Width of QRS complex should not exceed 110 mS
– QRS complex should be dominantly upright in leads I and II
– QRS and T waves tend to have the same general direction in the limb leads
– All waves are negative in lead aVR

I Lateral	aVR None	V1 Septal	V4 Anterior
II Inferior	aVL Lateral	V2 Septal	V5 Lateral
III Inferior	aVF Inferior	V3 Anterior	V6 Lateral

Fig. 3. Contiguous leads.

- R wave should grow from V1 to V4 (atleast) and S wave should grow from V1 to V3 and should disappear in V6
- P wave should be upright in I, II and V2 to V6
- There should be no Q wave or less Q wave in I, II, V2 TO V6
- T wave must be upright in I, II, V2 to V6

A sample ECG is shown in Fig. 4

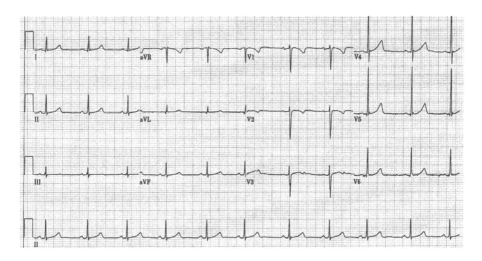

Fig. 4. Sample ECG.

2   Literature Survey

The analysis and prediction using ECG is always a tedious task even for many experienced consultants. The automated system that is proposed here will check the logical background and comes up with accurate prediction. As the proposed system shows high accuracy in prediction, it can be even used by non-medical people to analyse their ECG. They just need to give the ECG as an input to the image segmentation module. The entire system is developed to an app, which can be easily used by others.

There have been various attempts before also to create an algorithm or design a system which can be used to analyze ECG Shape Matching Algorithm:- The paper proposed an algorithm for capturing the shape similarity of ECG waveforms by using shape matching with dynamic time warping method. Overall shape similarity is measured using combination of shape similarity of each recording channel [1]. Thus predictions were made out of the captured shape.

Lightweight Algorithm: - This paper proposed a lightweight algorithm for rapid detection of ECG signal. To determine occurrence of the R-peak it uses the slopes between adjacent signals [2].

P wave Detection Algorithm:- Another paper presented an algorithm for efficient p wave detection which was based on the morphological characteristics of arrhythmia by using correlation and regression in ECG signal. The correlation coefficient then indicates the kind of arrhythmia disease. The algorithm was tested using database called MIT-BIH arrhythmia database where every p wave was classified appropriately. The results were presented in terms of the correlation coefficient [3].

3   Proposed System

Now a day there exists number of methodologies for ECG analysis [4]. All these process requires large storage space and extensive manual effort. Therefore the objective of this work is to build a platform, which accepts user ECG and help to detect variations and predict the disease. It can be used as a doctor's guide or as a customer product, which helps them to monitor the disease at a very early stage

3.1   Design and Implementation

The electrocardiogram is the simplest and oldest method for cardiac investigations, yet this provides a wealth of useful information and also remains as an essential part of the assessment of cardiac patients.

The ECG sample is given as the input to processing phase. In the processing phase the grid removal occurs and we will get the exact signal of the ECG. The processed image gets compared with the original ECG and the variations are identified in the evaluation phase. The variations predict problem of the patient. In the prediction phase the value gets stored and comparison of the peaks occurs and the output gets predicted. Figure 5 shows the system architecture for ECG reading.

The major steps of the process comprises of

- Pre-processing
- Lead segmentation
- ECG Waveform Extraction

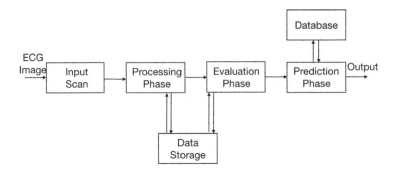

Fig. 5. System architecture.

Pre-processing: ECG images are scanned at a high resolution and then wave-form trace is captured at a good thickness. The foreground can be easily segmented from the background grid due to the limited number of colors in the image. This operation will retain both the ECG waveforms and the lead annotation (Figs. 6 and 7).

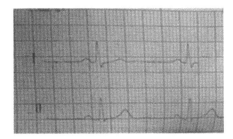

Fig. 6. Before preprocessing

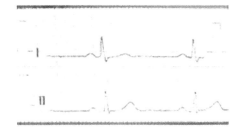

Fig. 7. After preprocessing.

During preprocessing and classification of ECG, noise sources can play a crucial role in reliable measurement. Although there are methods for detection the rates, most methods are too complicated for real-time implementation. Therefore, a low- complexity algorithm for automatically detecting and classifying the ECG noises including flat line (FL), time-varying noise (TVN), baseline wander (BW), abrupt change (ABC), muscle artifact (MA) and power line interference (PLI) is incorporated in the study.

The method is based on the moving average (MAv) and derivative filters and the five temporal features including turning points, global and local amplitude estimates, zero crossing and autocorrelation peak values. The method is tested and validated using a wide variety of clean and noisy ECG signals taken from the MIT-BIH arrhythmia database and Physionet Challenge database.

The method achieves an average sensitivity (Se) of 98.55%, positive productivity (+P) of 95.27% and overall accuracy (OA) of 94.19% for classifying the noises. Thus preprocessing and extraction of waves can be done will almost zero error factor.

Lead Segmentation: The ECG image records all lead recordings or 12 channels by interlacing three-second intervals from combinations of the leads. The standard 12-lead electrocardiogram is a representation of the heart's electrical activity recorded from electrodes on the body surface Isolate the different leads from the ECG image as different diseases are manifested differently in each of the lead. Using a Gaussian pyramid we smooth the background subtracted image as the scanned images are large and since grey-level correlation is robust than binary correlation. The below figure depicts lead segmentation. Leads aVR, aVL, and aVF are the augmented limb leads. They are derived from the same three electrodes as leads I, II, and III, but they use Goldberger's central terminal as their negative pole which is a combination of inputs from other two limb electrodes (Fig. 8).

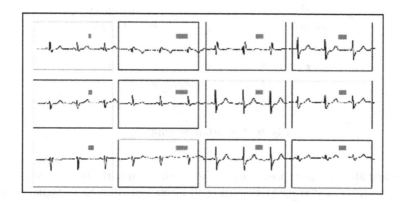

Fig. 8. ECG lead segmentation.

ECG Waveform Extraction: After the segmentation of each lead position, the ECG waveform is made out of it. These recordings are actually the time series or functions of lead-voltage vs. time. This output can be used later for disease prediction by comparing the peak values. Peak value comparison method and the shape algorithm used is discussed in next session.

4 Results and Discussion

Each ECG wave includes 12 leads and each lead has different function. The muscles actions are considered in each lead. These leads are taken together for comparison. The diseased samples and the normal ECGs get compared and variations are found and considering these variations predictions are made. Shape algorithm is used for detecting the various waveforms. The algorithm is performed in two steps; the first step is devoted to signal filtering and to calculate a set of parameters, which will be used in the second step to actually detect waveforms. A comprehensive description of the algorithm is shown in Fig. 9.

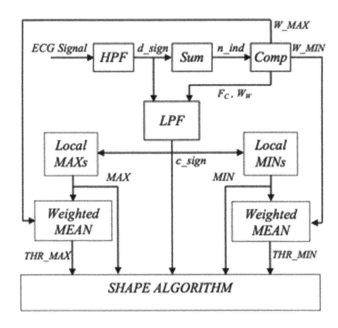

Fig. 9. Algorithm description.

The first step of signal processing is the elimination of the baseline variability. In particular, the first stage (HPF in Fig. 9) filters the input ECG Signal exploiting a high-pass FIR filter, designed to keep the computational complexity as low as possible; the width of the moving window of the filter is composed of only two unitary points, reducing the number of sums and avoiding the multiplications. This filter represents an approximation of the differential calculus and its aim is to highlight the slope steepness of the original signal. For the subsequent parts of the algorithm description, the output of this stage will be called differentiated signal (d_sign).

The second stage (Sum in Fig. 9) returns the indicator n_ind related to the noise level of the input signal. In order to reduce the number of operations and avoid multiplications, a row estimation of the noise at high frequencies, obtained by summing the absolute values of d_sign, is provided. In the stage Comp, the noise level indicator n_ind is compared with a set of thresholds tuned according to the characteristics of the acquisition system. This step produces two filter parameters (FC and WW) and two weight coefficients (W_MAX, W_MIN) used in the subsequent stages. A low-pass FIR filter (LPF) is then used to process the differentiated signal d_sign to obtain the so-called conditioned signal c_sign; the cut-off frequency FC and the width WW of the moving windows are reckoned in the Comp stage. The filter tuning is effective because the signal under elaboration is few tens of seconds long, thus the algorithm is able to dynamically deal with the time-variant noise. The conditioned signal c_sign is then processed by Local MAXs and Local MINs blocks, to obtain two vectors, MAX and MIN, containing the local maximums and minimums, respectively. The means of the MAX and MIN are then weighted by the two Weighted MEAN blocks using the coefficients W_MAX and W_MIN calculated in Comp [8].

In this way the two scalar values THR_MAX and THR_MIN are obtained. The conditioned signal c_sign, its local maximum and minimum vectors MAX and MIN as well as the weighted means THR_MAX and THR_MIN are the parameters provided to the stage devoted to the actual detection of the waveforms. Amplitude difference between samples is evaluated with respect to the previously defined threshold THR_MAX or THR_MAX_NB, according to the waveform under investigation.

Figure 10 show the peak detection in a QRS complex using the shape detection algorithm.

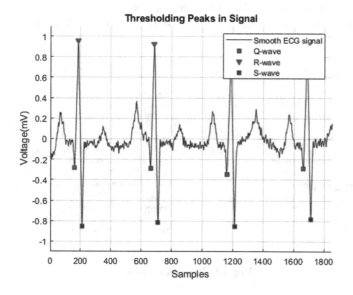

Fig. 10. Waveform detection using shape algorithm.

The algorithm is run over the entire set of inputs and the corresponding peaks are calculated which are given to the shape algorithm for prediction. Table 2 shows the comparison between the results obtained by the proposed scheme and the performance of the other traditional algorithms. The proposed scheme shows slightly equal performance with respect to the other algorithms of the same class. However, it should be underlined that, when an algorithm is designed to be run in real-time by devices with limited computational capacity, the obtained performance can be considered very much effective, thus validating the effectiveness of the proposed method.

Table 2. Performance comparison.

Algorithm	Precision (%)	Recall (%)
Pan-Tompkins [9]	99.56	98.76
Wavelet [11]	99.68	98.90
Empirical mode decomposition [10]	99.76	98.88
Area over the curve [12]	99.30	98.60
Proposed scheme	99.87	99.21

The proposed work has also been implemented as an android application (CardiaCare) where a QR code scanner helps in scanning the ECG input graphs. The processing module of the application does the graph extraction and comparison. An artificial neural network model is being developed for the comparison of the threshold, which helps in finding the difference in peak waves between the user inputs and the standard ECG graph values.

The final obtained predicted disease list is given in the Fig. 11.

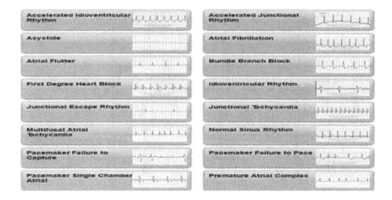

Fig. 11. Results obtained.

5   Conclusion

A decisive method for extraction and digitization of ECG signal from various sources such as thermal ECG printouts, scanned ECG and captured ECG images from devices is proposed. The technique produces an accurate waveform that is free from printed character and other noises. The extracted waveform is fed to a shape algorithm, which compares the threshold with the baseline graphs. Almost all anomalies can be effectively spotted and predicted using this method. Comparative study with similar algorithms has also been performed and the results have been analyzed. It was quite evident that for a real time processing system the accuracy obtained was much above the expected range. An android platform has also been build to the non-medicos people to have a routine check on the ECG and to have the analysis. Various improvements like, in-build lead sorting, muscle anomalies prediction can be seen as a scope for future enhancement.

References

1. Mahmood, T.S., Beymer, D., Wang, F.: Shape-based matching of ECG recordings. Healthcare Informatics Group, IBM Almaden Research Center (2013)
2. Adeluyi, O., Lee, J.-A.: R-READER: a lightweight algorithm for rapid detection of ECG signal R-peaks. In: IEEE International Conference (2011)
3. Joshi, A.K., Tomar, A., Tomar, M.: A review paper on analysis of electrocardiograph (ECG) signal for the detection of arrhythmia abnormalities. Int. J. Adv. Res. Electr. Electron. Instrum. Energy 3(10), 12466–12475 (2014)
4. Dhir, J.S., Panag, N.K.: ECG analysis and R peak detection using filters and wavelet transform. Int. J. Innov. Res. Comput. Commun. Eng. 2(2) (2014)
5. Shinwari, M.F., Ahmed, N., Humayun, H., Haq, I., Haider, S., Anam, A.: Classification algorithm for feature extraction using linear discriminant analysis and cross-correlation on ECG signals. Int. J. Adv. Sci. Technol. 48, 149–162 (2012)
6. Islam, M.K., Hague, A.N.M.M., Tangim, G., Ahammad, T., Khondokar, M.R.H.: Study and analysis of ECG signal using MATLAB LABVIEW as effective tools. Int. J. Comput. Electr. Eng. 4(3) (2012)
7. Messaoud, M.B., Kheli, B., Kachouri, A.: Analysis and parameter extraction of P wave using correlation method. Int. Arab J. Inf. Technol. 6(1), 40–46 (2009)
8. Safdarian, N., Maghooli, K., Dabanloo, N.J.: Classification of cardiac arrhythmias with TSK fuzzy system using genetic algorithm. Int. J. Signal Process. Image Process. Pattern Recognit. 5(2), 89–100 (2012)
9. Crema, C., Depari, A., Flammini, A., Vezzoli, A.: Efficient R-peak detection algorithm for real-time analysis of ECG in portable devices. IEEE Instrum. Meas. Soc. (2016)
10. Pan, J., Tompkins, W.J.: A real-time QRS detection algorithm. IEEE Trans. Biomed. Eng. 32, 230–236 (1985)
11. Satija, U., Ramkumar, B., Manikandan, M.S.: Low-complexity detection and classification of ECG noises for automated ECG analysis system. IEEE (2016)
12. Rooijakkers, M., Rabotti, C., Bennebroek, M., van Meerbergen, J., Mischi, M.: Low-complexity R-peak detection in ECG signals: a preliminary step towards ambulatory fetal monitoring. In: Annual International Conference of the IEEE, pp. 1761–1764 (2011)

Customer Retention Prediction with CNN

Yen Huei Ko[1], Ping Yu Hsu[1], Ming Shien Cheng[2(✉)],
Yang Ruei Jheng[1], and Zhi Chao Luo[1]

[1] Department of Business Administration, National Central University,
No. 300, Jhongda Road, Jhongli City 32001, Taoyuan County, Taiwan (R.O.C.)
yensintong@gmail.com
[2] Department of Industrial Engineering and Management,
Ming Chi University of Technology, No. 84, Gongzhuan Road, Taishan Dist.,
New Taipei City 24301, Taiwan (R.O.C.)
mscheng@mail.mcut.edu.tw

Abstract. The prediction of customer retention provides competitive advantage to enterprises. When customer purchases more with satisfaction, it will increase customer retention. Customer repurchase behavior represents customer with satisfaction on enterprise to buy again resulting in customer retention. In e-commerce and telemarketing, trust and loyalty are key factors influencing customer repurchase behavior. In the past, most researches were relied on questionnaire survey to collect data. The drawbacks of such approach are those participants may not be willing to fill the lengthy questions which caused low data collection rate and even low quality data being collected. This study is to apply data driven techniques to extract information from transaction logs in ERP system utilized to compute trust and loyalty based on the verified formulation. The values of defined trust and loyalty are treated as independent variables to predict customer repurchase behavior by Convolutional Neural Networks (CNNs). The prediction accuracy reaches 84%.

Keywords: Data driven technique · Trust · Loyalty ·
Customer repurchase behavior · Customer retention · CNN

1 Introduction

As company starts to focus on customer demands and strives to fulfill customer expectations, the implementation of customer relationship management (CRM) driving the continuous business growth and financial gain becomes the essential topic in E-commerce and Telemarketing industries. Customer retention depends on the interactions between parties and is defined as the continuous business relationship of customers toward with the company. Customer retention is the strategic purpose in customer relationship development and the driving force to sustain a long-term profitability. Customer retention is explored in the form of customer purchase again with satisfaction, and Eriksson and Vaghult [5] also demonstrated that the more actual purchase the higher customer retention. Zeithaml [16] depicted that loyal customers are

© Springer Nature Singapore Pte Ltd. 2019
Y. Tan and Y. Shi (Eds.): DMBD 2019, CCIS 1071, pp. 104–113, 2019.
https://doi.org/10.1007/978-981-32-9563-6_11

with higher customer retention rates. How to enhance customer loyalty and manage customer relationship starts to be important for enterprises. By extending relationships with the loyal customers and increasing sales with existing customers, companies thereby increase the profitability in business.

Trust and loyalty are the cornerstones to enhance customer relationship and improve the competitive advantages for customer retention [12]. Gefen [6] stated that trust is the prime factor to strength the willingness to buy. Although some researchers applied trust to speculate on repurchase intentions [4], repurchase intentions can not exactly represent the repurchase behaviors, only the actual repurchase behavior can significantly contribute the company's revenue.

In the past, the most of researches applied questionnaire survey to collect data. There are some drawbacks to collect data for researches, namely, time consuming, low efficient response, and low intentions for participants to fill the survey. In E-commerce blooming era, data scientists start to utilize data driven techniques to extract the valuable information from transaction logs. In this study, we propose to utilize data driven technique to extract information from customer transaction records in a call center stored in ERP system instead of applying questionnaire survey to collect data.

In recent years, a variety of neural network techniques have been employed commercially. CNN has been widely used in different aspects recently. Kalchbrenner et al. [10] employed Dynamic Convolutional Neural Network (DCNN) for semantic modeling of sentences. The model induces a feature graph over the sentence and the network is capable to capture the word relations of varying size. Abdel-Hamid et al. [1] proposed a modified CNN with inputs using frequency and time in both axis for speech recognition and described how to organize speech features into feature maps for CNN processing. In this study, by the inspiration from the aforementioned literatures, we propose operational definitions of trust and loyalty organized as feature maps fed into CNN to predict customer repurchase behavior. The prediction accuracy of customer repurchase behavior reaches 84%.

The objectives of this study are to verify the relationships among trust, loyalty and customer repurchase behavior, understand the constructing factors of trust and loyalty, verify the correlation of trust and loyalty, and confirm the impacts of trust and loyalty on customer repurchase behavior. From managerial consideration, through the prediction of customer purchase behavior, enterprises can strategically modify the marketing activities to enhance the levels of trust and loyalty in order to increase customer repurchase behavior and customer retention resulting in the growth of business.

This paper organized as follow: (1) Introduction: This part explains the research background and objectives. (2) Related Works: This part contains a review of scholars' researches. (3) Research Methodology: This part describes the research framework, operational definition of variables, and CNN algorithm. (4) Experiments and Results: This part describes and discusses the experimental results (5) Conclusions: We present the contributions and suggestions for future studies.

2 Related Work

2.1 Trust

McKnight and Chervany [14] stated four dimensions of trust, namely, benevolence, integrity, competence, and predictability. Gefen and Straub [7] suggests that customer trust is related to benevolence, integrity, competence and predictability.

Benevolence: In the trading process, enterprises should look at things from consumers' perspectives, prioritize the greatest benefits of consumers, provide consumers with the most needed assistance, and make consumers feel positive [14].

Integrity: Mayer et al. [13] proposed the principle of integrity is to give customer a reliable and sincere feeling. Integrity is a sense from customer feeling enterprise honest. Integrity represents the party is honest, fulfilling its commitment and treating others in accordance with the due moral standards [7, 14].

Competence: Competence refers to a person's performance in a special field that people agree with [13]. McKnight and Chervany [14] stated that competence means that companies can accurately understand customer needs and provide the most suitable services and products in the field of E-commerce.

Predictability: Gefen and Straub [7] pointed out that predictability refers to providing the right product or service at the time when the customer feels appropriate or needs service. Therefore, the customer likely intends to purchase the product or accept the service. McKnight and Chervany [14] proposes that the high predictability represents strong probability which a consumer's expectation for service or product will be satisfied at a certain point in time or situation.

Based on the conclusions of the aforementioned researches, we apply four dimensions of trust defined by McKnight and Chervany [14] which most of scholars have adopted.

2.2 Trust, Loyalty, and Repurchase Behavior

There are several advantages to companies with the high level of customer loyalty, including price insensitivity and product or brand recognition. Customer loyalty results in the company's benefits, including revenues and profits. Recency (R), Frequency (F) and Monetary (M) have been widely used as dimensions of customer loyalty [2]. Customer loyalty is derived from customer trust and contributes to generate company profit. Morgan and Hunt [15] suggested that trust is considered to be the essential factors for business growth and trust leads to a high degree of loyalty. Idrees et al. [9] claimed that consumer loyalty can be achieved through consumer trust and trust had a direct impact on loyalty.

Loyalty is the essential factor to retain customer and keep enterprise growth. Guenzi and Pelloni [8] suggested that loyalty result in repurchase intention and repurchase behavior. Dixon et al. [3] proposed that loyalty influences the customer purchase intention and store selection for future purchase. Kincade et al. [11] believed that customer's purchase behavior was resulting from the demand, and the repurchase behavior was driven by customer's satisfaction. According to the aforementioned researches, the satisfied customer will stay in the relationship with enterprise, keep the good faiths on enterprise, and purchase products again.

3 Research Methodology

3.1 Research Design

The dimensions of trust proposed by McKnight and Chervany [14], namely, Benevolence, Integrity, Competence, and Predictability (abbreviated as BICP), are utilized as the independent variables in this study. The dimensions of loyalty, namely, Recency (R), Frequency (F) and Monetary (M), are widely used in the aforementioned researches. The relation between trust and loyalty is derived from the aforementioned papers which trust leads to loyalty. In this study, we will apply multiple regression algorithm to empirically confirm the relation of trust and loyalty and verify the operational definitions of trust and loyalty. Furthermore, the values of dimensions of loyalty are extensively calculated by multiple regression equations and fed as the input data in CNN algorithm to predict customer repurchase behavior. The architecture diagram is shown in Fig. 1.

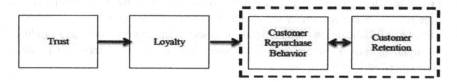

Fig. 1. Architecture of research model

3.2 Operational Definitions of Variables in Mathematical Representation

The operational definitions of trust, loyalty and repurchase behavior are described by mathematical formula shown below, respectively.

Trust. We adopt data driven technique to extract information from the transaction logs to construct the formulation for each dimension. The operational definitions of trust shown as follows,

1. Benevolence: According to literatures, benevolence is the activities that call agents care about customers' usage of products and periodically make phone call to comfort customers regarding what customers feel about services and products. We assume the total number of phone calls which call agent made to customer in a period of time represents the level of benevolence.
2. Integrity: When customer feels call agent reliable and honest, customer believes the call agent stands on the side of customer and accepts the new items which meet with

customer expectations. The definition of integrity is the sum of values from the transactions of new items divided by the number of days between the date of purchasing new product and the first date of caring call. The smaller number of days representing customer accepts promotion much easier, the higher level of integrity perceived by customer.

3. Competence: The well-trained call agent promotes the product bought by customer and met with customer's demand. The definition of competence is the sum of values from the transaction of the bought items in the past divided by the number of days between the date of purchasing the product and the first date of caring call. The smaller number of days, the higher level of competence perceived by customer.

4. Predictability: The call agent predicts when the product will be used up and executes the promotion calling. The definition of predictability is proposed as the number of confirmed orders including purchasing new and bought items in the past divided by the number of calling within time window of the research.

Loyalty. The operational definitions are described, respectively, as follows:

1. Recency: The number of days between the date of last day in investigating time window and the day of the last purchasing product represent recency. The smaller number of days, the higher possibility of customer to accept the next marketing campaign.

2. Frequency: It represents the total number of orders made by the customer over a time period. A larger number represents the greater customer loyalty.

3. Monetary: It indicates the total amount of money spending on purchasing products. A larger amount of money stands customer willing to spend more on products recommended by agent.

Repurchase Behavior. In Kincade et al. [11], the definition of repurchase behavior is adopted by questionnaire methodology. According to the real transaction information stored in ERP system, we can classify the accomplished order assigned as 1 and 0 otherwise.

3.3 CNNs Application

According to the operational definitions of trust and loyalty, the dimensions of trust and loyalty, namely, BICP and RFM, respectively, are calculated in each investigating time window. The relation between trust and loyalty is empirically demonstrated by multiple regression algorithm. Furthermore, the values of R, F, and M extensively calculated are utilized as the input data for CNN algorithm.

In CNN, each customer's transaction records are organized and simulated as a two-dimensional image. Regarding horizontal axis, it represents loyalty, values of RFM, derived from trust by using BICP as independent variables in multiple regression formula and normalized by taking nature log. Regarding vertical axis, each column represents the investigating time window.

4   Experiments and Results

In this study, the customer transaction data were retrieved from a call center in a pharmaceutical company in Taiwan from July 1, 2013 to June 30, 2016. The objective is aiming at investigating customer repurchase behavior deriving from trust and loyalty. By the aforementioned literatures, customer repurchase behavior represents loyalty building and customer retention.

In the selling processes of call center, there are three sequential stages of phone call, namely, call for caring, promotion and ordering, and call for after service. There are 19,033 valid calls are selected from 115,094 call records. In addition, there are 58% of customers repurchase happened in every six months. Therefore, we adopt every six-month as a time window, such as the first time window starting from July 1 to Dec. 31, 2013, to investigate customer repurchase behavior.

4.1   Experimental Results

Co-linearity Analysis. The co-linearity analysis needs to check among variables by using R programming language. The VIF (Variance Inflation Factor) values of defined trust and loyalty are shown in Tables 1 and 2, respectively. The values are less than 10 and there is no co-linearity trend.

Table 1. The VIF values of defined trust

Dimensions	VIF
Benevolence	1.795160
Integrity	1.567091
Competence	1.600038
Predictability	1.204697

Table 2. The VIF values of defined loyalty

Dimensions	VIF
Recency	1.046624
Frequency	1.604584
Monetary	1.588884

Relations of Trust and Loyalty. The relations of trust and loyalty are analyzed through multiple regressions. The results are shown in Tables 3, 4, and 5. In Table 3, the dimensions of trust are significantly related to recency.

Table 3. The multiple regression analysis of trust to recency

	Estimate	Std. Error	t value	P	
B(benevolence)	−0.079342	0.025202	−8.777	<2e−16	***
I(integrity)	0.043215	0.009040	4.721	2.47e−06	***
C(competence)	0.035198	0.009562	3.681	0.000237	***
P(predictability)	−0.336797	0.039716	−8.480	<2e−16	***

In Table 4, there is the positive relation between dimensions of trust and frequency of customer consumption.

Table 4. The multiple regression analysis of trust to frequency

	Estimate	Std. Error	t value	P	
B(benevolence)	0.124180	0.001772	70.067	<2e−16	***
I(integrity)	0.014891	0.001795	8.297	<2e−16	***
C(competence)	0.023231	0.001875	12.392	<2e−16	***
P(predictability)	0.758786	0.007786	97.451	<2e−16	***

In Table 5, the results show those 4 dimensions of trust have significant impact on monetary.

Table 5. The multiple regression analysis of trust to monetary

	Estimate	Std. Error	t value	P	
B(benevolence)	0.121536	0.005431	22.38	<2e−16	***
I(integrity)	0.060849	0.005500	11.06	<2e−16	***
C(competence)	0.095254	0.005745	16.58	<2e−16	***
P(predictability)	0.547109	0.023861	22.93	<2e−16	***

Consequently, results shown in Table 3, 4, and 5, the relations of trust and loyalty are confirmed which trust leads to loyalty and the operational definitions of trust and loyalty are verified. The relations of trust and loyalty are demonstrated by multiple regression equations shown below. According to multiple regression equations, the derivatives of RFM, namely, R^*, F^*, and M^*, respectively are obtained.

$$R^* = -0.079342 \times B + 0.043215 \times I + 0.035198 \times C - 0.336797 \times P + 1.728153$$

(1)

$$F^* = 0.124180 \times B + 0.014891 \times I + 0.023231 \times C + 0.758786 \times P - 0.123080$$

(2)

$$M^* = 0.121536 \times B + 0.060849 \times I + 0.095254 \times C + 0.547109 \times P + 3.069598$$

$$(3)$$

Prediction of Customer Repurchase Behavior. The derivatives, R^*, F^*, and M^* are adopted to investigate the effect of loyalty on customer repurchase behavior through multiple logistic regression analysis. The results are shown in Table 6.

Table 6. The multiple logistic regression analysis of derived loyalty to customer repurchase behavior

	Estimate	Std. Error	t value	P	
R^*(Recency)	0.03145	0.01911	1.646	0.0999	
F^* (Frequency)	0.32451	0.04953	6.552	6.82e−11	***
M^* (Monetary)	0.13787	0.03181	4.334	1.52e−05	***

In Table 6, the derived recency shows no significant influence on repurchase behavior. We propose to remove the derived recency dimension of loyalty to format a new input matrix for CNN algorithm.

CNN Algorithm. In this study, as the dataset is rather small, we adopt the low dimension model with the supervised approach to predict customer repurchase behavior. For each customer, the first four consecutive time windows from July 1, 2013 to June 30, 2015 in the vertical axis corresponding to values of F^* and M^* in horizontal axis are utilized for training and modeling in CNN algorithm. In the predicting stage, the testing dataset from January 1, 2014 to December 31, 2015 is adopted to predict customer repurchase behavior within January 1 and June 30, 2016.

There are 1922 consumer dataset as training each time and 823 consumer dataset as validation. The results of validation are shown in Fig. 2. Consequently, the training epoch is at 30 performing the best accuracy in this study.

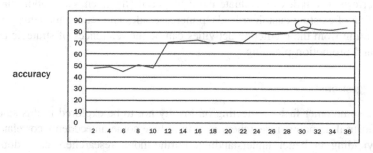

Fig. 2. The results of validation with various epoch

In predicting stage, the confusion matrix is shown in Table 7 and the accuracy yields 84%.

Table 7. The confusion matrix for prediction

Actual	Prediction	
	0	1
0	929	406
1	0	1146

5 Conclusions

5.1 Academic and Industrial Contributions

This research utilizes the transaction data from the call center in a pharmaceutical company in Taiwan to predict the customer repurchase behavior. The data driven technique is adopted to extract data directly from the ERP system and prevents some drawbacks from the traditional survey for data collection. According to aforementioned researches, we propose the research model that trust leads to loyalty, loyalty influences on customer repurchase behavior, and customer repurchase behavior represents customer retention. The relation of trust and loyalty is confirmed through multiple regression analysis and the operational definitions of dimensions for trust and loyalty are verified in this study. The prediction of customer repurchase behavior is performed by CNN algorithm and the accuracy achieves 84%. In this study, we have described how to apply CNN to predict customer repurchase behavior in a novel approach. In CNN's image structure, we are not only to accommodate the defined features, loyalty, but also to involve the temporal differences of transaction.

This study considers the situation in which the actual transaction interactivities are carried out through call center. In the past, it is very difficult to foresee what levels of trust and loyalty are from customer's perspectives, whether customer will visit again, and whether customer will be retained or not. According to results of this study, for industrial application, enterprise can have better understandings on customers' behaviors through the analysis of past transaction records to continuously strengthen service expertise and marketing planning. Secondly, manager can establish the system of key performance index to evaluate the managerial effectiveness. In addition, due to the retention of existing customer leads to increase sales and reduce marketing cost, the facts resulting from the prediction activities can be the references of strategic decision making and promotion planning.

5.2 Suggestions

There are some other factors affecting on loyalty not to be explored in this study. For other industrial domains, researchers may introduce some antecedents correlated with loyalty yielding in better understanding. Furthermore, researchers can adopt more control variables such as age, income, and genders for the studies of consumer behavior.

For future application of CNN algorithm, researchers may consider several aspects. First of all, introducing a variety of parameters combining with the domain knowledge will generate a larger image size and yield a better accuracy. Secondly, comparing design settings in terms of optimal epoch and neuron units will result in higher performance in CNN algorithm. The last, adopting more data for training the prediction model will offer the better accuracy.

References

1. Abdel-Hamid, O., Mohamed, A., Jiang, H., Deng, L., Penn, G., Yu, D.: Convolutional neural networks for speech recognition. In: 2014 IEEE/ACM Transactions on Audio, Speech, and Language Processing, pp. 1533–1545. IEEE, New Jersey (2014)
2. Bult, J.R., Wansbeek, T.: Optimal selection for direct mail. Market. Sci. **14**, 378–394 (1995)
3. Dixon, J., Bridson, K., Evans, J., Morrison, M.: An alternative perspective on relationships, loyalty and future store choice. Int. Rev. Retail Distrib. Consumer Res. **15**, 351–374 (2005)
4. Erciş, A., Ünal, S., Candan, F.B., Yıldırım, H.: The effect of brand satisfaction, trust and brand commitment on loyalty and repurchase intentions. Procedia – Soc. Behav. Sci. **58**, 1395–1404 (2012)
5. Eriksson, K., Vaghult, A.L.: Customer retention, purchasing behavior and relationship substance in professional services. Ind. Market. Manag. **29**, 363–372 (2000)
6. Gefen, D.: E-commerce: the role of familiarity and trust. Omega **28**, 725–737 (2000)
7. Gefen, D., Straub, D.W.: Consumer trust in B2C e-Commerce and the importance of social presence: experiments in e-Products and e-Services. Omega **32**, 407–424 (2004)
8. Guenzi, P., Pelloni, O.: The impact of interpersonal relationships on customer satisfaction and loyalty to the service provider. Int. J. Serv. Ind. Manag. **15**, 365–384 (2004)
9. Idrees, Z., Xinping, X., Shafi, K., Hua, L., Nazeer, A.: Consumer's brand trust and its link to brand loyalty. Am. J. Bus. Econ. Manag. **3**, 34–39 (2015)
10. Kalchbrenner, N., Grefenstette, E., Blunsom, P.: A convolutional neural network for modelling sentences. In: 52nd Annual Meeting of the Association for Computational Linguistics, pp. 655–665. Maryland (2014)
11. Kincade, D.H., Giddings, V.L., Chen-Yu, H.J.: Impact of product-specific variables on consumers post-consumption behaviour for apparel products: USA. J. Consumer Stud. Home Econ. **22**, 81–90 (1998)
12. Lee-Kelley, L., Gilbert, D., Mannicom, R.: How e-CRM can enhance customer loyalty. Market. Intell. Plan. **21**, 239–248 (2003)
13. Mayer, R.C., Davis, J.H., Schoorman, F.D.: An integrative model of organizational trust. Acad. Manag. Rev. **20**, 709–734 (1995)
14. Mcknight, D.H., Chervany, N.L.: What trust means in e-commerce customer relationships: an interdisciplinary conceptual typology. Int. J. Electron. Commerce **6**, 35–59 (2001)
15. Morgan, R.M., Hunt, S.D.: The commitment-trust theory of relationship marketing. J. Market. **58**, 20–38 (1994)
16. Zeithaml, V.A.: Service quality, profitability, and the economic worth of customers: what we know and what we need to learn. J. Acad. Market. Sci. **28**(1), 67–85 (2000)

Analysis of Deep Neural Networks
for Automobile Insurance Claim Prediction

Aditya Rizki Saputro$^{(\boxtimes)}$, Hendri Murfi, and Siti Nurrohmah

Department of Mathematics, Universitas Indonesia, Depok 16424, Indonesia
{aditya.rizki41,hendri,snurrohmah}@sci.ui.ac.id

Abstract. Claim prediction is an important process in an automobile insurance industry to prepare the right type of insurance policy for each potential policyholder. The volume of available data to construct the model of the claim prediction is usually large. Nowadays, deep neural networks (DNN) becomes more popular in the machine learning field especially for unstructured data likes image, text, or signal. The DNN model integrates the feature selection into the model in the form of some additional hidden layers. Moreover, DNN is suitable for the large volume of data because of its incremental learning. In this paper, we apply and analyze the accuracy of DNN for the problem of claim prediction which has structured data. First, we show the sensitivity of the hyperparameters on the accuracy of DNN and compare the performance of DNN with standard neural networks. Our simulation shows that the accuracy of DNN is slightly better than the standard neural networks in term of normalized Gini.

Keywords: Claim prediction · Structured data · Big data · Deep learning · Deep neural network

1 Introduction

An automobile insurance claim is a request from the policyholder for financial coverage caused by automobile-related loss. Automobile insurance provides bodily injury, property damage, comprehensive physical damage, and collision. The process of filling an automobile claim can be very different depending on the cause of damage, profile of the policyholder, and other factors[1].

Claim prediction is an important process in the automobile insurance industry because the insurance company can prepare the right type of insurance policy for each potential policyholder. Inaccuracies in automobile insurance company's claim predictions raise the cost of insurance for good drivers and reduce the price for bad ones. The more accurate prediction will allow them to adjust pricing better.

From a machine learning point of view, claim prediction can be categorized as a classification problem. Given the historical claim data, machine learning built a model that predicts if a policyholder will initiate an automobile insurance claim in the next year or not. Many machine learning models have been used for the claim prediction

---

[1] http://www.claimsjournal.com/news/national/2013/11/21/240353.htm

© Springer Nature Singapore Pte Ltd. 2019
Y. Tan and Y. Shi (Eds.): DMBD 2019, CCIS 1071, pp. 114–123, 2019.
https://doi.org/10.1007/978-981-32-9563-6_12

problem [1]. The volume of available data to construct the model of claim prediction is usually large. Therefore, we also need models that can handle a large volume of data [2]. Nowadays, deep learning becomes more popular in the machine learning field especially for unstructured data likes image, text, or signal [3, 4]. Deep neural networks (DNN) is a neural network model consisting of several hidden layers with many hyperparameters that need to be adjusted to form an optimum accuracy. The DNN model integrates the feature selection into the model in the form of some additional hidden layers. Moreover, DNN is suitable for the large volume of data because of its incremental learning.

In this paper, we apply and analyze the accuracy of DNN for the problem of claim prediction which has structured data. First, we show the sensitivity of the hyperparameters on the accuracy of DNN and compare the performance of DNN with standard neural networks. Our simulation shows that the accuracies of DNN are slightly better than the standard neural networks in term of normalized Gini.

The rest of this paper is organized as follows: In Sect. 2, the review of DNN. Section 3 describes the experiment. In Sect. 4, we discuss the results of the simulations. Finally, we give a conclusion in Sect. 5.

2 Deep Neural Networks

Deep Neural Networks (DNN) is a type of feed-forward neural networks. The 'deep' indicates that the neural networks have several layers. Generally, DNN has one input layer, two or more hidden layers, and one output layer.

Originally, DNN is used to process unstructured data such as images, text, and sounds [3, 4]. However, DNN can also be used to process structured data by modifying its layers. The DNN structure is modified by adding the entity embedding process after the input layer and the dropout in each hidden layer. Figure 1 is the DNN structure we use in this paper. We use three hidden layers and use entity embedding to convert category features into a vector. We also use dropout for eliminating some neurons in each hidden layer.

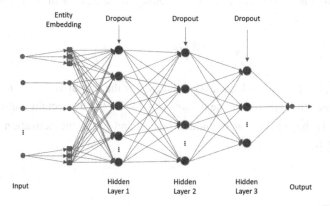

Fig. 1. DNN structure with three hidden layers

2.1 Entity Embedding

In machine learning, embedding is a process of projecting an input into another more convenient representation. Word embedding is a popular embedding method in DNN where words or phrases from the vocabulary are mapped to vectors of real numbers. There are several word embedding methods, one of which is one-hot encoding. For example, the word embedding of "Deep learning is deep" using the one-hot encoding is as follows: categories the phrase such as word_deep, word_learning, and word_is, for each category then creates a vector where its elements are one for the intended category and zero for the other categories.

In DNN, there is an embedding method that is suitable for processing problems that have structured data with categorical features called the entity embedding. The entity embedding is mapping the category features into a vector by adding linear neurons after one-hot encoding [5]. Given x, an input of a categorical feature that has m categorical values. Next, x is mapped with one-hot encoding to a vector $\boldsymbol{\delta} = (\delta_1, \delta_2, \ldots, \delta_m)^T$ of length m where the indices indicate the categorical features. Given a $d \times m$ matrix of embedding weights W, the output of the entity embedding layer is $x = W\boldsymbol{\delta}$. The d-length vector x is the result of the entity embedding process for xx. This d is commonly called as the size of embedding. The size of embedding needs to be determined in advance and can affect the model accuracy.

Until now there is still no rule to determine the exact size of the entity embedding. The only rule is to choose the size of embedding in the range 1 to $c - 1$ where c is the number of the category in the categorical features. However, some machine learning practitioners try to construct the rules for determining the size of entity embedding. One of them is Jeremy Howard who built a solution to the problem of Rossman store sales. According to Howard, the embedding size is $d = (c + 1)div2$ where max$d = 50^2$.

2.2 Hidden Layers

The hidden layer is the intermediate layer between the input layer and an output layer. In the hidden layer, neurons receive as input vectors from the previous layer and perform a simple computation, i.e., a weighted sum of the input vectors followed by mapping with a nonlinear function [4]. Generally, the data flow in the l-hidden layer is as follows:

$$y^{(l)} = f^{(l)}\left(W^{(l)}y^{(l-1)} + w_0^{(l)}\right) \tag{1}$$

where $l = 1, \ldots, L$ is the hidden layer index, $y^{(l)}$ is the output vector of hidden layer l which becomes the input of the next layer $(l + 1)$, $W^{(l)}$ is the weight matrix of hidden layer l, $w_0^{(l)}$ is the bias vector of hidden layer l and $f^{(l)}$ is the activation function of hidden layer l.

---

[2] http://towardsdatascience.com/deep-learning-structured-data-8d6a278f3088.

Two hyperparameters can be adjusted in hidden layers, e.g., the number of hidden layers and the number of neurons in hidden layers. We must adjust these hyperparameters to obtain optimum accuracy of the model. There are no exact rules to determine the number of hidden layers, and we use three hidden layers in our simulations. To determine the number of neurons, we use the heuristic and rule of thumb method [6]. The rule of thumb method is described as follows:

- The number of neurons should be in the range between the size of the input layer and the size of the output layer
- The number of neurons should be 2/3 of the input layer size, plus the size of the output layer
- The number of neurons should be less than twice the input layer size.

2.3 Dropout

Dropout is the process of eliminating neurons in input or hidden layers. By eliminating some neurons, the dropout process prevents overfitting which is a situation where the DNN fits all training data well. However, it gives worse accuracy in testing data. The dropout is also useful for reducing the running time of DNN.

Each input vector of a layer l with a dropout rate p is multiplied by $r_j^{(l)}$ which is a random variable with Bernoulli distribution and $p \in (0, 1)$ [7]. A random variable with Bernoulli distribution is used because the dropout process is like the Bernoulli experiment which produces one of two possible outcomes namely 'yes' or 'no'. In the dropout process, a random variable with Bernoulli distribution is used to determine whether a neuron is still being used or not.

The parameter p or dropout rate is one of hyperparameter on DNN that indicates the probability of the layer are dropped out. We can use a heuristic approach to determine the dropout rate. However, Srivastava stated that in general the optimal value of p is 0.5 [7].

2.4 Normalized Gini Coefficient

To evaluate the model, we use the normalized Gini coefficient as a model evaluation metric. Originally, the Gini coefficient is used to calculate a nation's inequality of income distribution. Next, it is adapted into a model evaluation metric for classification problems. One method to calculate normalized Gini coefficient for a binary classification problem is described below.

Given a target vector $y = (y_1, y_2, \ldots, y_n)^T$ where $y_i \in \{0, 1\}, 1 \le i \le n$ is the correct class of the i-th observation and a predicted probability vector $f(x) = (f(x_1), f(x_2), \ldots, f(x_n))^T$ where $f(x_i), 1 \le i \le n$ is the predicted probability that the i-th observation will fall into the class '1' obtained from a classification algorithm. First, sort the $f(x)$ elements in large to small order, then use a sorting method to change the order of the initial vector elements y into a sequence of elements y corresponding to the element $f(x)$ that has been sorted before. After that, the normalization coefficient value can be calculated by:

$$Y = \frac{cl_0 * cl_1}{2} \qquad (2)$$

$$nG = \frac{Y - X}{Y} \qquad (3)$$

where X is the number of iterations needed for sorting y elements, cl_i is the number of observations in y which have class classification 'i', and nG is the normalized Gini coefficient. The value of the normalized Gini coefficient is $0 \leq nG \leq 1$. The nG value that is getting closer to 1, the prediction for each x_i is more accurate [2].

3   Experiment

3.1   Datasets

To build and evaluate the model of claim prediction, we use the datasets from Porto Seguro's Safe Driver Prediction[3]. There are two datasets, i.e., a training dataset to build the model and a testing dataset to evaluate the accuracy of the model. The two datasets have the following characteristics:

- There were 595,212 observations on training data and 892,816 observations on testing data
- There are 57 features about the individual driver, vehicle, the operational area, and features containing previously calculated values. The features consist of binary and categorical features
- The labels of the dataset are '0' for 'no claim' and '1' for 'submit a claim'
- Training data is not balanced because it only has about 6% for the target label '1'
- There are around 2.49% missing values in training data and testing data.

3.2   Data Preprocessing and Entity Embedding

The available datasets cannot directly be used in this study. We need to preprocess the datasets before we use them in DNN model. The data preprocessing in this study are as follows:

- The missing value is changed to a new category value that is -1
- Selecting features by removing 20 features in the features containing previously calculated values that have low feature-target correlation values
- Perform an entity embedding process with four dimensions of entity embedding, as follows:
  - Size A: $d = (c + 1)div2$ with max$d = 50$
  - Size B: $d = (c + 1)div2$ with max$d = 30$
  - Size C: $d = (c + 1)div2$ with max$d = 10$
  - Size D: $d = (c + 1)div2$ with max$d = 5$.

---

[3] https://www.kaggle.com/c/porto-seguro-safe-driver-prediction/data.

3.3 Hyperparameters Tuning

In this study, three hyperparameters must be optimized in the model selection, i.e., the number of neuron units in the hidden layers or neurons size, the dropout rates and the dimensions of the entity embedding or embedding size. Table 1 gives the candidate values for each hyperparameter.

Table 1. Hyperparameter

Neuron sizes	Dropout rates	Entity embedding size
40 – 25 – 15	0,1	Size A
60 – 40 – 15	0,2	Size B
80 – 50 – 30	0,3	Size C
100 – 40 – 15	0,4	Size D
120 – 80 – 40	0,5	
140 – 100 – 60		

We select the optimal hyperparameters from 120 combinations of six neuron sizes, five dropout rates, and four entity embedding sizes. We conduct three simulations for a hyperparameter combination to get the average accuracy of the DNN model for the hyperparameter combination. The combination of the hyperparameters which give the most accurate model is selected as the final hyperparameters for the final DNN model.

3.4 Evaluation

To calculate the accuracy of a model, we submit the prediction of the model for testing data to the owner of dataset[4]. We get two values from the submission, i.e., the private score and public score. The private score is the normalized Gini coefficient value for 70% of the testing dataset. The public score is the normalized Gini coefficient value for 30% of the testing dataset. In this study, we use the private score to analyze the accuracy of a model. The similar mechanism is applied to calculate the accuracy of the standard neural network model.

4 Result

4.1 Effect of Dropout Rate

First, we analyze the sensitivity of each hyperparameter from 120 combinations. The accuracy scores of the testing dataset are given in Table 2. In Table 2, there are some accuracy scores with underline font showing the optimal scores for each combination of embedding sizes and neuron sizes. Mostly, these underline accuracy scores are from the simulations with a dropout rate of 0.1. We also can say that in general, the optimum dropout rate is 0.1.

---

[4] https://www.kaggle.com/c/porto-seguro-safe-driver-prediction/submit.

4.2   Effect of Embedding Size

From Table 2, we can calculate the average accuracy score for the simulations with the same embedding size and the same neuron sizes as shown in Figs. 2 and 3. From Fig. 2, we see that the accuracy scores of embedding sizes tend to increase with the neuron size of 120-80-40 and 140-100-60. For the other neuron sizes, the accuracy scores of embedding sizes tend to decrease. Overall, we reach the optimum accuracy score at the C-size embedding. In general, from Fig. 3 we can see that the embedding sizes do not have a significant effect on the accuracy scores.

Table 2. The accuracy scores of the DNN model.

Neuron Sizes	Dropout Rates	Embedding Sizes			
		A	B	C	D
40-25-15	0.1	0.27839	0.27853	0.27760	0.27679
	0.2	0.27607	0.27621	0.27749	0.27431
	0.3	0.27626	0.27636	0.27496	0.27495
	0.4	0.27545	0.27429	0.27467	0.27476
	0.5	0.27306	0.27414	0.27290	0.27259
60-40-15	0.1	0.27963	0.27990	0.27935	0.27908
	0.2	0.27797	0.27827	0.27904	0.27330
	0.3	0.27795	0.27911	0.27744	0.27810
	0.4	0.27765	0.27690	0.27529	0.27691
	0.5	0.27606	0.27594	0.27529	0.27530
80-50-30	0.1	0.27963	0.28049	0.28070	0.28103
	0.2	0.27968	0.27989	0.27971	0.28011
	0.3	0.27999	0.27927	0.27914	0.27837
	0.4	0.27917	0.27889	0.27867	0.27821
	0.5	0.27723	0.27801	0.27693	0.27667
100-40-15	0.1	0.28037	0.28088	0.28123	0.28078
	0.2	0.28060	0.27973	0.28082	0.27944
	0.3	0.28043	0.28079	0.27951	0.27936
	0.4	0.27976	0.28010	0.27852	0.27853
	0.5	0.27792	0.27694	0.27730	0.27700
120-80-40	0.1	0.28027	0.28092	0.28182	0.28136
	0.2	0.28010	0.28070	0.28104	0.28045
	0.3	0.28078	0.28039	0.28084	0.27989
	0.4	0.27979	0.28021	0.27951	0.27900
	0.5	0.27570	0.27892	0.27837	0.27817
140-100-60	0.1	0.27990	0.28016	**0.28231**	0.28172
	0.2	0.28018	0.28056	0.28119	0.28132
	0.3	0.27959	0.28055	0.28102	0.28069
	0.4	0.27979	0.27985	0.27986	0.27971
	0.5	0.27930	0.27927	0.27877	0.27861

4.3   Effect of Neurons Size

Using Fig. 3, we also can see the sensitivity of neurons sizes to the accuracy of the DNN model. Figure 3 shows that the accuracy scores tend to grow quadratically as the neuron sizes get bigger. On average, the neuron size which gives the optimal accuracy scores is 140-100-60.

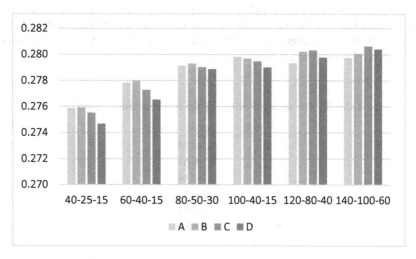

Fig. 2. The sensitivity of embedding sizes to the accuracy of the DNN model

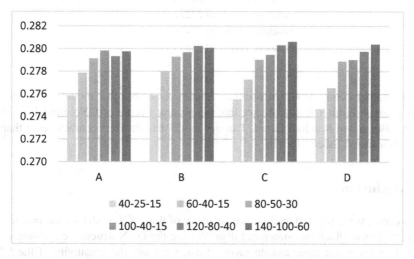

Fig. 3. The sensitivity of neurons sizes to the accuracy of the DNN model

4.4   Comparison

There is an accuracy score with bold font in Table 2. This accuracy is the optimum accuracy score among all the combinations. The optimum accuracy score is 0.28231 which is the accuracy score for the DNN model with the dropout rate of 0.1, the C-size embedding and the size of the neuron of 140-100-60.

For comparison purpose, we use the standard neural networks with one hidden layer. We select the optimal number of neurons from six candidates as shown in Table 3. From Table 3, we see that the optimum accuracy score of the standard neural networks is 0.26214. It means that the standard neural networks give a slightly smaller accuracy score than of the DNN model.

Table 3. The accuracy scores of the standard neural networks.

Neurons sizes	Accuracy scores
40	**0,26214**
60	0,26153
80	**0,26214**
100	0,25881
120	0,25682
140	0,25965

Table 4. The accuracy scores of other models.

Models	Accuracy scores
XGBoost	0,28518
AdaBoost	0,27271
Stochastic GB	0,28499
Random Forest	0,25933

For another comparison, we use the accuracy score of other models from Fauzan [2] in Table 4. From Table 4, we see that the accuracy score of DNN model is bigger than AdaBoost and Random Forest. But our model has smaller accuracy score than of the XGBoost and Stochastic GB.

5   Conclusion

In this paper, we apply and analyze the accuracy of the DNN model for the problem of claim prediction which has structured data. We use the DNN structure consisting of an embedding layer and three hidden layers. First, we show the sensitivity of the DNN hyperparameters to the accuracy of the DNN model. We get the optimal DNN model with the neuron size of 140-100-60, the C-size embedding, and the dropout rate of 0.1.

Moreover, our simulations show that the DNN model gives slightly more accurate than the standard neural networks in term of normalized Gini score.

Acknowledgements. This work was supported by Universitas Indonesia under PIT 9 2019 grant. Any opinions, findings, and conclusions or recommendations are the authors' and do not necessarily reflect those of the sponsor.

References

1. Weerasinghe, K.P.M.L.P., Wijegunasekara, M.C.: A comparative study of data mining in the prediction of auto insurance claims. European Int. J. Sci. Technol. **5**(1), 47–54 (2016)
2. Fauzan, M.A., Murfi, H.: The accuracy of XGBoost for insurance claim prediction. Int. J. Adv. Soft Comput. Appl. **10**(2) (2018)
3. LeCun, Y., Bengio, Y., Hinton, G.: Deep learning. Nature **521**, 436–444 (2015)
4. Montavon, G., Samek, W., Muller, K.R.: Methods for interpreting and understanding deep neural networks. Digit. Signal Process. **73**, 1–15 (2018)
5. Guo, C., Berkhahn, F.: Entity embeddings of categorical variables. arXiv preprint arXiv:1604. 06737 (2016)
6. Panchal, F.S., Panchal, M.: Review on methods of selecting a number of hidden nodes in the artificial neural network. Int. J. Comput. Sci. Mob. Comput. **3**(11), 455–464 (2014)
7. Srivastava, N., Hinton, G., Krizhevsky, A., Sutskever, I., Salakhutdinov, R.: Dropout: a simple way to prevent neural networks from overfitting. J. Mach. Learn. Res. **15**(1), 1929–1958 (2014)

The Performance of One Dimensional Naïve Bayes Classifier for Feature Selection in Predicting Prospective Car Insurance Buyers

Dilla Fadlillah Salma$^{(\boxtimes)}$, Hendri Murfi$^{(\boxtimes)}$, and Devvi Sarwinda$^{(\boxtimes)}$

Department of Mathematics, Universitas Indonesia, Depok 16424, Indonesia
{dilla.fadlillah,hendri,devvi}@sci.ui.ac.id

Abstract. One of the products sold by insurance companies is car insurance. To offer this product, one of the techniques used by the company is cold calling. This method often decreases the sellers' mentalities because they face many rejections when offering insurance products. This problem can be reduced by classifying prospective buyers' data first. The data can be classified as customers with the potential to buy insurance and customers who have no potential to buy insurance. From the obtained data, there are certainly many features that support the classification process. However, not all features contributed to improving classification accuracy. Machine learning especially the method of feature selection helps to reduce dimensions and to improve classification accuracy. In this paper, we examine One-Dimensional Naïve Bayes Classifier (1-DBC) as a feature selection method that is applied to two classifier methods, i.e., Support Vector Machine and Logistic Regression. Our simulations show that the two classifiers can use fewer features to produce comparable accuracies in classifying prospective car insurance buyers.

Keywords: Car insurance · Feature selection · Logistic regression · One Dimensional Naïve Bayes Classifier · Support vector machine

1 Introduction

Car transportation is a technology in the field of transportation that is widely used by people all over the world today. From 2006 to 2015, there were 1.2 billion car users worldwide[1]. By using car transportation, there are many possible negative risks for its users and owners, such as accidents and theft. According to WHO data in 2017, there are more than 1.25 million cases of traffic accidents every year[2]. The high risk provides a good opportunity for insurance companies to offer insurance products, one of which is selling a car insurance product. To sell car insurance, companies need to do a presentation to buyers that also requires meeting agreements, and to obtain such meeting agreements; you require cold calling [1].

---

[1] https://www.statista.com/statistics/281134/number-of-vehicles-in-use-worldwide/.

[2] http://www.who.int/mediacentre/factsheets/fs358/en/.

© Springer Nature Singapore Pte Ltd. 2019
Y. Tan and Y. Shi (Eds.): DMBD 2019, CCIS 1071, pp. 124–132, 2019.
https://doi.org/10.1007/978-981-32-9563-6_13

Cold calling is one of the methods that can be used to offer car insurance services. Cold calling is a sales technique with a very high rejection rate. This fact causes a decrease in the sellers' mentalities to make the best offer. Thus, offering car insurance with the cold calling technique will be more effective if the salesperson has the data of the customers that are classified whether customers have the potential to buy insurance or not. By doing the classification, it will reduce the problem of rejection and increase the sales percentage.

From a machine learning point of view, the problem of determining prospective buyers of car insurances is a classification problem of supervised learning because the form of the output value is class, which is the chance to either buy car insurance or not. Machine learning makes the problem easy for humans and produces more accurate results than what humans do [2, 3]. In the classification process, features are used as references to predict the classification. So, feature selection is an important part of optimizing the performance of the classifier [4].

The feature selection method becomes very important to get the best features with high accuracy of the classification. There are many feature selection methods, and the recent one is One-Dimensional Naive Bayes Classifier (1-DBC) [5]. 1-DBC is a feature selection method that uses Naive Bayes to classify data by considering one feature only. Then, we use the accuracy of the classification to score the feature. The feature selection method is applied to all scored features to select the most important ones. In this paper, we apply the 1-DBC as a feature selection method to order the features of a classification problem in predicting prospective car insurance buyers. We examine the performance of the feature selection method based on the accuracy of two classifiers, e.g., SVM and logistic regression. Our simulations show that the two classifiers can use a smaller number of features to produce comparable accuracies.

The outline of this paper is as follows: In Sect. 2, we present the reviews of related methods, i.e., Naive Bayes, SVM, logistic regression, and 1-DBC for feature scoring. Section 3 describes the results of our simulations and discussions. Finally, a general conclusion about the results is presented in Sect. 4.

2 Methods

There are various kinds of feature selection methods that have been used by previous researchers. Examples of the types of features that can be used for this problem include Sequential Feature Selection, Mutual Information (MI), and Information Gain (IG) [6, 7]. Jalil et al. present a comparative study on correlation and information gain algorithms to evaluate and to produce a subset of crime features. With the selected features, their accuracy becomes 96.94% [8]. Another almost similar study is the prediction of customer attrition of commercial banks using an SVM model. In this study, He et al. reach the prediction accuracy by 98.95% [9].

In 2017, Cinelli et al. used the 1-DBC method to improve the accuracy of the SVM on microarray data. After the implementation of the 1-DBC, SVM reached the accuracy of more than 90% [5]. This result is not so different from the two previous studies. However, the 1-DBC method is easier and simpler to apply to data with fewer features. 1-DBC used by Cinelli et al. is the application of Bayes rules to calculate the probability log ratio of features that are included in two classes, which is different with 1-DBC used by us that calculate the score of each feature using the Naive Bayes Classifier.

SVM has been claimed as one of the most popular and convenient techniques and logistic regression that widely used in many areas of scientific research [10, 11]. Therefore, to analyze the performance of 1-DBC, we use those two classifiers in predicting prospective buyers for solving problems related to the classification of data Before fitting the customer data, we conduct a preprocessing to make the data suitable for each classifier.

2.1 Naïve Bayes

Naive Bayes is a classification algorithm that is very fast and simple. Naïve Bayes is a classification technique based on the Bayes theorem assuming independence between its features. This method is the basis of the 1-DBC method. Let $x = (x_1, x_2, \ldots x_M)^T$ be a data vector where i-th features expressed by x_i, and C_k be the target class. Based on the Bayes theorem, the Naïve Bayes equation can be stated as follows [12]:

$$P(C_k|x_i) = \max_{k \in \{1,\ldots,K\}} P(C_k) \prod_{i=1}^{M} P(x_i|C_k) \tag{1}$$

where $P(C_k|x_i)$ is the posterior probability of class k, $P(C_k)$ is the prior probability of class k, and $P(x_i|C_k)$ is the probability of the i-th feature given with class k.

2.2 Support Vector Machine

Support vector machine is one of a supervised learning model for the problem of classification. The main purpose of the SVM model is to determine a hyperplane that can maximize margins so that it can classify the data into the correct class. The general form of the hyperplane for the classification problem is as follows [12]:

$$y(x) = w^T x + b \tag{2}$$

where x is the input vector, w is the weight vector, and b is the bias parameter. The hyperplane works well for linearly separable data. However, the data that exist in the real world are often not linearly separable data. Therefore, we need a method to overcome the nonlinear data by implementing soft margin and kernel functions. The kernel function for SVM that will be used in this study is the RBF Gaussian kernel function with the following forms:

$$K(x, x_j) = exp(-\gamma \| x - x_j \|^2) \tag{3}$$

where $\gamma \geq 0$ is a parameter of the kernel functions.

2.3   Logistic Regression

Logistic regression is one of the machine learning model developed with a probability approach to solving the classification problem. This method uses a generalized linear model function to solve the two-class classification problems. The general form of logistic regression is as follows [12]:

$$y(x) = (w^T x) \tag{4}$$

where σ is a sigmoid logistical function in the form of

$$\sigma(a) = \frac{1}{1 + e^{-a}} \tag{5}$$

Given a training data $\{x_n, t_n\}, n = 1, 2, \ldots, N$, where $t_n \in \{0,1\}$. Then, $p(t_n = 1|w) = \sigma(w^T x_n) = y_n$ and the joint probability $t = (t_1, \ldots, t_N)^T$ is:

$$p(\mathbf{t}|\mathbf{w}) = \prod_{n=1}^{N} y_n^{t_n} \{1 - y_n\}^{1-t_n} \tag{6}$$

2.4   1-DBC Based Feature Selection

Before feature selection, we need to determine the distribution of the feature. According to this distribution, there are three types of Naive Bayes models that can be used, i.e., Gaussian Naive Bayes, Multinomial Naive Bayes, and Bernoulli Naive Bayes [13]. Gaussian Naive Bayes is a feature selection method where the features are normally distributed. Because some features are not binary, we need to normalize the features to match the Gaussian models. Multinomial Naive Bayes is most appropriate for features with categorical data, and Bernoulli Naive Bayes is suitable if data of a feature is binary.

The Naive Bayes is a classification method, and the 1-DBC method is the Naive Bayes method using one feature only. It means that the method classifies the data by considering one feature only. The accuracy of the method is used to score the corresponding feature. The accuracy is an average accuracy by using 3-fold cross-validation. Finally, all features are ordered by their scores to be used in feature selection.

3   Results and Discussion

In this section, we describe the performance of the 1-DBC method as a feature selection method for two classifiers, i.e., SVM, and logistic regression. For this purpose, we use the accuracy measures called true positive and true negative for all cases.

3.1   Dataset

We use a dataset from a bank in the United States to construct the classification model[3]. The bank offers car insurance products using the cold calling technique. To increase sales of car insurance services, the company constructs machine learning models to classify potential customers. The statistics of the dataset are as follow:

1. The number of customers is 4000 customers
2. There are 18 binary features
3. Each customer has a label of '0' for customers who do not buy insurance products and '1' for customers who buy car insurance products. 60% of customers have labels of '0' and 40% of customers have labels of '1'. This means that the data is good enough to describe each category.
4. There are some features that have NaN values in the dataset. Job feature has 19 NaN values, Communication feature has 902 NaN values, Education feature has 169 NaN values, and Outcome feature has the highest NaN values of 3042. The missing values are filled by the average value of corresponding features.

3.2   Computer Software

The machines and computer software used to run the feature selection algorithm and classification of prospective car insurance buyers in this study have the following specifications:

1. Operating system: Windows 10 64-bit
2. Processor: Intel (R) Core (TM) i7-6500U CPU @ 2.50 GHz (4 CPUs), ~2.6 GHz
3. Software: Python 3.6.4 64 bits, Qt 5.6.2, PyQt5 5.6 on Windows.

3.3   Preprocessing

We need to preprocess the dataset because there are some missing values and some categorical values. These missing values and categorical values make the classification process hampered or even completely impossible. We perform three preprocessing processes including overcoming the missing values, changing the categorical values, and normalizing data [14].

There are several approaches that can be used to overcome the problem of missing values. Such approaches include removing columns with missing values, replacing the missing values with the average values of the corresponding features, and predicting the missing values. In this study, we replace the missing values with the average values of the corresponding features. This method is suitable for data that are not too large[4].

We need to transform the categorical values because some classification methods do not support to process data in the form of categorical data. Therefore, we convert categorical data into numerical data using a label encoder method. The label encoder

---

[3] https://www.kaggle.com/emmaren/cold-calls-data-mining-and-modelselection/data.

[4] https://www.analyticsindiamag.com/5-ways-handle-missing-values-machine-learning-datasets.

method is a method used to convert a categorical feature into a numeric-value vector. The size of the vector equals the number of categories in the categorical feature[5].

The last preprocessing is normalization. We need to normalize the dataset because the dataset has a different scale, some are large, and some are small. The preprocessing method used to normalize the dataset is the Z-score method. The Z-Score method is one of the normalization methods that use the mean and standard deviation to normalize the dataset.

3.4   Feature Scoring

First, we show the feature rank in Fig. 1. The score of a feature is the accuracy of the 1-DBC method when the method considers only the feature to classify the dataset. From Fig. 1, we see that the highest-score feature is the Calllength feature that results in the accuracy of 72.19%, while the lowest-score feature is the Balance feature which only results in the accuracy of 58.5%.

3.5   Feature Selection

We show the performance of the 1-DBC method for feature selection in this section. To examine its performance, we use two classifiers, i.e., SVM and logistic regression. According to [5], the 1-DBC feature selection method can improve the accuracy of the SVM for classification of CDR3 repertoires. Also, we consider logistic regression because these models give good classification accuracies for a similar dataset. Both models produce the accuracies of more than 80%.

According to the feature ranking, we select the top n features to classify the dataset using SVM, and logistic regression, where n is 4, 5, 6, 8, 10, 12, 15, 16, 17, and 18. It means that we do not use a feature selection method when we set n equals to 18.

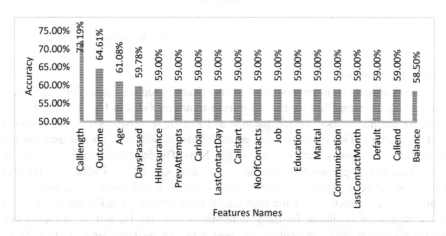

Fig. 1.   The rank of features producing by the 1-DBC feature selection method

[5] https://www.analyticsvidhya.com/blog/2015/11/easy-methods-deal-categorical-variables-predictive-modeling/.

Support Vector Machine

From the previous section, it was mentioned that the kernel used in SVM is the RBF kernel which is the most widely used kernel in SVM for nonlinear data. Therefore, we need to optimize two parameters of the SVM, i.e., parameter trade-off C and parameter *gamma* of the RBF kernel. The candidates of parameters C are 0.1, 1, 3 and 5, while the candidates of parameter *gamma* are 0,001, 0.01, and 0.005. From these candidate parameters, we choose one that produces the highest accuracy for the final SVM model. Figure 2 shows the accuracies of the SVM model for n-top features.

From Fig. 2, we see that the accuracy of the SVM is 78.91% if we use four-top features. The 5th features improve the accuracy of the SVM significantly. The next six features also improve the accuracy of the support vector machine. The accuracy of the SVM is similar for the feature sizes of 15, 16, 17, and 18. It means that an SVM can achieve the best accuracy by considering only 83% of features.

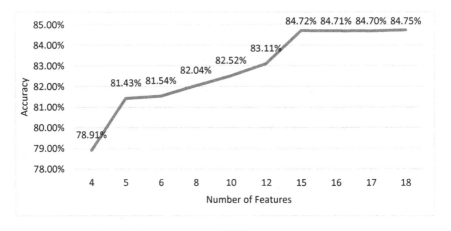

Fig. 2. The accuracies of SVM for the top n features

Logistic Regression

The second classification method used in this study is logistic regression. The accuracies of logistic regression for the top n features are provided in Fig. 3. From Figs. 2 and 3, we see that support vector machine and logistic regression give a quite similar trend of accuracy. Logistic regression gives the lowest accuracy when it uses the top four features. The lowest accuracy is 76.58%. Next, The 5th features also improve the previous accuracy significantly achieving the accuracy of 78.42%. The next six features give a quite different trend for support vector machine and logistic regression. The accuracy tends to fluctuate for the addition of these features. Finally, the accuracy of logistic regression is similar for the feature sizes of 15, 16, 17, and 18. It means that logistic regression also can achieve the best accuracy by considering only 83% of features.

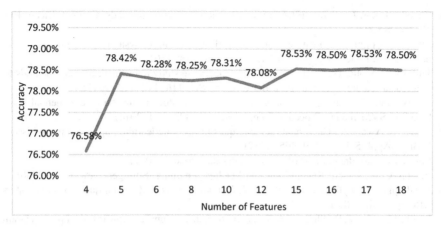

Fig. 3. The accuracies of logistic regression for the top *n* features

4   Conclusions

In this paper, we consider the use of the 1-DBC method for feature selection. This 1-DBC feature selection method score each feature using the accuracy of each feature to classify the dataset. Using this 1-DBC feature selection method, the rank of features from the most to the less influential features are calllength, outcome, age, dayspassed, HH insurance, prevattempts, carloan, lastcontactday, callstart, noofcontacts, job, education, marital, communication, and lastcontactmonth. Next, we implement the 1-DBC feature selection method with two classifiers, i.e., SVM, and logistic regression. Our simulations show that the two classifiers can use a smaller number of features to produce comparable accuracies.

Acknowledgment. This work was supported by Universitas Indonesia under PITTA 2019 grant. Any opinions, findings, and conclusions or recommendations are the authors' and do not necessarily reflect those of the sponsor.

References

1. Schiffman, S.: Cold Calling Techniques: (That Really Work!). Adams Media, Avon (2014)
2. Mitchell, T.M.: Machine learning and data mining. Commun. ACM **42**(11), 30–36 (1999)
3. Mitchell, T.M.: The Discipline of Machine Learning. Carnegie Mellon University, School of Computer Science, Machine Learning Department, Pittsburgh (2006)
4. Wang, S., Li, D., Song, X., Wei, Y., Li, H.: A feature selection method based on improved fisher's discriminant ratio for text sentiment classification. Expert Syst. Appl. **38**, 8696–8702 (2011)
5. Cinelli, M., et al.: Feature selection using a one dimensional naïve Bayes' classifier increases the accuracy of support vector machine classification of CDR3 repertoires. Bioinformatics **33**, 951–955 (2017)

6. Chandrashekar, G., Sahin, F.: A survey on feature selection methods. Comput. Electr. Eng. **40**, 16–28 (2014)
7. Jain, D., Singh, V.: Feature selection and classification systems for chronic disease prediction: a review. Egyptian Inform. J. **19**, 179–189 (2018)
8. Jalil, M.A., Mohd, F., Noor, N.M.M.: A comparative study to evaluate filtering methods for crime data feature selection. Procedia Comput. Sci. **116**, 113–120 (2017)
9. He, B., Shi, Y., Wan, Q., Zhao, X.: Prediction of customer attrition of commercial banks based on SVM model. Procedia Comput. Sci. **31**, 423–430 (2014)
10. Soofi, A., Awan, A.: Classification techniques in machine learning: applications and issues. J. Basic Appl. Sci. **13**, 459–465 (2017)
11. Liu, Y.: On goodness-of-fit of logistic regression model (2007)
12. Bishop, C.M.: Pattern Recognition and Machine Learning. Springer, New York (2006)
13. Vanderplas, J.: Python Data Science Handbook: Tools and Techniques for Developers. OReilly, Beijing (2016)
14. Malik, J.S., Goyal, P., Sharma, A.K.: A comprehensive approach towards data preprocessing techniques & association rules. In: Proceedings of the 4th National Conference (2010)

Clustering

Clustering Study of Crowdsourced Test Report with Multi-source Heterogeneous Information

Yan Yang[1], Xiangjuan Yao[1(✉)], and Dunwei Gong[2]

[1] School of Mathematics, China University of Mining and Technology,
Xuzhou 221116, China
yaoxj@cumt.edu.cn
[2] School of Information and Control Engineering,
China University of Mining and Technology, Xuzhou 221116, China

Abstract. Crowdsourced testing is an emerging testing method in the field of software testing and industrial practice. Crowdsourced testing can provide a more realistic user experience. But crowdsourced workers are independent of each other, they may submit test reports for the same issue, resulting in highly redundant test reports submitted. In addition, crowdsourced test reports with multi-source heterogeneous information tend to have short text descriptions, but the screenshots are rich, and using only text information can lead to information bias in test reports. In view of this, this paper attempts to use the screenshot information in the crowdsourced test report to assist the text information to cluster the crowdsourced test report. Firstly, the text similarity and screenshot similarity of crowdsourced test reports are calculated respectively, then the similarity between crowdsourced test reports is weighted. Finally, test reports are grouped by clustering algorithm based on similarity measure. Testers only need to audit the test report as the representative, which greatly reduces the pressure of the tester's report audit. The final experimental results show that the effective use of the screenshot information in the test report can achieve higher clustering accuracy.

Keywords: Crowdsourced test report · Redundancy · Clustering

1 Introduction

In crowdsourced software testing, a large number of online workers involve in the completion of test tasks, It has been proved can shorten the test cycle and reduce the test cost. unlike traditional software testing [1,2], crowdsourced testing outsources testing tasks to non-specific groups. These non-specific groups, i.e. crowdsourced workers, are characterized by diversity and richness. They can

Supported by National Natural Science Foundation of China (No. 61573362, 61773384, and 61502212), and National Key Research and Development Program of China (No. 2018YFB1003802-01).

provide real feedback from users in different test environments, discover software bugs or improve the software quality based on users functional requirements.But the workers are independent of each other and may submit test reports of the same problem. Coupled with the inherent characteristics of crowdsourced testing, there are a lot of redundant test reports on the crowdsourcing platform. A large number of redundant test reports have greatly increased the review pressure of report reviewers. Therefore, it is urgent to automatically detect redundant crowdsourced test reports.

In crowdsourced testing, crowdsourced workers usually perform a series of operations on mobile phones and submitting test reports [3]. Because it is difficult for mobile devices to write long descriptive reports, it is relatively easy to get other multimedia information such as screenshots or short video operations. As a result, test reports submitted by crowdsourced workers are often multi-source heterogeneous. They prefer a small amount of textual descriptions and supplement the description with multimedia information. The method for distinguishing traditional redundant test reports can no longer meet the needs of crowdsourced test reports with multi-source heterogeneous information. How to effectively utilize the screenshot information and solve the problems of information missing and ambiguity caused by the simplicity of text description is a challenge for crowdsourced test report clustering. Jiang et al. [4] used textual information to perform fuzzy clustering on crowdsourced test reports, and divided a small number of test reports with multiple defect descriptions into different sets. Analysis from the submitted test report data shows that the number of reports containing multiple defect tests is small, because crowdsourced workers are more inclined to submit multiple short single defect test reports within the specified time.

In view of this, this paper proposes a clustering method for crowdsourced test reports by integrating text description and screenshot information. Firstly, the text feature vectors and image feature histogram of reports are obtained by text mining technology and image recognition, and the similarities between test reports are calculated. Secondly, the test report is clustered by clustering algorithm. If the similarity reaches the given threshold, the test reports will be aggregated into the same cluster. Test reports in the same cluster are considered redundant. The experimental results show that high clustering accuracy can be obtained by using the screenshot information to assist text information.

The structure of this paper is organized as follows: Sect. 2 introduces the related work of this paper; The method of processing the crowdsourced test report is proposed in Sect. 3; Sect. 4 presents the algorithm of clustering the crowdsourced test report; Experiments are listed in Sect. 5. Finally, Sect. 6 concludes this paper.

2   Related Work

2.1   Crowdsourced Software Testing

Traditional software testing typically employs professional personnel to operate the program under specified conditions, design test cases, identify program errors, measure software quality, and evaluate whether it meets design requirements [5–7].

Crowdsourced technology has now been applied to different areas of software testing. Howe [8] first proposed the concept of "crowdsourced". Liu et al. [9] studied the application of crowdsourced technology in usability testing. Pastore et al. [10] used crowdsourced technology to generate test input to solve test oracle problems. Dolstra et al. [11] studied the application of crowdsourced technology in GUI test tasks. Musson et al. [12] used the crowd to measure the actual performance of the software. Nebeling et al. [13] proved that the crowdsourced test is an effective method to verify the Web interface.

2.2   Detection of Duplicate Defect Reports

There are a lot of duplicate reports in the bug tracking system, and effective detection of duplicate reports can save developerstime and effort. Runeson et al. [14] used natural language processing techniques to sort similar error reports. Wang et al. [15] used text information and execution information to detect duplicate defect reports. Jalbert and Weimer [16] introduced a classifier that comprehensively uses the surface features of the report and textual semantic similarity to identify duplicate defect reports. Tian et al. [17] extended Jalbert and Weimer's work, and used the support vector machine (SVM) to identify duplicate reports. Sun et al. [18] used the SVM to construct a discriminant model to calculate the possibility of two test reports as duplicate reports. Subsequently, in a later article [19], they extended the BM25 algorithm in which new similarity measures are used to identify duplicate defect reports.

For the crowdsourced test report, Wang et al. [20] proposed a clustering-based crowdsourced test report classification method by using the text information. Subsequently, Wang et al. [21] proposed a crowdsourced test report classification method based on local active learning by asking the user report label. Feng et al. [22] prioritize test reports based on textual application of diversity strategies and risk strategies. Feng et al. [3] calculated the text distance and screenshot distance between test reports, and finally sorted the test reports based on the equilibrium distance between the two.

3   Information Processing of Crowdsourced Test Report

Highly redundant problem with test reports submitted for crowdsourcing platforms, this chapter proposes a clustering method of crowdsourced test reports, which integrates text and image information. The clustering framework of crowdsourcing test reports fusing text and image information is shown in Fig. 1.

Fig. 1. Crowdsourced test report automatic clustering framework

3.1   Test Text Information Processing

The test text description information and the screenshot information are respectively extracted and processed.

Extracting Text Feature Vectors: For the text description information in the extracted crowdsourced test report, we use natural language processing (NLP) technology to process the test text information. First, the test report text information is processed by word segmentation. Word segmentation is carried out using the widely used Language Technology Platform (LTP) [23]. Secondly, using the part-of-speech tag service provided by LTP, the verbs and nouns in the test report are selected to construct the keyword set. Finally, the vector space model is used to convert the text information in the test report into a feature vector.

Calculating Text Similarity: After obtaining the text feature vectors of crowdsourced test reports, we use cosine similarity to calculate the text similarity between test reports. Many experimental results show that it is very effective to use cosine similarity to calculate document similarity. The formula for calculating text similarity of test reports and is as follows:

$$ST(R_i, R_j) = \frac{\sum\limits_{k=1}^{n} W_{ik} \times W_{jk}}{\sqrt{\sum\limits_{k=1}^{n} W_{ik}^2} \times \sqrt{\sum\limits_{k=1}^{n} W_{jk}^2}}$$

From the formula, it can be seen that the greater the value of $ST(R_i, R_j)$, the higher the similarity of test reports R_i and R_j.

3.2   Screenshot Information Processing

The research and application of image recognition technology in the field of software engineering are relatively few. In this paper, We utilize spatial pyramid matching algorithm (SPM) to process images.

Extracting Feature Histogram of Screenshots: Screenshots not only provide a specific screen interface for real application scenarios, but also provide a view of scenarios where defects occur during testing [3]. However, reports submitted by different workers have different resolution and background complexity. Therefore, we use spatial pyramid matching (SPM) algorithm to recognize screenshots and extract feature histograms.

Computing the Similarity of Screenshots: Here we calculate the similarity of screenshots by using the Bhattacharyya coefficient.The feature histograms of screenshot of test reports R_i and R_j are $H_i(l)$ and $H_j(l)$ respectively. The formulas for calculating the similarity of screenshot of test reports R_i and R_j are as follows [24]:

$$SS(R_i, R_j) = \sum_{l=1}^{L} \sqrt{H_i(l)H_j(l)}$$

It can be seen from the formula that the more similar the screenshot feature histograms of test reports R_i and R_j, the larger the value of $SS(R_i, R_j)$.

Some test reports may have multiple screenshots. For the case of multiple screenshots, the similarity between different screenshots of two reports is calculated, and then the average value is obtained.

3.3   Calculate the Similarity Between Crowdsourced Test Reports

After calculating the text similarity and screenshot similarity between the crowdsourced test reports, we combine these values by experimenting with $(0, 1]$ selecting the appropriate weighted ω within the range. We define the weighted similarity of a pair of crowdsourced test reports as:

$$Sim(R_i, R_j) = ST(R_i, R_j)\omega + SS(R_i, R_j)(1 - \omega)$$

At that time, $\omega = 1$ indicated that the test report did not contain screenshot information, and only needed to calculate text similarity. The weight ω can be adjusted by experiment.

4   Clustering Algorithm

The clustering algorithm used in this paper is an improvement to the algorithm used by Gopalan et al. [25]. Unlike the algorithm proposed by them, after we have obtained the similarity value between the crowdsourced test reports, we first choose the two similarity values. The test report is represented as two different clusters. In the remaining test reports, the first test report is compared with the existing two cluster representatives. If the similarity value is greater than or equal to the specified threshold δ, the test report is inserted. In the cluster with the highest similarity value, if the similarity value is less than the specified threshold δ, it is added as a new cluster and is represented as a new cluster; the next test report is similar to the representative of the existing three clusters. The comparison of the value and the threshold, if the similarity value is higher than the threshold, is inserted into the cluster with the highest similarity, otherwise as the representative of the new cluster, and so on, until all the crowdsourced test reports are traversed. The concrete steps of the algorithm are shown in Fig. 2.

Input: p crowdsourcing test reports $R = \{R_1 R_2, \cdots, R_i, \cdots, R_p\}$ and crowdsourcing test report weighted similarity $sim(R_i, R_j)$ matrix SIM

Output: t clusters

Step 1: Initialize the basic parameters of the algorithm. Setting the weighted value m, threshold δ, traversing the crowdsourcing test report number $\alpha = 0$, $t = 0$;

Step 2: Find $Min\{sim(R_i, R_j)\}$ from SIM, and take R_i and R_j in $Min\{sim(R_i, R_j)\}$ as representatives of two clusters, namely $\{R_i\}$ and $\{R_j\}$, at this time, $\alpha = 2$, $t = t + 2$;

Step 3: $\alpha = \alpha + 1$, take the smallest subscript R_s in set $\{R - R_i - R_j\}$ and calculate $Max\{sim(R_s, R_i), sim(R_s, R_j)\}$. If $Max\{sim(R_s, R_i), sim(R_s, R_j)\} \geq \delta$, R_s is added to the cluster corresponding to $Max\{sim(R_s, R_i), sim(R_s, R_j)\}$; otherwise $t = t + 1$, R_s is the representative of the new cluster, namely $\{R_s\}$.

Step 4: Determine whether $\alpha \leq p$ is satisfied, if the condition is satisfied, then go to step 3; otherwise output t, and end the algorithm.

Fig. 2. Clustering algorithm of crowdsourced test reports based on similarity

5   Experiments

5.1   Data Acquisition

The crowdsourced test report we obtained was the test report used by Feng et al. [3], the crowdsourced test process of five applications simulated by three companies and more than 300 students, and the corresponding test report was obtained. The five applications are:

- CloudMusic: a music player made by NetEase;
- SE-1800: intelligent substation monitoring application produced by Shuneng Technology;
- Ishopping: Taobao shopping app made by Alibaba;
- Justforfun: picture expression sharing application;
- UBook: An online education app developed by New Oriental.

From the statistical analysis of the obtained test report data, it can be found that there are three application reports containing screenshots more than half of the total number of test reports, and the other two applications containing screenshots of the test report, although not more than half of the total number of test reports, but the proportion Close to 50%, so the crowdsourced test report with screenshots is ubiquitous. We can use the screenshot information in the crowdsourced test report to assist in supplementing the text information to cluster the test report.

In addition, it can be seen from Table 1 that a small number of test reports with multiple defect descriptions are available. In CloudMusic application test report set, the number of multi-defect test reports is only 3.36%; in Justforfun application test report set, the number of multi-defect test reports is more, but only 18.90%. So the method of this chapter only clustered the test report once.

Table 1. Distribution of crowdsourced test report screenshots and multi-defect test reports for five applications.

	CloudMusic	SE-1800	Ishopping	Justforfun	UBook
The total number of test report	238	348	408	291	443
Number of test reports with graphs (proportion)	116 (48.74%)	277 (79.60%)	276 (67.65%)	128 (43.99%)	286 (64.56%)
Number of multi-defect test reports (proportion)	8 (3.36%)	36 (10.34%)	28 (6.86%)	55 (18.90%)	57 (12.87%)

5.2   Evaluation Indicators

To measure the effectiveness of clustering, we used F-measure to evaluate clustering effects. Among them, some indicators are defined as follows in Table 2.

Table 2. Distribution of crowdsourced test reports with screenshots.

		Test report category predicted by clustering algorithm	
		Redundancy test report	Non-redundant test report
Test report actual classification	Redundancy test report	TP	FN
	Non-redundant test report	FP	TN

Firstly, the clustering results of each cluster are evaluated, and then the average value is calculated to evaluate the overall clustering effect. The formulas for defining the indicators for evaluating the clustering effect are as follows:

Average accuracy is defined as:

$$\overline{Pr} = \frac{\sum_{\kappa=1}^{t} TP_{\kappa}}{\sum_{\kappa=1}^{t} (TP_{\kappa} + FP_{\kappa})}$$

The average recall rate is defined as:

$$\overline{Re} = \frac{\sum_{\kappa=1}^{t} TP_{\kappa}}{\sum_{\kappa=1}^{t} (TP_{\kappa} + FN_{\kappa})}$$

Finally, the harmonic mean value between the two is taken as the comprehensive evaluation of clustering algorithm, that is, to evaluate the clustering algorithm.

$$\overline{F} = \frac{2 * \overline{Pr} * \overline{Re}}{\overline{Pr} + \overline{Re}}$$

From the above formula, we can see that the larger the $\overline{F}$ value, the better the clustering effect.

5.3 Experimental Results and Analysis

By observing the data set, we found that a small number of test reports did not contain any descriptive information. Therefore, in order to eliminate the impact of these invalid test reports on clustering, the test reports with zero feature vectors after text processing are deleted and the remaining test reports are clustered.

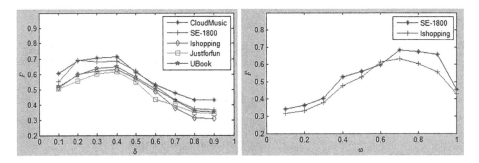

Fig. 3. Crowdsourced test reports $\overline{F}$ values at different thresholds.

Fig. 4. Crowdsourced test reports $\overline{F}$ values at different weights.

(1) **The first group of experimental results and analysis.** Firstly, we analyze the effect of different thresholds on clustering results. In this group of experiments, all test reports of different applications are clustered using the method proposed in this paper. The thresholds δ are 0.1-0.9 and the weights ω are 0.7. The $\overline{F}$ value of each application test report set is calculated, calculation results show in Fig. 3.

As can be seen from Fig. 3, given different thresholds, different $\overline{F}$ values will be obtained. The $\overline{F}$ value of test report set of CloudMusic, UBook, Justforfun and ishopping application gradually increases from 0.1 to 0.4. When the threshold is 0.4, the $\overline{F}$ value reaches its peak, then it gradually declined and stabilized. For SE-1800, the change trend of $\overline{F}$ value is a little more complicated. Also, we can see that when the threshold value of SE-1800 application test report set is 0.2, the $\overline{F}$ value is 0.6921; when the threshold value is 0.4, the $\overline{F}$ value is 0.6836, the difference between the two is small. Therefore, we believe that when the given threshold is 0.4, the clustering effect of crowdsourced test reports of these five applications is better.

(2) **The second group of experimental results and analysis.** The impact of different weights on clustering results is analyzed. In this group of experiments, the weight is 0.1-1, and the threshold is set to 0.4. We selected SE-1800 and Ishopping applications with more images for experiments. Calculate

the $\overline{F}$ value of each application test report. The experimental results are shown in Fig. 4.

As can be seen from Fig. 4, setting different weights will affect the clustering effect of test reports. When ω is 0.7, the $\overline{F}$ value of test report set of SE-1800 and Ishopping application reaches the highest, which is 0.6836 and 0.6318. When ω is 0.1, the test report clustering is mainly based on screenshot information, and $\overline{F}$ value is low. This is because not every test report contains screenshots. If only the screenshots are used as the main part, the impact of text information in the test report will be neglected. In addition, SPM algorithm only focuses on the texture features of the image. Therefore, only screenshots are the main information, and the clustering effect is poor. When the weight ω is 1, the test report clustering only uses text information. The $\overline{F}$ values of test reports for SE-1800 and Ishopping applications are only 0.4557 and 0.4312, because the text information in test reports is too brief.

(3) **The third group of experimental results and analysis.** Finally, using this method and text information only, the clustering effect of test reports is compared experimentally. The method of this article is to fuse text and image information with weight $\omega = 0.7$. Using text information only means weight $\omega = 1$. The thresholds were optimized by the previous experiments, $\delta = 0.4$. The results are listed in Table 3.

Table 3. Different methods of contrast experiment.

Application	Fusion of text and image information $\overline{F}$	Using only text information $\overline{F}$
CloudMusic	0.7123	0.5743
SE-1S00	0.6836	0.4557
Ishopping	0.6318	0.4312
Justforfun	0.6171	0.5003
UBook	0.6484	0.5276

From Table 3, it can be found that the $\overline{F}$ value obtained by the method proposed in this article is higher than that obtained by clustering only text information. Because only using text information, text information is simple, and different applications have specific operating interfaces, using screenshots can reduce the semantic deviation caused by short text appropriately. Therefore, clustering analysis of crowdsourced test reports is effective by fusing text and image information in test reports.

6   Summary

This paper proposes a clustering method for crowdsourced test reports with multi-source heterogeneous information. According to the characteristics of crowdsourced test reports, this paper effectively utilizes the screenshot information in test reports and reduces the diversity of text semantics.

In order to reduce the test review pressure of testers and save a lot of manpower and material resources, this paper uses the combination of screenshot information and text information in the test report to cluster test reports. By calculating the similarity of the test reports, a clustering algorithm is used to cluster the test reports based on the similarity. Redundant test reports are clustered into the same cluster. Finally, the experimental results demonstrate the effectiveness of the proposed method.

It should be noted that although this method can improve the clustering accuracy of crowdsourced test reports, it also has shortcomings. When there are multiple screenshots in a test report, taking the vector average of multiple screenshots, without considering that multiple screenshots may describe different defects, simply obtaining the vector average will affect the clustering effect. How to effectively solve this problem requires further research and discussion.

References

1. Yao, X.J., Gong, D.W., Zhang, G.J.: Constrained multi-objective test data generation based on set evolution. IET Softw. **9**(4), 103–108 (2015)
2. Yao, X.J., Gong, D.W., Wang, W.L.: Test data generation for multiple paths based on local evolution. Chin. J. Electron. **24**, 46–51 (2015)
3. Feng, Y., Jones, J.A., Chen, Z.Y., Fang, C.R.: Multi-objective test report prioritization using image understanding. In: ACM International Conference on Automated Software Engineering, pp. 202–213. ACM, Singapore (2016)
4. Jiang, H., Chen, X., He, T., Chen, Z., Li, X.: Fuzzy clustering of crowdsourced test reports for apps. ACM Trans. Internet Technol. **18**(2), 1–28 (2018)
5. Gong, D.W., Zhang, W.Q., Yao, X.J.: Evolutionary generation of test data for many paths coverage based on grouping. J. Syst. Softw. **84**(12), 2222–2233 (2018)
6. Yao, X.J., Harman, M., Jia, Y.: A study of equivalent and stubborn mutation operators using human analysis of equivalence. In: Proceedings of 36th International Conference on Software Engineering, ICSE, Hyderabad, pp. 919–930 (2014)
7. Gong, D.W., Yao, X.J.: Testability transformation based on equivalence of target statements. Neural Comput. Appl. **21**, 1871–1882 (2012)
8. Howe, J.: The rise of crowdsourcing. Wired Mag. **14**(6), 1–4 (2006)
9. Liu, D., Bias, R.G., Lease, M., Kuipers, R.: Crowdsourcing for usability testing. Proc. Am. Soc. Inf. Sci. Technol. **49**(1), 1–10 (2013)
10. Pastore, F., Mariani, L., Fraser, G.: CrowdOracles: can the crowd solve the oracle problem? In: IEEE 6th International Conference on Software Testing. Verification and Validation, pp. 342–351. IEEE Computer Society, Luxembourg (2013)
11. Dolstra, E., Vliegendhart, R., Pouwelse, J.: Crowdsourcing gui tests. In: 2013 IEEE 6th International Conference on Software Testing. Verification and Validation, pp. 332–341. IEEE Computer Society, Luxembourg (2013)
12. Musson, R., Richards, J., Fisher, D., Bird, C., Bussone, B., Ganguly, S.: Leveraging the crowd: how 48,000 users helped improve lync performance. IEEE Softw. **30**(4), 38–45 (2013)

13. Nebeling, M., Speicher, M., Grossniklaus, M., Norrie, M.C.: Crowdsourced web site evaluation with crowdstudy. In: Brambilla, M., Tokuda, T., Tolksdorf, R. (eds.) ICWE 2012. LNCS, vol. 7387, pp. 494–497. Springer, Heidelberg (2012). https://doi.org/10.1007/978-3-642-31753-8_52

14. Runeson, P., Alexandersson, M., Nyholm, O.: Detection of duplicate defect reports using natural language processing. In: International Conference on Software Engineering, pp. 499–510. IEEE Computer Society, Minneapolis (2007)

15. Wang, X., Zhang, L., Xie, T., Anvik, J., Sun, J.: An approach to detecting duplicate bug reports using natural language and execution information. In: International Conference on Software Engineering, pp. 461–470. IEEE Computer Society, Leipzig (2008)

16. Jalbert, N., Weimer, W.: Automated duplicate detection for bug tracking systems. In: IEEE International Conference on Dependable Systems and Networks with Ftcs and DCC, pp. 52–61. IEEE, Anchorage (2008)

17. Tian, Y., Sun, C., Lo, D.: Improved duplicate bug report identification. Softw. Maint. Reengineering **94**(3), 385–390 (2012)

18. Sun, C., Lo, D., Wang, X., Jiang, J., Khoo, S.C.: A discriminative model approach for accurate duplicate bug report retrieval. In: Proceedings of the IEEE International Conference on Software Engineering, pp. 45–54. IEEE Computer Society, Cape Town (2010)

19. Sun, C., Lo, D., Khoo, S.C., Jiang, J.: Towards more accurate retrieval of duplicate bug reports. In: International Conference on Automated Software Engineering, ASE, USA, pp. 253–262 (2011)

20. Wang, J., Cui, Q., Wang, Q, Wang, S.: Towards effectively test report classification to assist crowdsourced testing. In: ACM International Symposium on Empirical Software Engineering and Measurement. ACM, Spain (2016)

21. Wang, J., Wang, S., Cui, Q., Wang, Q.: Local-based active classification of test report to assist crowdsourced testing. In: ACM International Conference on Automated Software Engineering, pp. 190–201. ACM, Singapore (2016)

22. Feng, Y., Chen, Z., Jones, J.A., Fang, C.R., Xu, B.: Test report prioritization to assist crowdsourced testing. In: Proceeding of the 2015 10th Joint Meeting on Foundations of Software Engineering, pp. 225–236. ACM, Bergamo (2015)

23. Che, W., Li, Z., Liu, T.: LTP: a Chinese language technology platform. In: International Conference on Computational Linguistics. Demonstrations, pp. 23–27. Chinese Information Processing Society of China, Beijing (2010)

24. Bhattacharyya, A.: On a measure of divergence between two statistical populations defined by their probability distributions. Bull. Calcutta Math. Soc. **35**, 99–109 (1943)

25. Gopalan, R.P., Krishna, A.: Duplicate bug report detection using clustering. In: International Conference on Software Engineering, pp. 104–109. IEEE Computer Society, Hyderabad (2014)

Strategy for the Selection of Reactive Power in an Industrial Installation Using K-Means Clustering

Fredy Martínez$^{(\boxtimes)}$ ⓘ, Fernando Marínez ⓘ, and Edwar Jacinto ⓘ

Universidad Distrital Francisco José de Caldas, Bogotá D.C., Colombia
{fhmartinezs,fmartinezs,ejacintog}@udistrital.edu.co
http://www.udistrital.edu.co

Abstract. This paper uses K-means clustering as a strategy to determine the minimum, average and maximum reactive power ranges required by an industrial plant for the design of its power factor (PF) solution. The strategy calculates three cluster points in the sensed power data set. The analysis is carried out by comparing the consumption of reactive power against active power during one week of plant operation. This data is used to analyze the behavior of the power factor, and to specify the reactive power for a 6-step mixed compensation bank. The strategy was successfully used to design a power factor corrector in which the specified capacitor bank demonstrated high performance.

Keywords: Compensation bank · Industrial · K-means clustering ·
Power factor · Reactive power · Unsupervised learning

1 Introduction

Clustering is a type of unsupervised learning, is widely used to classify data that do not have labels [5,11]. K-Means Clustering is one of the popular clustering algorithms [8,16]. The job of the K-Means Clustering is to establish k nodes, each one representing the center of a cluster of data. For each node desired then, the algorithm positions that center (the centroid) at the point where the distance between it and the nearest points is on average smaller than the distance between those points and the next node. By optimizing this process it is possible to sort the data around the centroid, and finally, to apply labels to each subset of data.

Grouping and segmentation of data into a large data set is an analysis strategy used in many applications [2,4,10]. This tool allows you to identify characteristics by groups of data, which can solve a problem by fragmenting it. In image processing, for example, it enables the segmentation of an image according to different criteria to determine specific areas or filter out non-relevant information [3,19]. This solution can be useful for the diagnosis of a disease or for the navigation of an autonomous robot [12,15].

© Springer Nature Singapore Pte Ltd. 2019
Y. Tan and Y. Shi (Eds.): DMBD 2019, CCIS 1071, pp. 146–153, 2019.
https://doi.org/10.1007/978-981-32-9563-6_15

We are applying data segmentation as a support in the design of a low power factor (PF) solution for an industrial plant. The PF in an electrical installation depends on the electrical load installed there. The excess of inductive loads (motors) in this type of places leads to a low PF, which translates into low energy efficiency (due to the high currents present), bad voltage regulation, reduction of the installed electrical capacity (also due to the large currents in the conductors), potential damage to the electrical equipment if not corrected correctly, problems with generators and transformers in the power grid [7], and additional costs by the electric utility company [13].

Traditionally, this problem is tackled by injecting reactive power into the plant's electrical grid, usually with a capacitor bank [6,18]. The use of capacitors allows to improve the FP of industrial plants, also reduces billing penalties imposed by the electric power utility, and decreases system losses by improving electrical efficiency. This solution is classic and relatively inexpensive but requires consideration of other possible problems related to the connection of the capacitors [1,18]. In today's electrical installations it is important to consider the proportion of non-linear loads, and therefore the harmonic distortion of the system [14,17]. There is always the possibility of resonance of the capacitor bank with some harmonic of the system, which can be catastrophic.

In industrial plants, motors are widely used in machinery, elevators, fans, pumps, etc. The problem with the intense use of these machines is the reactive power they take from the system's power supply, which always leads to a lagging PF [9]. This requirement for reactive power results in higher grid losses and reduced voltage levels, which in turn increases the cost of electrical energy. This is why it is always necessary to enhance the FP.

Another important consideration in PF correction is that the electrical load in industrial plants is not constant. The design of the solution must be economical, allowing the installation and connection of the appropriate number of capacitors depending on the performance of the plant. The solution then contemplates the switching (through a PLC) of static shunt capacitor of suitable value according to the power profile of the plant [9].

The paper is organized as follows. The Sect. 2 presents a description of the problem. Section 3 describes the proposed strategy and a general design outline. Section 4 introduces some results obtained with the proposed strategy. Finally, conclusion and discussion are presented in Sect. 5.

2   Problem Statement

When talking about FP there are two problems to consider: displacement PF and total PF. The first case corresponds to the classical conception of PF (also known as $cos\varphi$), in which the electrical installation has mainly linear loads (resistive, inductive and/or capacitive loads), and therefore the apparent power (S) is different from the real power (P) due to the appearance of reactive power (Q) required by reactive loads (inductive and/or capacitive loads), although the problem is usually generated by the excess of inductive reactive load. This phenomenon is best understood in the classic triangle of powers (Fig. 1).

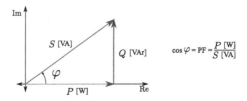

Fig. 1. Power triangle

Total PF is the relationship between real power (P) and apparent power (S) when the distortion power caused by harmonic currents is included. Non-linear loads such as electronic type loads produce these harmonics in the load currents. However, in the problem studied in this research, the majority of the loads are linear and particularly inductive. This is why all the discussion below relates to displacement PF.

When in an electrical installation the apparent power is equal to the real power (or at least its difference is very small), then the angle φ is equal to zero, and $cos\varphi = 1$. In this case, the system supplies the minimum amount of current required to produce the same amount of work when operating the load, and therefore from the electrical point of view operates at its highest performance point. On the other hand, if there is any appreciable value of reactive power, then the apparent power will be greater than the real power, and the current required to move the same load will be greater. In cases where reactive power is present, the angle φ will be different from zero, and the $cos\varphi$ will be less than one. The increase in current reduces the installed capacity of the electrical network, since the reactive current occupies the conductors, but also increases the losses of the system quadratically $\left(i^2R\right)$. This kind of electricity consumption is also penalized by the electric utility company.

Consequently, due to the costs associated with low PF, electrical installations must have PF correction equipment. In general, there are two techniques used in industrial plants: shunt capacitors and motor over excitation (caused by capacitors). The first strategy corresponds to the most economical and efficient technique for correcting displacement PF. Shunt capacitors are used to provide an additional main capacitive load. This counteracts the inductive delay load of the plant and improves the power factor to acceptable limits. However, it must be considered that industrial plants do not always operate with a constant load, which is why it is necessary to switch the number of capacitors according to the point of operation. The second technique is a variant and consists of connecting high capacity capacitors to the motor terminals. The idea is that the current of the capacitor is equal to the magnetizing current of the motor, creating resonance and over-voltages at the motor terminals. However, the oversized capacitor may cause damaging over-voltages on the conductor insulation or excessive currents.

The solution required by the industrial plant considers the switching of shunt capacitors. For the selection of the capacitors and the switching characteristics, the power consumption of the plant must be characterized. After having the power profile of the plant, appropriate values must be selected for the minimum, average and maximum reactive power required by the system.

3   Methodology

We record the real, apparent and reactive power of the plant for one week, taking data every minute. The grouping and analysis of data were carried out comparing the consumption of reactive power versus real power day by day, that is to say, every 1440 data. In total, the data set consists of 10080 records. The recorded behavior of the reactive power is shown in Fig. 2.

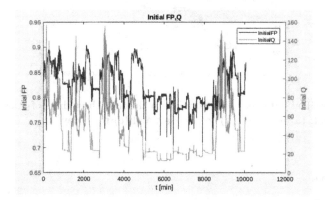

Fig. 2. Reactive power recorded for one week (green). From the power readings the PF is calculated in the same time interval (blue) (Color figure online)

In this way, the data set is divided into seven data groups, one group per day. For each day, data is grouped around three points using K-means clustering. Each of these points will become characteristic of the consumption profile defining the minimum, average and maximum reactive power values per day. Figure 3 shows the behavior of the data without grouping for the first three days.

The three reactive power data per day are then used to define the sizes and quantities of the capacitors to be purchased for installation in the industrial plant, after analyzing the possible switching sequences. This final selection also meets the commercially available values (from 3 kVAr to 51 kVAr, and in 3 kVAr increments).

The switching sequence depends on the measured reactive power and the PF present in the system. The selection is made using a table containing all the possible combinations of these variables.

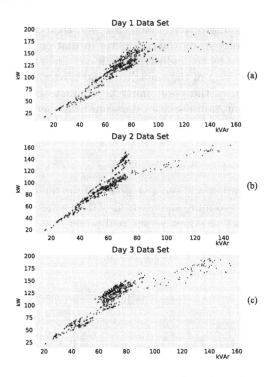

Fig. 3. Unclassified data set corresponding to the first three days of operation of the industrial plant: (a) first day, (b) second day, and (c) third day

4   Results

We apply the strategy to determine the three representative points of each subgroup of data (per day). Figure 4 shows the results for the first three days (black stars). The data that are represented by the cluster points are identified by a single color (red, blue and green). Due to the dependence of the data (higher real power means more reactive power) on the data, the process always gives a value for the minimum, average and maximum reactive power.

The results for all data subgroups (every day) are:

- Day 1: 24.6 kVAr, 54.4 kVAr and 77.6 kVAr
- Day 2: 25.6 kVAr, 54.4 kVAr and 74.9 kVAr
- Day 3: 36.6 kVAr, 71.6 kVAr and 129.4 kVAr
- Day 4: 20.3 kVAr, 50.6 kVAr and 73.9 kVAr
- Day 5: 13.4 kVAr, 20.6 kVAr and 22.8 kVAr
- Day 6: 21.2 kVAr, 35.6 kVAr and 55.6 kVAr
- Day 7: 24.0 kVAr, 50.1 kVAr and 80.7 kVAr

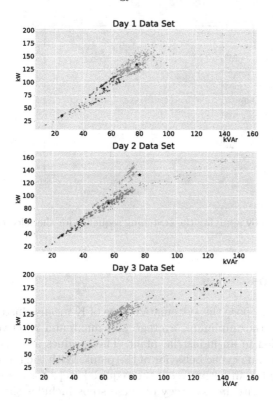

Fig. 4. Classified data set corresponding to the first three days of operation of the industrial plant: (a) first day, (b) second day, and (c) third day (Color figure online)

According to this, the values selected for the six steps are:

- 12 kVAr
- 15 kVAr
- 21 kVAr
- 24 kVAr
- 30 kVAr
- 36 kVAr

By applying the switching rules for the selection of the reactive power, the appropriate step is selected. With this value, the final behavior of the corrected PF can be calculated (Fig. 5).

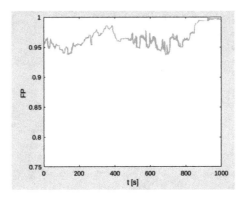

Fig. 5. PF corrected by applying the switching sequence for the selected six-step mixed compensation bank

5 Conclusions

In this paper, we propose the systematized use of K-means clustering as a strategy to determine the minimum, average and maximum reactive power values from data recorded in an industrial plant. These values are calculated at one-day intervals assuming cyclic behavior of the plant. With the representative data of reactive power for an entire week, the required steps for a six-step corrector are selected. The switching strategy of these steps includes the reactive power and PF values for the selection of the capacitor to be connected. Evaluating the performance of the strategy from the test data shows how it is possible to maintain an average FP above 0.95.

Acknowledgments. This work was supported by the Universidad Distrital Francisco José de Caldas, in part through CIDC, and partly by the Technological Faculty. The views expressed in this paper are not necessarily endorsed by Universidad Distrital. The authors thank the research group ARMOS for the evaluation carried out on prototypes of ideas and strategies.

References

1. Andrews, D., Bishop, M., Witte, J.: Harmonic measurements, analysis, and power factor correction in a modern steel manufacturing facility. IEEE Trans. Ind. Appl. **32**(3), 617–624 (1996)
2. Edelkamp, S., Pomarlan, M., Plaku, E.: Multiregion inspection by combining clustered traveling salesman tours with sampling-based motion planning. IEEE Robot. Autom. Lett. **2**(2), 428–435 (2017)
3. Ghita, O., Dietlmeier, J., Whelan, P.: Automatic segmentation of mitochondria in EM data using pairwise affinity factorization and graph-based contour searching. IEEE Trans. Image Proces. **23**(10), 4576–4586 (2014)
4. González, O., Acero, D.: Robot semi-autónomo para el transporte de pacientes con movilidad reducida. Tekhnê **13**(2), 49–63 (2016)

5. Gorzalczany, M., Rudzinski, F.: Generalized self-organizing maps for automatic determination of the number of clusters and their multiprototypes in cluster analysis. IEEE Trans. Neural Netw. Learn. Syst. **29**(7), 2833–2845 (2018)
6. Grebe, T.: Application of distribution system capacitor banks and their impact on power quality. IEEE Trans. Ind. Appl. **32**(3), 714–719 (1996)
7. Heger, C., Sen, P., Morroni, A.: Power factor correction - a fresh look into today's electrical systems. In: IEEE-IAS PCA 54th Cement Industry Technical Conference, pp. 1–13 (2012)
8. Hong, J., Yiu-Ming, C.: Subspace clustering of categorical and numerical data with an unknown number of clusters. IEEE Trans. Neural Netw. Learn. Syst. **29**(8), 3308–3325 (2018)
9. Jain, R., Sharma, S., Sreejeth, M., Singh, M.: PLC based power factor correction of 3-phase induction motor. In: IEEE International Conference on Power Electronics. Intelligent Control and Energy Systems (ICPEICES-2016), pp. 1–5 (2016)
10. Kamper, H., Jansen, A., Goldwater, S.: Unsupervised word segmentation and lexicon discovery using acoustic word embeddings. IEEE/ACM Trans. Audio Speech Lang. Process. **24**(4), 669–679 (2016)
11. Lawal, I., Poiesi, F., Anguita, D., Cavallaro, A.: Support vector motion clustering. IEEE Trans. Circ. Syst. Video Technol. **27**(11), 2395–2408 (2017)
12. Martínez, F., Acero, D.: Robótica Autónoma: Acercamientos a algunos problemas centrales. Distrital University Francisco José de Caldas, CIDC (2015). ISBN 9789588897561
13. Nieto, A., Brinez, D., Lopez, J., Marin, P., Cabrera, S., Paya, D.: Electrical cost optimization for electric submersible pumps: systematic integration of current conditions and future expectations. In: SPE Middle East Oil & Gas Show and Conference, pp. 1–15 (2017)
14. Romero, M., Gallego, L., Muller, S., Meyer, J.: Characterization of non-linear household loads for frequency domain modeling. Ingeniería e Investigación **35**(1), 65–72 (2015)
15. Santiago, C., Nascimento, J., Marques, J.: Automatic 3-D segmentation of endocardial border of the left ventricle from ultrasound images. IEEE J. Biomed. Health Inform. **19**(1), 339–348 (2015)
16. Sarkar, A., Maulik, U.: Rough based symmetrical clustering for gene expression profile analysis. IEEE Trans. NanoBioscience **14**(4), 360–367 (2015)
17. Seme, S., Lukac, N., Stumberger, B., Hadziselimovic, M.: Power quality experimental analysis of grid-connected photovoltaic systems in urban distribution networks. Energy **139**, 1261–1266 (2017)
18. Shwedhi, M., Sultan, M.: Power factor correction capacitors: essentials and cautions. In: 2000 Power Engineering Society Summer Meeting, pp. 1317–1322 (2000)
19. Wangsheng, Y., Zhiqiang, H., Peng, W., Xianxiang, Q., Liguang, W., Huanyu, L.: Weakly supervised foreground segmentation based on superpixel grouping. IEEE Access **6**, 12269–12279 (2018)

A Collaborative Filtering System Using Clustering and Genetic Algorithms

Soojung Lee[✉]

Gyeongin National University of Education,
155 Sammak-ro, Anyang 13910, Republic of Korea
sjlee@gin.ac.kr

Abstract. Recommender systems have been essential these days to assist online customers to acquire useful information. However, one of the popular types of the systems called memory-based collaborative filtering suffers from several fundamental problems in spite of its main advantages such as simplicity and efficiency. This study addresses the scalability problem which is one of major problems of the system. We employ a clustering technique to handle the problem and propose a novel idea using the genetic algorithm to enhance the performance of the system in terms of prediction accuracy, not to mention scalability. Experimental results demonstrated successful performance achievements of the proposed method under various data conditions.

Keywords: Collaborative filtering · Recommender system ·
Clustering · Genetic algorithm

1   Introduction

Recommender systems are very useful for people to obtain information on various types of products such as movies, books, music, etc. There have been developed several types of implementation for recommender systems [1,14]. Content-based filtering (CBF) makes use of user features to recommend items with similar features. Hence, this method has a main drawback that it can not recommend items with features distinct from the user's. Another implementation type named collaborative filtering (CF) tackles this problem of CBF by referring to the history of items preferred by other users to make recommendation. Thus, it does not need to analyze the contents of user or item features, but only maintains the preference history of users. Preference information is obtained either explicitly from the user ratings data or implicitly from product purchases, viewing time, or click-throughs [14].

This paper focuses on collaborative filtering for its wide and successful usage in practical recommender systems. It is usually classified into memory-based and

This research was supported by Basic Science Research Program through the National Research Foundation of Korea (NRF) funded by the Ministry of Education (2017R1D1A1B03029576).

model-based based on its implementation techniques [1]. Memory-based systems are simple to implement and known to be efficient, but they have major problems such as data sparsity, cold-start, graph-sheep, and scalability [14]. These problems are mainly caused due to their underlying principle, i.e., manipulating the user-item ratings matrix. On the contrary, instead of maintaining the matrix, model-based techniques attempt to learn a model from rating data through various data mining or machine learning algorithms. Some of the developed famous models include matrix factorization, support vector regression, latent sematic models, Bayesian probabilistic models, and clustering [14].

This paper addresses clustering-based CF. It is recognized as a useful tool to reduce the scalability problem of CF, which is very critical to online performance of the system, while providing comparable recommendation quality as traditional CF [12]. We discuss the most widely used clustering method, K-means algorithm, and suggest a novel idea to enhance the performance of K-means clustering CF by the use of genetic algorithms.

The remainder of this paper is organized as follows: In the next section, we discuss previous works related to this study. Section 3 describes the proposed method, followed by performance experiments in Sect. 4. Section 5 concludes this paper.

2 Background

2.1 Previous Works on Clustering Collaborative Filtering

K-means clustering methods have been employed by several studies to improve the performance of collaborative filtering. Xue et al. suggests a smoothing-based method on the basis of clusters that combines the advantages of both memory-based and model-based CF approaches [16]. The study proposed in [5] applied genetic algorithms to K-means clustering to find the relevant clusters more effectively. Gong proposed a personalized recommendation approach that joins the user clustering and item clustering techniques to solve the problems of scalability and sparsity in collaborative filtering [4]. [15] used cluster ensembles of self-organizing maps and K-means. They experimented with three ensemble methods using the Movielens dataset to show that cluster ensembles can provide better recommendation performance than single clustering techniques. The study presented in [8] assumed a different environment of multi-criteria user ratings. Their work showed regression-based techniques and automatically detected customer segments through clustering.

Although K-means algorithm is simple and easy, it has some drawbacks [5]: First, the resulting clusters may depend on the initial centers; Second, the algorithm may fall into a local optima because it uses a hill-climbing strategy. In order to overcome drawbacks of K-means algorithm, a self-constructing clustering algorithm is proposed in [7] to form clusters automatically as well as to reduce the dimensionality related to the number of products.

Self-organizing map (SOM) is another well-known clustering method. It uses a neural-network based unsupervised learning mechanism for clustering. It has been widely used due to its good performance. In [6], a recommender system combining collaborative filtering with SOM is proposed. All users are segmented by demographic characteristics and users in each segment are clustered according to the preference of items using the SOM network. Roh et al. proposed a three-step model for CF recommendation, which consists of profiling, inferring, and predicting steps [11]. Their model combines a CF algorithm with two machine learning processes, SOM and case based reasoning. The study in [17] used the association rules mining to fill the vacant ratings and employed SOM clustering to make prediction. Purbey et al. discussed the SOM algorithm used for CF in detail and evaluated its performance [9].

In spite of several efficient applications of SOM, SOM has some limitations that may degrade its performance. One major drawback is that SOM has no mechanism to determine the number of clusters, as K-means. In addition, it is hard to determine initial weights and stopping conditions as other neural network-based algorithms.

3   Proposed Algorithm

The recommendation process of the traditional CF system works as follows. In order to decide whether to recommend an item for a target user, the system first finds neighbor users of the target user and computes a predicted rating from the ratings given by the neighbors to the item. Neighbors consist of those who have shown similar rating behavior to items in the past. This process may significantly degrade the online performance of the system due to the computational load to find similar users, i.e., neighbors. By forming clusters, similarity computation and rating prediction are made only within the cluster to which the target user belongs, thus enabling to lessen the scalability problem of CF.

In this paper, we use K-means algorithm to cluster users, but overcome its drawbacks by employing genetic algorithms, namely, the difficulty of determining the number of clusters. The basic idea of K-means algorithm is to have any two users within a cluster show higher similarity than those belonging to different clusters. A genetic algorithm is an evolutionary algorithm that simulates the process of natural evolution. It is famous for finding optimal solutions through the principles of natural genetics [2,3]. The algorithm maintains a population of solutions which are evolved from generation to generation through the genetic operators until an optimal solution satisfying a given criteria called fitness is found.

In our GA operation, we aim to find the optimal number of clusters. In addition, we have another objective of finding the optimal size of each cluster. Hence, a solution s is defined as $s = (K, N(C_1), N(C_2), \ldots, N(C_K))$, where K represents the number of clusters and $N(C_i)$ the size of cluster i. Obviously, there are restrictions on the range of gene values, i.e., $K << |U|$, $\sum_i N(C_i) = m|U|, m \geq 1$, where $|U|$ is the total number of users in the system. Different from K-means algorithm, we allow a user to belong to several clusters, where m is the mean number of clusters containing the user. Our intention behind this strategy is to have more neighbors in a cluster to consult ratings than when using K-means, as our algorithm yields a cluster of the larger size when $m > 1$. We maintain the population size constant throughout the generations.

- P: population of solutions
- PS: population size
- P_{new}:   population of new solutions
- N_{gens}:   number of generations
- f_{th} :   fitness threshold
- f_i, f_{best}, f_{worst} :   fitness of solution i, best, and worst fitness among all solutions at the current generation, respectively.
- $Prob_{Si}$:   selection probability of solution i
- $Prob_C$:   crossover probability
- $Prob_M$:   mutation probability

1.   Initialize solutions in P.
2.   **For** each solution i in P
2.1   Run the clustering algorithm to cluster users based on solution i.
2.2   Evaluate the clustering CF system to compute the fitness f_i.
2.3   Update f_{best} and f_{worst}.
3.   **while** the number of generations< N_{gens} and f_{best}>f_{th} **do**
3.1   Calculate $Prob_{Si}$ for each solution i in P.
3.2   Select two solutions s_1 and s_2 in P with the selection probability.
3.3   With the probability $Prob_C$, apply the one-point crossover operator to s_1 and s_2.
3.3.1   If s_1 and s_2 are crossed over, insert the crossed-over offsprings into P_{new}.
3.3.2   else insert s_1 and s_2 into P_{new}.
3.4   With the probability $Prob_M$, flip a random bit of the solutions inserted at Step 3.3.
3.5   If $|P_{new}|$<PS, then go to Step 3.2.
3.6   Let $P=P_{new}$ and P_{new}={}.
3.7   **For** each solution i in P
3.7.1   Run the clustering algorithm to cluster users based on solution i
3.7.2   Evaluate the clustering CF system to compute the fitness f_i.
3.7.3   Update f_{best} and f_{worst}.
4.   **end while**
5.   Return the clustering CF system with the solution yielding the best fitness f_{best}.

Fig. 1. The proposed clustering CF system utilizing the genetic algorithm

Three genetic operators are used in our algorithm: selection, crossover, and mutation. All of them are very typical and mostly used in the genetic algorithms. Figure 1 describes the detailed steps of the algorithm. For a fitness function, we choose MAE (Mean Absolute Error) for measuring performance of collaborative filtering systems, which is representative metric for prediction quality.

At each generation, two solutions with the better fitnesses are selected with probability. These solutions are then crossed over with the predetermined probability to produce two new offsprings. Each of these offsprings is to undergo the mutation step where a random bit is flipped with a given mutation probability. These three genetic operations are repeated until the number of new offsprings reaches the given number of solutions, PS. With each of this new solution, the clustering algorithm is executed to build clusters of users based on the gene values of the solution, i.e., the number of clusters and each cluster size. Our proposed clustering algorithm is based on K-means and described in Fig. 2. The CF system with the clusters resulting from the proposed clustering algorithm is then evaluated to obtain fitnesses. The algorithm terminates if it reaches a given number of generations or if the best fitness of solutions in the population is less than a given threshold.

Input parameters:
- K: number of clusters.
- N(C$_i$): Size of cluster C$_i$, i=1, 2, ..., K.   ΣN(C$_i$)=m x total number of users
- R: user-item rating matrix
- sim(u, v): similarity value between users u and v
- N: threshold for the number of iterations

begin
1. Select a user randomly as a center of each cluster C$_i$.
2. **while** (iter <=N and any center of a cluster is changed) **do**
2.1   **for** each cluster C$_i$ **do**
2.2   Compute similarity between the center and all users.
2.3   Sort all users according to the computed similarity.
2.4   Determine N(C$_i$) number of users in decreasing order of similarity to belong to C$_i$.
2.5   Compute a new center for C$_i$.
2.6   **end for**
2.7   iter = iter + 1
3. **end while**
end

Fig. 2. The proposed clustering algorithm

Our strategy of the genetic algorithm is considered unique and novel compared to the state-of-the-art methods, in that not only the targets for optimization are more than one, but also the fitness concerns the performance quality of the CF system, instead of similarity as in the studies presented in [3,13,18]. Like the previous researches, the proposed optimization approach is better to run offline, not to hinder the online CF performance, as the clustering and the genetic algorithm take unignorable time, as reported in the literature [3,14].

4   Performance Evaluation

4.1   Experimental Background

We conducted experiments with two datasets publicly open and popular in the related research area, MovieLens 1M (http://www.grouplens.org) and Jester (http://goldberg.berkeley.edu/jester-data). They are chosen due to their distinct characteristics, which would help to reveal the performance of the system in more details. MovieLens has an integer rating scale of one to five and very few user ratings data with the sparsity level of 0.9607 in our experiment, whereas Jester provides a real scale from −10 to +10 with the sparsity level of 0.2936, thus much denser than MovieLens. Detailed description on each dataset can be found on the corresponding website.

The baseline similarity measure is the mean squared differences (MSD) [14]. We used this measure for both K-means algorithm and our proposed clustering algorithm. Two different experiments of K-means algorithm are conducted, one with the number of clusters of five and the other with that of ten, using the legends of KM-C5 and KM-C10, respectively. The proposed clustering CF system is denoted as CL-GA in the legend. Among the whole ratings data, we used 80% of them for training and the rest to produce performance results.

To evaluate performance, we measured prediction quality of the system using a well-known metric of MAE (Mean Absolute Error). It is defined as the mean difference between the predicted rating of an unrated item and its corresponding real rating. To reflect different rating scales of datasets used in our experiment, we normalized MAE to obtain values within zero to one range, namely NMAE. We made the rating prediction using a weighted sum formula which is usually used to accumulate the ratings by similar users [10].

As for the parameters used by the proposed algorithms, we use the values in Table 1. The selection probability of each solution i, $Prob_{S_i}$, is set to the inverse of MAE and normalized into $[0, 1]$.

Table 1. Parameter values for the genetic and the clustering algorithms

PS	N_{gens}	f_{th} (MovieLens/Jester)	$Prob_C$	$Prob_M$	N	m
60	30	0.65/2.5	0.5	0.3	10	3

4.2   Results

Figure 3 shows NMAE results with varying number of the nearest neighbors using the datasets. In Fig. 3(a), it is observed that NMAE performance of all methods experimented is gradually improved to a steady state, as more similar users are consulted using MovieLens. It is also found that K-means clustering based CF turns out to yield much poorer results than the others. This result is supported by [14] stating that there is a tradeoff between scalability and prediction performance. In particular, with more clusters, KM performs even worse in terms of NMAE. This is surely because the size of the cluster would be normally smaller with a larger number of total clusters and so there would be less number of neighbors in a cluster to consult the ratings. Nevertheless, it is noted that the proposed method (CL-GA) significantly overcomes such disadvantage of the normal clustering CF. CL-GA showed still better performance than MSD with the help of the genetic algorithm, which implies that the optimized assignment of a user to clusters is more effective than the number of referenced neighbors on NMAE.

The methods show different behavior using Jester as shown in Fig. 3(b). MSD is defeated by the two K-means CF systems, KM-C5 and KM-C10, as well as by CL-GA. The reason comes from the different data environment that Jester has much denser ratings than MovieLens. Hence, the clusters resulting from K-means are assumed to contain neighbors more helpful for accurate rating predictions in this dense dataset.

It is found that the construction of more clusters is still advantageous as seen in the results of KM-C10 compared to those of KM-C5. However, when we experimented with 15 clusters using K-means CF, we obtained worse results than KM-C5, implying that there should be a proper number of clusters for best performance of CF using K-means. As seen in the figure, the optimized number of clusters pursued by CL-GA gives a significant positive effect on NMAE performance. That effect should be further supported by the optimal size of each cluster. However, the clustering based CF systems demonstrate the diminishing effect on the performance as more neighbors in a cluster are referred to, since similarities with them should become lower and thus inaccuracies on rating predictions calculated from them would increase accordingly.

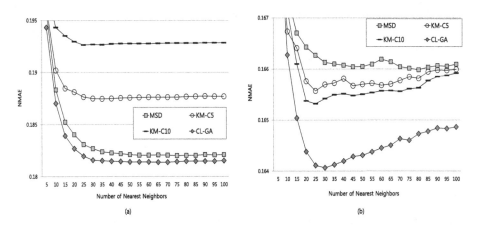

Fig. 3. Mean absolute error using (a) MovieLens and (b) Jester datasets

5   Conclusion

This study proposes a clustering based collaborative filtering system to reduce the scalability problem of the system. We employed K-means algorithm for clustering but overcomed its drawbacks by adopting the genetic algorithm. The number of clusters and the size of each cluster are both optimized through our approach. Experimental studies are conducted using datasets with different characteristics to investigate the performance of the systems on various data environment. Besides lessening the scalability problem, the proposed method made significant improvement of prediction accuracy, as found in the experimentation results. In future research works, we plan to explore further optimization of the system as well as thorough investigation of the performance of the developed system in various aspects.

References

1. Aamir, M., Bhusry, M.: Recommendation system: state of the art approach. Int. J. Comput. Appl. **120**(12), 25–32 (2015)
2. Alhijawi, B., Kilani, Y.: Using genetic algorithms for measuring the similarity values between users in collaborative filtering recommender systems. In: The 15th IEEE/ACIS International Conference on Computer and Information Science, pp. 1–6 (2016)

3. Bobadilla, J., Ortega, F., Hernando, A., Alcal, J.: Improving collaborative filtering recommender system results and performance using genetic algorithms. Knowl.-Based Syst. **24**(8), 1310–1316 (2011)
4. Gong, S.: A collaborative filtering recommendation algorithm based on user clustering and item clustering. J. Sofw. **5**(7), 745–752 (2010)
5. Kim, K.J., Ahn, H.: A recommender systems using GA K-means clustering in an online shopping market. Expert Syst. Appl. **34**, 1200–1209 (2008)
6. Lee, M., Choi, P., Woo, Y.: A hybrid recommender system combining collaborative filtering with neural network. In: De Bra, P., Brusilovsky, P., Conejo, R. (eds.) AH 2002. LNCS, vol. 2347, pp. 531–534. Springer, Heidelberg (2002). https://doi.org/10.1007/3-540-47952-X_77
7. Liao, C.L., Lee, S.J.: A clustering based approach to improving the efficiency of collaborative filtering recommendation. Electron. Commerce Res. Appl. **18**, 1–9 (2016)
8. Nilashi, M., Jannach, D., bin Ibrahim, O., Ithnin, N.: Clustering- and regression-based multi-criteria collaborative filtering with incremental updates. Inf. Sci. **293**, 235–250 (2015)
9. Purbey, N., Pawde, K., Gangan, S., Karani, R.: Using self-organizing maps for recommender systems. Int. J. Soft Comput. Eng. **4**(5), 47–50 (2014)
10. Resnick, P., Lakovou, N., Sushak, M., Bergstrom, P., Riedl, J.: Grouplens: an open architecture for collaborative filtering of netnews. In: Proceedings of the ACM Conference on Computer Supported Cooperative Work, pp. 175–186. ACM Press (1994)
11. Roh, T.H., Oh, K.J., Han, I.: The collaborative filtering recommendation based on SOM cluster-indexing CBR. Expert Syst. Appl. **25**(3), 413–423 (2003)
12. Sarwar, B., Karypis, G., Konstan, J., Riedl, J.: Recommender systems for large-scale e-commerce: scalable neighborhood formation using clustering. In: The Fifth International Conference on Computer and Information Technology (2002)
13. Shivhare, H., Gupta, A., Sharma, S.: Recommender system using fuzzy C-means clustering and genetic algorithm based weighted similarity measure. In: International Conference on Computer, Communication and Control (IC4), pp. 1–8. IEEE (2015)
14. Su, X., Khoshgoftaar, T.M.: A survey of collaborative filtering techniques. Adv. Artif. Intell. **2009** (2009)
15. Tsai, C.F., Hung, C.: Cluster ensembles in collaborative filtering recommendation. Appl. Soft Comput. **12**, 1417–1425 (2012)
16. Xue, G., Lin, C., Yang, Q., Xi, W., Zeng, H., Yu, Y.: Scalable collaborative filtering using cluster-based smoothing. In: Proceedings International Conference on Special Interest Group on Information Retrieval, pp. 114–121. ACM (2005)
17. Ye, H.: A personalized collaborative filtering recommendation using association rules mining and self-organizing map. J. Softw. **6**(4), 732–739 (2011)
18. Zhang, F., Chang, H.Y.: A collaborative filtering algorithm employing genetic clustering to ameliorate the scalability issue. In: IEEE International Conference on e-Business Engineering, pp. 331–338 (2006)

Using K-Means Algorithm for Description Analysis of Text in RSS News Format

Paola Ariza-Colpas[1,2(✉)], Ana Isabel Oviedo-Carrascal[2],
and Emiro De-la-hoz-Franco[1]

[1] Universidad de La Costa, CUC, Barranquilla, Colombia
{parizal,edelahoz}@cuc.edu.co
[2] Universidad Pontificia Bolivariana, Medellín, Colombia
ana.oviedo@upb.edu.co

Abstract. This article shows the use of different techniques for the extraction of information through text mining. Through this implementation, the performance of each of the techniques in the dataset analysis process can be identified, which allows the reader to recommend the most appropriate technique for the processing of this type of data. This article shows the implementation of the K-means algorithm to determine the location of the news described in RSS format and the results of this type of grouping through a descriptive analysis of the resulting clusters.

Keywords: RSS news's format · Simple K-means · Bag of words · Stopwords · Text mining

1 Introduction

The purpose of text mining is the extraction of relevant information that is contained in the different formats in which you can find this type of data as they are: journal articles, web pages, among others [1, 2]. Text mining currently has several applications. In research, it is normally used in various sectors of application such as medicine to make various discoveries and advances in medical care, allowing health professionals to have faster and more efficient information about resources [3, 4]. Organizations use this type of text analysis to support real-time decision-making and effective management of different types of consultations carried out by clients [5, 6].

In terms of information security, text mining allows relevant analyzes to be carried out on texts that lead to terrorist actions and crimes that may be generated by browsing different Web sites [7, 8]. In social networks, this type of analysis is normally used to identify trends, including the analysis of the sentiment of different users on the Web, of places or places, which allows the establishment of a recommendation scale for products and services [9, 10].

The mining of texts in terms of texts of news origin allows to clarify the existing reliance between the news that is being raised around a particular region to be able to determine concurrence in relevant aspects and that require attention by the control organisms.

© Springer Nature Singapore Pte Ltd. 2019
Y. Tan and Y. Shi (Eds.): DMBD 2019, CCIS 1071, pp. 162–169, 2019.
https://doi.org/10.1007/978-981-32-9563-6_17

This article is divided into the following sections: in the first section of the report the different research works that have been developed around the themes, in the second section the structure of the dataset and the methods used for the analysis of the data, in the third section is defined the methodology developed to perform the experimentation process, in the fourth section the results of the experimentation are shown and finally the results of the implementation of the algorithms are shown.

2 Brief Review of Literature

Text mining is mainly focused on being able to identify or derive new information from texts, the main relevance of this type of applications is the process of indexing large volumes of data. That is why a large number of authors have studied different ways of performing analyzes on this type of data [11, 12], identifies that the problems of text classification and the representation of a document have a strong impact on the performance of the systems Learning. The high dimensionality of classical structured representations can lead to cumbersome calculations due to the large size of real-world data. Consequently, there is a need to reduce the amount of information that is handled to improve the classification process.

On the other hand, Wu [13, 14], proposes a method of text detection that consists of two main steps: connected components (CC) of extraction and non-textual filtering. For the extraction of the CC, an adaptive approach of grouping of color of multiple scale is realized, that can extract text of images in different complexities of color and is robust to contrast the variation. For non-textual filtering, the text covariance descriptor (TCD) was combined with the gradient oriented histogram (HOG) to construct feature vectors and used to distinguish text from the background at the character level and text line.

Regarding the detection of text in several languages for the indexing of Aradhya video, proposes a method to detect text located in a greater or lesser proportion in the complex background of images/video. This approach consists of four stages: In the first stage, the wavelet transform is applied and a Gabor filter is applied. The performance of the approach is exhibited by presenting promising experimental results for 101 video images, using publicly available data ICDAR 2003 and ICDAR 2013 [15, 16].

Regarding the use of clustering to evaluate the progression of the difficulty of the text, Cheng-Hsien took as a basis different series of textbooks in Taiwan, to be able to identify the level of complexity of the text in the context. In particular, based on two methodological emphases. First, the difficulty of the text is analyzed quantitatively by the frequency lists based on the BNC corpus, taking into account both the vocabulary and the structure of the complexity of the texts; Second, the Clustering algorithm was used to identify text difficulty levels on a predefined basis. This corpus-based calculation method determines not only objectively the developmental gaps in a series of textbooks, but also identifies progress by evolving into the more complex vocabulary and structures. This rigorous evaluation of textbooks provides a common framework for evaluation in the increasing difficulty of teaching text of different languages [17, 18].

There are different solutions to analyze different characteristics of the text, Aliem [19, 20] shows how modularity characteristics can serve as a useful criterion to determine the clusters in the matrices that are generated in the analysis of documents. In this paper we present the performance of CoClus, a new algorithm effective in the analysis of diagonal blocks that manages to minimize the characteristics of modularity. Maximization is used as an iterative alternative for the optimization of procedures, in contrast with algorithms that use the spectral characteristics of discrete optimization problems. After many experiments in several document dataset they show that the method is very efficient, stable and outperforms other algorithms of co-diagonal clustering by blocks dedicated to the same task. Another important advantage of the use of modularity in the co-clustering context is that it provides a simple new way of determining the appropriate number of co-clusters.

Additionally, it has been identified that text classification can help users effectively manage and exploit useful information hidden in large-scale documents. However, the scarcity of data and the sensitivity to the semantic context often hinder the performance of classification in short texts. In order to overcome the weakness, this work proposes a unified framework to expand short texts based on the grouping of embedded plugs and neural networks (CNN). Empirically, semantically related words are usually found close to each other divided by spaces. Therefore, the semantically related words are the first ones that are identified through the general application of clustering. Wang, represents in multiscale units the semantic units represented in the short texts. In the incorporation of spaces, the semantic units are chosen to constitute extended matrices. Finally, for a short text, the projected Matrix and the expanded matrices are combined and entered into the CNN algorithm in parallel. The experimental results in two reference points validate the efficacy of the proposed method [21, 22].

The use of learning algorithms for text classification assumes the availability of a large number of documents that have been organized and labeled correctly by human experts for use in the training phase. Unless the text documents in question have been in existence for some time, the use of an expert system is inevitable due to the tediousness of manually organizing and labeling thousands of groups of text documents. Therefore, unsupervised learning systems are needed to automatically group data in the training phase. In addition, even when there is knowledge, the variation is high when the subject that is being classified depends on personal opinions and is open to different interpretations. Shafiabady, [23–25], describes a methodology that uses Self Organizing Maps (SOM) and, alternatively, does the automatic grouping using the correlation coefficient (CorrCoef). Consequently, clusters are used as labels to train the vector machine (SVM). Experiments and results are presented based on the application of the methodology for some sets of standard text data in order to verify the accuracy of the proposed system. The current results are used to evaluate the effect that the reduction of dimensionality and the changes that in the clustering schemes have on the accuracy of the SVM. The results show that the proposed combination has a higher precision compared to the training of the learning machine through expert knowledge.

3   Materials and Methods

3.1   Description and Preparation of Dataset

For the development of this experimentation was built a dataset built from the RSS format of news from different parts of the world, which is made up of 3 instances and 100 attributes that are described below:

- Title: enunciation of the title of the news.
- Description: body of the news.

First, the file containing the dataset of the documents is created, then the processing is done using the stopofwords filter that contains the weka tool to be later analyzed by the classification and clustering techniques (Fig. 1).

Fig. 1. Dataset preprocessing.

Being the whole $X = (x_i)$, $i = 1,...., n$ of n dimensional points that are grouped into a set of groupings K, C_k, $k = 1,...., K$.

The K-means algorithm finds a partition such that the quadratic error between the empirical mean of a cluster and the points in the cluster is minimized. Allowing it U_k to be the cluster average C_k. The quadratic error between U_k and points in cluster C_k is defined as: [17]

$$J(C_k) = \sum_{x_i \in C_k} \|X_i - \mu_k\|^2 \tag{1}$$

The goal of K-means is to minimize the sum of the squared error over all K groups.

$$J(C) = \sum_{k=1}^{K} \sum_{x_i \in C_k} \|X_i - \mu_k\|^2 \tag{2}$$

Minimizing this objective function is known to be a NP-hard problem (even for K = 2) [18].

Thus, K-means, is an algorithm that can only converge to a local minimum, although recent study has shown with a high probability K-means could converge to the global optimum when the groupings are well separated [19].

K-means starts with an initial partition with K groups and assigns groups of patterns in order to reduce the squared error. Since the quadratic error always decreases with an increase in the number of K groups (with J(C) = 0 when K = n), it can be minimized only for a fixed number of clusters.

The main steps of K- means algorithm are the following [20].

- Select an initial partition with K groupings; Repeat steps 2 and 3 until the cluster membership stabilizes.
- Generate a new partition by assigning each pattern to its nearest cluster center.
- Calculate the new cluster centers.

4   Experimentation

Taking into account the dataset of news descriptions, the process of preparing the data was carried out in order to carry out the training of the method.

For the execution of the algorithm WEKA software was used, taking into account the terms and license of use.

The K-means method was chosen for the implementation, based on the need to be able to group the news events located in certain regions. The results can be seen in Table 1

Table 1. Data aggregation results with the chosen clustering

Cluster	Numbers of record
Cluster 0	3 (3%)
Cluster 1	11 (11%)
Cluster 2	28 (28%)
Cluster 3	33 (33%)
Cluster 4	25 (25%)

The data distribution for the training and testing process was made using crossed validation, the tool selects a percentage of the data for training and another for testing with the proposed method. Below is the descriptive analysis after the implementation of the method.

5   Results

The word bag concept was applied with the English descriptions and the respective stopwords, making use of the simple k-means algorithm with the configuration of 5 clusters obtaining the following results.

In cluster 0, where 3% of the records are located. Group the description of the news that was developed in Colombia (Table 2).

Table 2. Cluster 0 description

Attribute	Cluster 0
Country	66.7%
Located	66.7%
Region	66.7%
Colombia	66.7%

In cluster 1, where 11% of the records are located. Group the description of the news that was developed in United States (Table 3).

Table 3. Cluster 1 description

Attribute	Cluster 1
States	100%
United	100%
Country	63.6%
Located	72.7%

In cluster 2, where 28% of the records are located. Group the description of the news that was developed in department and municipality of Colombia (Table 4).

Table 4. Cluster 2 description

Attribute	Cluster 2
Colombia	66.7%
Department	33.4%
Municipality	33.4%

In cluster 3, where 33% of the records are located. Group the description of the news that was developed in cities of Canada (Table 5).

Table 5. Cluster 3 description

Attribute	Cluster 2
City	90.9%
Located	75.7%
Canada	72.7%

In cluster 4, where 25% of the records are located. Group the description of the news that was developed county of the United States (Table 6).

Table 6. Cluster 4 description

Attribute	Cluster 2
County	52%
States	100%
United	100%
City	80%

6 Conclusions

It is demonstrated that the application of the K-means method allows to contribute to the descriptive analysis of the news in the RSS format, allowing characterization by location to identify the place where the news is being developed. After the description of the clusters, the compactness of the results can be identified, allowing interpreting in future works where certain events or natural phenomena take place in certain regions or cities of the world.

References

1. Palechor, F., De la hoz manotas, A., De la hoz franco, E., Colpas, P: Feature selection, learning metrics and dimension reduction in training and classification processes in intrusion detection systems. J. Theor. Appl. Inf. Technol. **82**(2) (2015)
2. Calabria-Sarmiento, J.C., et al.: Software applications to health sector: a systematic review of literature (2018)
3. Sen, T., Ali, M.R., Hoque, M.E., Epstein, R., Duberstein, P.: Modeling doctor-patient communication with affective text analysis. In: 2017 Seventh International Conference on Affective Computing and Intelligent Interaction (ACII), pp. 170–177. IEEE (2017)
4. Jeon, S.W., Lee, H.J., Cho, S.: Building industry network based on business text: corporate disclosures and news. In: 2017 IEEE International Conference on Big Data (Big Data), pp. 4696–4704. IEEE (2017)
5. Irfan, M., Zulfikar, W.B.: Implementation of fuzzy C-Means algorithm and TF-IDF on English journal summary. In: 2017 Second International Conference on Informatics and Computing (ICIC), pp. 1–5. IEEE (2017)
6. De-La-Hoz-Franco, E., Ariza-Colpas, P., Quero, J.M., Espinilla, M.: Sensor-based datasets for human activity recognition–a systematic review of literature. IEEE Access **6**, 59192–59210 (2018)
7. Zhang, X., Yu, Q.: Hotel reviews sentiment analysis based on word vector clustering. In: 2017 2nd IEEE International Conference on Computational Intelligence and Applications (ICCIA), pp. 260–264. IEEE (2017)
8. Vieira, A.S., Borrajo, L., Iglesias, E.L.: Improving the text classification using clustering and a novel HMM to reduce the dimensionality. Comput. Methods Programs Biomed. **136**, 119–130 (2016)
9. Wu, H., Zou, B., Zhao, Y.Q., Chen, Z., Zhu, C., Guo, J.: Natural scene text detection by multi-scale adaptive color clustering and non-text filtering. Neurocomputing **214**, 1011–1025 (2016)

10. Palechor, F.M., De la Hoz Manotas, A., Colpas, P.A., Ojeda, J.S., Ortega, R.M., Melo, M.P.: Cardiovascular disease analysis using supervised and unsupervised data mining techniques. JSW **12**(2), 81–90 (2017)
11. Aradhya, V.M., Pavithra, M.S.: A comprehensive of transforms, Gabor filter and k-means clustering for text detection in images and video. Appl. Comput. Inform. (2014)
12. Bharti, K.K., Singh, P.K.: Opposition chaotic fitness mutation based adaptive inertia weight BPSO for feature selection in text clustering. Appl. Soft Comput. **43**, 20–34 (2016)
13. Li, C.H.: Confirmatory factor analysis with ordinal data: comparing robust maximum likelihood and diagonally weighted least squares. Behav. Res. Methods **48**(3), 936–949 (2016)
14. Melissa, A., François, R., Mohamed, N.: Graph modularity maximization as an effective method for co-clustering text data. Knowl.-Based Syst. **109**(1), 160–173 (2016)
15. Mendoza-Palechor, F.E., Ariza-Colpas, P.P., Sepulveda-Ojeda, J.A., De-la-Hoz-Manotas, A., Piñeres Melo, M.: Fertility analysis method based on supervised and unsupervised data mining techniques (2016)
16. Wang, P., Xu, B., Xu, J., Tian, G., Liu, C.L., Hao, H.: Semantic expansion using word embedding clustering and convolutional neural network for improving short text classification. Neurocomputing **174**, 806–814 (2016)
17. Shafiabady, N., Lee, L.H., Rajkumar, R., Kallimani, V.P., Akram, N.A., Isa, D.: Using unsupervised clustering approach to train the Support Vector Machine for text classification. Neurocomputing **211**, 4–10 (2016)
18. Zhang, W., Tang, X., Yoshida, T.: Tesc: an approach to text classification using semi-supervised clustering. Knowl.-Based Syst. **75**, 152–160 (2015)
19. De França, F.O.: A hash-based co-clustering algorithm for categorical data. arXiv preprint arXiv:1407.7753 (2014)
20. Echeverri-Ocampo, I., Urina-Triana, M., Patricia Ariza, P., Mantilla, M.: El trabajo colaborativo entre ingenieros y personal de la salud para el desarrollo de proyectos en salud digital: una visión al futuro para lograr tener éxito (2018)
21. Jain, A.K.: Data clustering: 50 years beyond K-means. Pattern Recognit. Lett. **31**(8), 651–666 (2010)
22. Drineas, P., Frieze, A.M., Kannan, R., Vempala, S., Vinay, V.: Clustering in large graphs and matrices. In: SODA, vol. 99, pp. 291–299 (1999)
23. Meila, M., Shi, J.: Learning segmentation by random walks. In: NIPS, pp. 873–879 (2000)
24. Jain, A.K., Dubes, R.C.: Algorithms for clustering data (1988)
25. Guerrero Cuentas, H.R., Polo Mercado, S.S., Martinez Royert, J.C., Ariza Colpas, P.P.: Trabajo colaborativo como estrategia didáctica para el desarrollo del pensamiento crítico (2018)

RETRACTED CHAPTER: Differential Evolution Clustering and Data Mining for Determining Learning Routes in Moodle

Amelec Viloria[1](✉), Tito Crissien Borrero[1], Jesús Vargas Villa[1],
Maritza Torres[2], Jesús García Guiliany[3], Carlos Vargas Mercado[4],
Nataly Orellano Llinas[5], and Karina Batista Zea[4]

[1] Universidad de La Costa, St. 58 #66, Barranquilla, Atlántico, Colombia
{aviloria7, rectoria, jvargas41}@cuc.edu.co
[2] Universidad Centroccidental "Lisandro Alvarado", Barquisimeto, Venezuela
mtorres@ucla.edu.ve
[3] Universidad Simón Bolívar, Barranquilla, Colombia
jesus.garcia@unisimonbolivar.edu.c
[4] Corporación Universitaria Latinoamericana, Barraquilla, Colombia
carlosvargasmercado0103@gmail.com,
kbatistazea@hotmail.com
[5] Corporación Universitaria Minuto de Dios UNIMINUTO,
Barranquilla, Colombia
Nataly.Orellano@gmail.com

Abstract. Data mining techniques are being widely used in the field of education from the arise of e-learning platforms like Moodle, WebCT, Claroline, and others, and the virtual learning system they entail. Information systems store all activities in files or databases which, correctly processed, may offer relevant data to the teacher. This paper reports the use of data mining techniques and Differential Evolution Clustering for discovering learning routes frequently applied in the Moodle platform. Data were obtained form 4.115 university students monitored in an online course using Moodle 3.1. Firstly, students were grouped according to the data from a final qualifications report in a course. Secondly, the data of the Moodle logs about each cluster/group of students was used separately with the aim of obtaining more specific and precise models of the student behavior in the processes.

Keywords: K-means · Clustering · Differential evolution · Data mining · Moodle

1 Introduction

The new knowledge discovered by data mining techniques on e-learning information systems is one of the areas addressed by Educational Data Mining (EDM) [1]. This new knowledge may be useful for both teachers and students who follow recommended activities and resources to favor their learning, and teachers can obtain objective feedback about their teaching tasks. Teachers can evaluate the structure of the course

---

The original version of this chapter was retracted: The retraction note to this chapter is available at https://doi.org/10.1007/978-981-32-9563-6_35

and its effectiveness during the learning process, as well as classify the students into groups according to their orientation as a result of the monitoring actions [2, 3].

Process Mining (PM) is a technique for mining data on applications that generate event logging to identify possible processes in a variety of application domains. The application of process mining activities should result in models of business process flows and information on their historical use (more frequent routes, less performed activities, etc.) [4, 5].

This study aims to use Automatic Clustering Differential Evolution (ACDE) and data mining techniques to determine the number of clusters and learning routes of a group of university students in a Moodle virtual course with the purpose of facilitating the teaching-learning process.

2   Theoretical Review

2.1   Automatic Clustering Using Differential Evolution

Researches used automatic clustering methods for determining the number of clusters based on the Evolutionary Computation (EC) technique. K-means method has done a lot and has been published with different methods namely Automatic Clustering using Differential Evolution (ACDE), combining methods between PSO and k-means on Dynamic Clustering with Particle Swarm Optimization (DCPSO), and Genetic Clustering for unknown k clustering (GCUK) [8].

Automatic clustering methods have been used to determine the number of clusters in the k- means but are yet to achieve accurate cluster result. Therefore, it is necessary to improve the performance of automated grouping methods used for determining the number of clusters.

The ACDE method is the most popular EC technique which has effectively improved the performance of automatic clustering methods proposed by previous researchers. ACDE predicated on the differential evolution (DE) method is one of the strongest, fastest, and most efficient global search heuristics methods in the world that is very easy to use with high-dimensional data [9].

The ACDE was developed by [10] and the combination of ACDE and k-means methods was termed the automatic clustering approach based on the differential evolution method combined with k-means for crisp clustering method aimed at improving clustering performance in the k-means method (ACDE-k-means). The ACDE method computed the number of clusters automatically and is able to balance the evolutionary process of DE methods to achieve better partitions than the classic DE. However, the classic DE method still depends on user's considerations to determine the k activation threshold thereby affecting the performance of the DE method [11–13].

2.2   Moodle

Moodle is the result of the Thesis of Martin Dougiamas from the University of Perth, in Western Australia. This university teacher wanted a tool to facilitate the social constructivism and cooperative learning. Its name comes from the acronym of Modular Object oriented Dynamic Learning Environment, although other sources mention that it comes from the English verb moodle that describes the process of wandering lazily through something, and doing things when one decides. According to the author's words, he wanted: "A program that is easy to use and as intuitive as possible". Mastering Moodle is simple, it is hardly necessary to control an iconography composed of 15 fully significant symbols [14].

It is an e-learning tool that enables the non-face-to-face learning of the students, an aspect to be considered with many of the students who cannot attend classes because of their work or personal situations, which makes it necessary to have a tool that facilitates the virtuality, fundamental issue with the new format of tutorship that will force a greater organizational work, as well as the management of practices and the works derived from the integration of more active pedagogies according to the philosophy of the New School theory [15].

3   Materials and Methods

3.1   Database

The data used in this study was obtained from a Moodle 3.1 course used by 4,115 university students of Industrial Engineering from a private University in Colombia. The study was conducted during the 2017-2018 academic year. The research is carried out on a mandatory ninth-year subject matter.

Each unit is composed of three types of content:

- Level of declarative knowledge: theoretical content, information, and how to put the weekly learning-to-learn strategies into practice.
- Level of procedural knowledge: practical tasks where students have to put into practice their declarative knowledge.
- Level of conditional knowledge: discussion forums where the students have to deal with topics on how they can use strategies of the week in different contexts.

The mandatory tasks of each unit were: perform the practical task and publish at least one comment in each forum.

The suggested tasks of each unit were: understand the theoretical contents and put them into practice in the task, and share their experience on the topic of the week in the forum.

The system employs several different sources of information on which the data obtained from the work carried out by the students throughout the program are based.

On the one hand, Table 1 shows the variables that are taken into account, which determine the interaction that each student has on the Moodle platform.

These variables are calculated from the Moodle registry and different database tables.

These data obtained from Moodle and the different tables of databases are processed and converted into an .ARFF file to later apply a clustering algorithm.

Table 1. Variables that show the interaction of students in Moodle [16, 17].

Name	Description	Extraction method
Theory time	Total time spent on theoretical components of the contents	The sum of the time periods between resource view and the next different action
Tasks time	Total time spent on teaching tasks	The sum of the time periods between quiz view/quiz attempt/quiz continue attempt/quiz close attempt and the next different action
Forums time	Total time spent in forums review	The sum of the time periods between forum view and the next different action
Theory days	How many days, in a time period of 15 days, are left to check the content at least once (in days)	Resource view date since the content is available
Tasks days	How many days, in a time period of 15 days, are left to check the task at least once (in days)	Task view date since the content is available
Days of "Delivery"	How many days, in a time period of 15 days, are left to complete them (in days)	Task view date since the content is available
Words in the forums	Number of words published in forums	Extract the number of words in everything published by forum add discussion or forum add replay
Phrases in the forums	Number of sentences published in forums	Extract the number of sentences of everything published by forum add discussion or forum add replay

3.2   Methods

The research applies a methodology that uses clustering to group students by type and thus be able to improve the models extracted with mining of processes (see Fig. 1).

The proposal performs a pre-grouping for groups of students with similar characteristics. Subsequently, mining of processes is applied to discover specific models of students' behaviors. As shown in Fig. 1, the research proposes two different types of clustering/grouping [18, 19]:

- Manual: students are directly grouped using the final grade obtained in the course.
- Automatic: Students are grouped by applying clustering ACDE on information about the interaction they perform during the course in the Moodle platform.

There are two types of students in the Manual grouping:

1. Students whose final grade is less than 5 (suspended students)
2. Students whose final grade is greater than or equal to 5 (approved students)

For the automatic grouping, the variables used and its description to perform the clustering come from the interaction of the students in Moodle, as shown in Table 1.

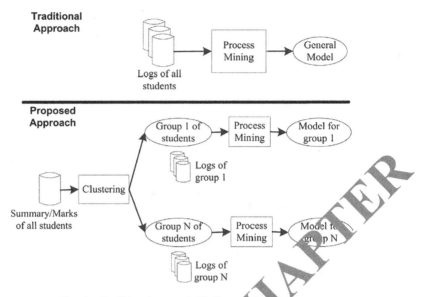

Fig. 1. Traditional research Vs Proposed research.

3.3   Data Mining

This research uses the ProM open source tool, which is a specific software for issues related to mining processes and the Heuristic miner algorithm that is based on frequency of patterns since it focuses in the event tracking [20, 21].

In addition, the heuristic Miner is a heuristic network drawn as a cyclic directed graph, which shows, in this case, the most frequent behavior of students in each data set used. Default parameters of the Heuristic Miner algorithm of ProM is used, and as a measure of quality, Adjustment or Fitness [22].

The Adjustment indicates the difference between the behavior actually observed in the registry and the behavior described by the process model. A sequence of activities that belong to the same case is called trace. The traces of the registry can be associated with execution paths specified by the process model. If the model has a low Adjustment, it indicates that the mining model of processes do not correctly parse most of the traces of record. This may be due to the presence of noise, as a result of activities which are not considered and connections that are missing [23, 24].

4   Results

The experiments were conducted using a computing platform with Intel Celeron 2.16 GHz CPU, 8 GB RAM and Microsoft Windows 10-bit Home 64 used as operating system and MATLAB version R2016a used as the data analytics tool. The resulting clusters are shown in Table 2 using ACDE.

Table 2. Dataset Descriptions

Clusters	Description
Base 1	2,300 students (2,200 approved and 100 suspended)
Base 2	1,204 students (1,154 approved and 50 suspended)
Base 3	611 students (601 suspended and 10 approved)

Using the Heuristic Miner, the obtained results are shown in Table 3.

Table 3. Results of the Adjustment value of the different models

The Data Set	Adjustment
All students	0.8333
Approved	0.9117
Suspended	0.9375
Students Cluster 0	0.9130
Students Cluster 1	0.9024
Students Cluster 2	0.900

Table 3 shows that the lowest value of Adjustment Measure was obtained by using all data of students jointly, in which 3,4 6 out of 4,115 students fit in with the obtained model, i.e. the 83% of all students. On the other hand, all other models (obtained using clustering both manually and by AODE) get an Adjustment value higher than 90% in all cases. The higher adjustment value was obtained when using data of suspended students, where 150 of the 160 students fit into the model obtained, i.e., the 93.75% of suspended students. Therefore, in this case, these specific models obtained using manual and automatic clustering represent/fit better than the general model obtained from all students.

Table 4 displays information about the level of complexity or size of each of the models obtained.

Two typical measures of graph theory have been used (the total number of nodes and the total number of links) in order to see the level of complexity of the obtained models.

Table 4. Complexity/size of the obtained models

The data set	N° Nodes	N° Links
All students	32	70
Approved students	113	244
Suspended Students	12	24
Students Cluster 0	61	121
Students Cluster 1	59	110
Students Cluster 2	38	84

Table 4 shows that the smallest model, and therefore more easily understandable, was obtained with the suspended students, followed by all students, and students in cluster 2. On the other hand, the other three models are much larger and complex. It is believed that the reasons could be:

- In the data set of all students, students exhibit different behaviors and show just some common actions because there are different types of students (approved and suspended).
- In the data set of students that suspended and cluster 2, students show only some common behavior patterns because this type of students participates/interacts little in the Moodle platform.
- In the data set of students who approved, cluster 0 and cluster 1, students show many more common behavior patterns because these students are more active users of Moodle.

Finally, the models with the best and worst Adjustment are shown. In our heuristics networks, boxes represent the events carried out by students when they interact with the Moodle platform and arches/links represent the relationship/dependencies between events. Then, Fig. 2 shows the heuristic network obtained when using students who approved in the course.

It is evident that these students have a greater number of associated subnets since the interaction with the platform is greater and, therefore, there is a greater diversity in terms of common events among these students.

On the other hand, Fig. 3 shows two subnets that are still most of the students who suspended in the course.

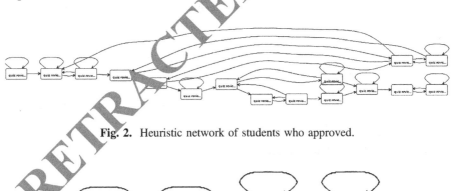

Fig. 2. Heuristic network of students who approved.

Fig. 3. Heuristic network of students who suspended.

From an educational and practical point of view, this information can be used to provide feedback to the teachers on student learning and could be easily used to draw new students at risk of suspend in the course. For example, teachers just must check if the new students follow the same specific behavior routes/patterns followed by heuristic network students who suspended, that is, if they visit the same pages, see the same examinations, in the same order of suspended students above.

5 Conclusions

This paper proposes the use of grouping or clustering to improve the mining of educational processes and, at the same time, optimize both performance/adjustment and comprehensibility/size of the obtained model. The comprehensibility of the obtained model is a basic objective in education, due to the transfer of basic knowledge that this concept entails.

Charting, models, or a more accessible visual representation for teachers and students make these results very useful for the learning process follow up and to provide feedback, being one of future challenges to do it in real time. In addition, Moodle does not provide specific visualization tools of data used by students to enable the different actors of the learning process understand these large amounts of "gross" data, and be aware of what is happening in distance education, in addition to extending the use of the results of adaptive hypermedia learning environment in which it is very useful to motivate students or recommend learning routes to enhance learning experiences in a more strategic way.

References

1. Viloria, A., Lis-Gutiérrez, J.P., Gaitan-Angulo, M., Godoy, A.R.M., Moreno, G.C., Kamatkar, S.J.: Methodology for the design of a student pattern recognition tool to facilitate the teaching - learning process through knowledge data discovery (big data). In: Tan, Y., Shi, Y., Tang, Q. (eds.) DMBD 2018. LNCS, vol. 10943, pp. 670–679. Springer, Cham (2018). https://doi.org/10.1007/978-3-319-93803-5_63
2. Ballesteros Román, A.: Minería de Datos Educativa Aplicada a la Investigación de Patrones de Aprendizaje del Estudiante en Ciencias. Centro de Investigación en Ciencia Aplicada y Tecnología Avanzada, Instituto Politécnico Nacional, México City (2012)
3. Ben Salem, S., Naouali, S., Chtourou, Z.: A fast and effective partitional clustering algorithm for large categorical datasets using a k-means based approach. Comput. Electr. Eng. 68, 463–483 (2018). https://doi.org/10.1016/j.compeleceng.2018.04.023
4. Chakraborty, S., Das, S.: Simultaneous variable weighting and determining the number of clusters—A weighted Gaussian means algorithm. Stat. Probab. Lett. 137, 148–156 (2018). https://doi.org/10.1016/j.spl.2018.01.015
5. Abdul Masud, M., Zhexue Huang, J., Wei, C., Wang, J., Khan, I., Zhong, M.: Inice: a new approach for Identifying the Number of Clusters and Initial Cluster Centres. Inf. Sci. (2018). https://doi.org/10.1016/j.ins.2018.07.034
6. Rahman, M.A., Islam, M.Z., Bossomaier, T.: ModEx and seed-detective: two novel techniques for high quality clustering by using good initial seeds in K-Means. J. King Saud Univ. - Comput. Inf. Sci. 27, 113–128 (2015). https://doi.org/10.1016/j.jksuci.2014.04.002
7. Rahman, M.A., Islam, M.Z.: A hybrid clustering technique combining a novel genetic algorithm with K-means. Knowl.-Based Syst. 71, 345–365 (2014). https://doi.org/10.1016/j.knosys.2014.08.011

8. Ramadas, M., Abraham, A., Kumar, S.: FSDE-forced strategy differential evolution used for data clustering. J. King Saud Univ. - Comput. Inf. Sci (2016). https://doi.org/10.1016/j.jksuci.2016.12.005

9. Yaqian, Z., Chai, Q.H., Boon, G.W.: Curvature-based method for determining the number of clusters. Inf. Sci. (2017). https://doi.org/10.1016/j.ins.2017.05.024

10. Tîrnăucă, C., Gómez-Pérez, D., Balcázar, J.L., Montaña, J.L.: Global optimality in k-means clustering. Inf. Sci. (Ny) **439–440**, 79–94 (2018). https://doi.org/10.1016/j.ins.2018.02.001

11. Xiang, W., Zhu, N., Ma, S., Meng, X., An, M.: A dynamic shuffled differential evolution algorithm for data clustering. Neurocomputing (2015). https://doi.org/10.1016/j.neucom.2015.01.058

12. Garcia, A.J., Flores, W.G.: Automatic clustering using nature-inspired metaheuristic: a survey. Appl. Soft Comput (2016). https://doi.org/10.1016/j.asoc.2015.12.001

13. Das, S., Abraham, A., Konar, A.: Automatic clustering using an improved differential evolution algorithm. IEEE Trans. Syst. Man, Cybern. - Part A Syst. Humans 38, 218–237 (2008). https://doi.org/10.1109/TSMCA.2007.909595

14. Costa, C., Alvelos, H., Teixeira, L.: The use of MOODLE e-learning platform, a study in a Portuguese University. Procedia Technology **5**, 334–343 (2012)

15. El-Bahsh, R., Daoud, M.: Evaluating the use of MOODLE to achieve effective and interactive learning: a case study at the German Jordanian University. In: Proceedings of the 35th Annual IEEE International Conference on Computer Communications, pp. 1–5 (2016)

16. Coll, S.D., Treagust, D.: Blended learning environment: approach to enhance student's learning experiences outside school (LEOS). MIER J. Educ. Stud. Trends Pract. **7**, 2 (2018)

17. Kuo, R., Suryani, E., Yasid, A.: Automatic clustering combining differential evolution algorithm and k-means algorithm. In: Lin, Y.K., Tsao, Y.C., Lin, S.W. (eds.) Proceedings of the Institute of Industrial Engineers Asian Conference 2013, pp. 1207–1215. Springer, Singapore (2013). https://doi.org/10.1007/978-981-4451-98-7_143

18. Piotrowski, A.P.: Review of differential evolution population size. Swarm Evol. Comput. **32**, 1–24 (2017). https://doi.org/10.1016/j.swevo.2016.05.003

19. Kaya, I.: A genetic algorithm approach to determine the sample size for attribute control charts. Inf. Sci. (Ny) **179**, 1552–1566 (2009). https://doi.org/10.1016/j.ins.2008.09.024

20. Dobbie, G., Sing, Y., Riddle, P., Ur, S.: Research on particle swarm optimization based clustering: a systematic review of literature and techniques. Swarm Evol. Comput. **17**, 1–13 (2014). https://doi.org/10.1016/j.swevo.2014.02.001

21. Departamento Administrativo Nacional de Estadística.: Página principal. Recuperado de: DANE (2018). http://www.dane.gov.co/

22. Torres-Samuel, M., Vásquez, C.L., Viloria, A., Varela, N., Hernández-Fernandez, L., Portillo-Medina, R.: Analysis of Patterns in the University World Rankings Webometrics, Shanghai, QS and SIR-SCimago: Case Latin America. In: Tan, Y., Shi, Y., Tang, Q. (eds.) DMBD 2018. LNCS, vol. 10943, pp. 188–199. Springer, Cham (2018). https://doi.org/10.1007/978-3-319-93803-5_18

23. Vásquez, C., Torres, M., Viloria, A.: Public policies in science and technology in Latin American countries with universities in the top 100 of web ranking. J. Eng. Appl. Sci. **12** (11), 2963–2965 (2017)

24. Torres-Samuel, M., et al.: Efficiency analysis of the visibility of Latin American Universities and their impact on the ranking web. In: Tan, Y., Shi, Y., Tang, Q. (eds.) DMBD 2018. LNCS, vol. 10943, pp. 235–243. Springer, Cham (2018). https://doi.org/10.1007/978-3-319-93803-5_18

Student Performance Assessment Using Clustering Techniques

Noel Varela[1(✉)], Edgardo Sánchez Montero[1], Carmen Vásquez[2],
Jesús García Guiliany[3], Carlos Vargas Mercado[4],
Nataly Orellano Llinas[5], Karina Batista Zea[4], and Pablo Palencia[5]

[1] Universidad de la Costa, St. 58 #66, Barranquilla, Atlántico, Colombia
{nvarela2, esanchez2}@cuc.edu.co
[2] Universidad Nacional Experimental Politécnica "Antonio José de Sucre",
Barquisimeto, Venezuela
cvasquez@unexpo.edu.ve
[3] Universidad Simón Bolívar, Barranquilla, Colombia
jesus.garcia@unisimonbolivar.edu.co
[4] Corporación Universitaria Latinoamericana, Barraquilla, Colombia
carlosvargasmercado0103@gmail.com,
kbatistazea@hotmail.com
[5] Corporación Universitaria Minuto de Dios – UNIMINUTO,
Barranquilla, Colombia
Nataly.Orellano@gmail.com,
pablo.palendia.d@uniminuto.edu.co

Abstract. The application of informatics in the university system management allows managers to count with a great amount of data which, rationally treated, can offer significant help for the student programming monitoring. This research proposes the use of clustering techniques as a useful tool of management strategy to evaluate the progression of the students' behavior by dividing the population into homogeneous groups according to their characteristics and skills. These applications can help both the teacher and the student to improve the quality of education. The selected method is the data grouping analysis by means of fuzzy logic using the Fuzzy C-means algorithm to achieve a standard indicator called Grade, through an expert system to enable segmentation.

Keywords: Clustering · Fuzzy C-means algorithm · Fuzzy logic ·
Expert system

1 Introduction

In Latin America, higher education public institutions currently face the challenge of improving their academic quality with scarce financial resources and, at the same time, coping with the demands of the new social and economic contexts in the global society [1, 2]. For this reason, there is an evident concern about developing processes and products at both academic and administrative levels and optimizing the use of available resources. An interest of great importance for university authorities is the student educational outcome whose study and analysis require the use of adequate tools for generating indicators that guide the decision-making processes at this educational level [3].

© Springer Nature Singapore Pte Ltd. 2019
Y. Tan and Y. Shi (Eds.): DMBD 2019, CCIS 1071, pp. 179–188, 2019.
https://doi.org/10.1007/978-981-32-9563-6_19

Regarding the issues mentioned above, several authors focus on the student performance as one of the most critical and urgent problems to face, since this issue is usually showing low academic results of students at Latin America universities. The impact of this problem is widely known in relation to the high drop-out rates, high repetition rates, high number of students with a lag in their studies, low grade averages, and low licensing rates [4–6].

In education, grading is the process of applying standardized measures of different levels to assess the performance of the students in a course. As a result, Grading can be used by potential employers or educational institutions to evaluate and compare the applicant's knowledge. On the other hand, data grouping is the process of extracting previously unknown, valid, useful, and positionally hidden patterns of large data sets. The main objective of clustering is to divide the students into homogeneous groups according to their characteristics and skills [7]. These applications can help both the teacher and the student to improve the quality of education. This research applies the cluster analysis to segment students into groups according to their characteristics [8].

Data Clustering is a statistical technique for data analysis without supervision. It is used for classification of homogeneous groups to discover patterns and hidden relationships that help make decisions in a quick and efficient way. By the use of this technique, a large data set is segmented into subsets called clusters (groups) [9–11]. Each cluster is a collection of similar objects, gathered into the same group, and different to objects in other clusters. Concepts of Fuzzy Logic are defined to perform the clustering, which finally leads to the determination of the Grade indicator to obtain the student performance.

2   Theoretical Review

2.1   Data Grouping (Clustering)

A clustering algorithm organizes items into groups based on similarity criteria. The Fuzzy C-means is a clustering algorithm where each item can belong to more than one group - hence the word Fuzzy - where the membership degree for each element is given by a probability distribution on the clusters [12, 13].

2.1.1   Fuzzy C-Means Algorithm (FCM)
FCM is a clustering algorithm developed by Dunn and subsequently improved by Bezdek. It is useful when the required number of clusters is predetermined. Therefore, the algorithm tries to put each data points in one of the clusters. What makes FCM different is that it does not decide the allocation of a data point to a certain group, instead, it calculates the probability of that point to belong to that cluster (degree of membership). Therefore, depending on the required accuracy of clustering, appropriate tolerance measures can be used. Given that the absolute composition is not calculated, FCM can be extremely fast because the number of iterations needed to achieve a specific grouping corresponds to the accuracy required [14, 15].

a.1 – Iterations. In each iteration of FCM algorithm, the following objective function J is minimized, Eq. 1 [16]:

$$J = \sum_{i=1}^{N} \sum_{j=1}^{C} \delta_{ij} \|x_i - c_j\|^2 \tag{1}$$

Where N is the number of data points, C is the number of needed clusters, cj is the vector of centers for the cluster j, and δ_{ij} is the degree of membership to the i-th data point xi on the cluster j. The norm $\| xi - cj \|$ measures the similarity (or closeness) of the data point xi to the vector of centers cj in the cluster j. Note that, in each iteration, the algorithm maintains a vector of centers for each cluster. These data points are calculated as their own weighted average, where the weights are given by the degrees of membership.

a.2 - Degrees of Membership. For a data point xi, the degree of membership in the cluster j is calculated as follows, Eq. 2 [17]:

$$\delta_{ij} = \frac{1}{\sum_{k=1}^{C} \left[\frac{\|x_i - C_j\|}{\|x_i - C_k\|} \right]^{\frac{2}{m-1}}} \tag{2}$$

Where, m is the fuzziness coefficient and the vector of centers cj is calculated as follows, Eq. 3 [17]:

$$c_j = \frac{\sum_{i=1}^{N} \delta_{ij}^m \cdot x_i}{\sum_{i=1}^{N} \delta_{ij}^m} \tag{3}$$

In the previous Eq. (3), δ_{ij} is the value of the degree of membership calculated in the previous iteration. Note that, at the beginning of the algorithm, the degree of membership for the data point i to the cluster j began with a random value θ_{ij}, $0 < \theta_{ij} < 1$, such as the Eq. 4 [16]:

$$\sum_{j}^{C} \delta_{ij} = 1 \tag{4}$$

a.3 - Fuzziness Coefficient. In Eqs. (2) and (3), the fuzziness coefficient m, with $1 < m < \infty$ measures the required clustering tolerance. This value determines how many clusters may overlap with each other. The higher the value of m, the greater the overlap between clusters. In other words, the higher the fuzziness coefficient used by the algorithm, a greater number of data points fall within a "fuzzy" band where the degree of membership is neither 0 nor 1 but will be somewhere in the middle [12, 18].

a.4 - Termination Condition. The required accuracy of the membership degree determines the number of iterations made by the FCM algorithm. This measure of accuracy is calculated using the degree of membership of an iteration to the next, taking the largest of these values in all the data points considering all groups. If the measure of accuracy between iteration k and $k + 1$ is represented with ε, its value is calculated in the following way, Eq. 5 [18]:

$$\varepsilon = \Delta_i^N \cdot \Delta_j^C \cdot |\delta_{ij}^{k+1} - \delta_{ij}^k|$$ (5)

Where, δ^k ij and δ^{k+1} ij are respectively the degrees of membership in the iteration k and $k + 1$, and the operator δ, when supplied with a vector of values, returns the largest value of that vector.

3 Method

3.1 Database

The data used in this study was obtained from a Moodle 3.1 course consisting of 4,115 university students of Industrial Engineering from a private university in Colombia. The study was conducted during the 2017–2018 academic term. The research is carried out on a mandatory ninth-year subject matter.

3.2 Methods

3.2.1 Fuzzy Logic

Fuzzy logic manages the imprecision and uncertainty in a natural way, where it is possible, to provide a representation of knowledge with human orientation [4]. In recent years, there are many research studies where fuzzy logic, neural networks, classic neural networks, and the fuzzy neural networks have been employed on student modeling systems. This research applies the Fuzzy C-means Clustering algorithm for the automatic generation of the membership function. The Rule Based Fuzzy Expert System is also applied to automatically convert the crisp data into a fuzzy set and calculate the total score of a student in the exams of the first term (parc-1), second term (parc-2), and third term (parc-3). This type of knowledge in the management system can improve policies and strategies for enhancing the system quality [19, 20].

(i). Crisp value: The crisp value is the grade obtained by the student in the exams constituting the crisp input. (ii). Fuzzification: Fuzzification means that the crisp value (student's grade) becomes the fuzzy input value with the help of the proper membership function. (iii). Inference Mechanism: Defines the different types of fuzzy rules ("If/Then Rule") to assess the academic performance of students. (iv). Fuzzy Output: Determines an output value of the membership function for each active rule ("If/Then Rule"). (v). Defuzzification (Performances): Defuzzification means calculating the final result (performance value) with the help of the proper defuzzification method. In this research, Center of Area (COA) is used for defuzzification (performance assessment) [21, 22].

The membership function is a graphical representation of the magnitude of participation of each entry. A chart that defines how the value of membership between 0 and 1 is assigned to each point in the input space. The input space is often referred to as the universe of discourse or universal set that contains all possible elements of particular interest. A "weight" is associated with each of the processed entries, the functional overlap between entries is defined, and finally, determines an output response. The rules use the input membership values as weighting factors to determine their influence on the fuzzy output sets of the conclusion in the final output. Once the functions are inferred, they are scaled and combined, and they are then defuzzified in a crisp output that activates the system [21, 23].

3.2.2 Fuzzification

Fuzzification consists on the process of transforming crisp values into degrees of membership for linguistic terms of fuzzy sets. The membership function is used to associate a score to each linguistic term. It also refers to the transformation of an objective term into a fuzzy concept. All this activity is performed by Matlab [5] assisted by the Toolbox of Fuzzy Logic, through the FCM function, which format is: [centers, U,objFun] = fcm(data,Nc,options) [24].

Tickets for this function are [23]:

- Data: Matrix with input data that should be grouped, specified as an array with Nd rows, where Nd is the number of data points. The number of columns of data is equal to the dimensionality of the data.
- Nc: Number of clusters chosen by the user.
- Options(1): Exponent of the matrix of fuzzy partition U, specified as a scalar greater than 1.0. This option controls the amount of fuzzy overlap between the groups, with higher values indicating a greater degree of overlap (default = 2).
- Options(2): Maximum number of iterations (default = 100).
- Options(3): Difference between variations of desired centroid (default = 1e−5).
- Options(4): Shows iterations (default = 1).

While the outputs are [24]:

- Center: Coordinates of the Centers of final clusters, returned as a matrix with Nc rows that contain the coordinates of each cluster. The number of columns in the centers is equal to the dimensionality of the data to be grouped together.
- U: Matrix of diffuse partition, returned as an array with Nc rows and Nd columns. The item $U(i, j)$ indicates the degree of membership of the data point j in the cluster i. For a given data point, the sum of the membership values for all groups is one.
- objFun: Values of the objective function for each iteration, returned as a vector.

4 Results

The data make up a numeric matrix (data) of dimension n × 3 where n = 4115 is the number of students assessed, and the first, second, and third columns show values obtained in parc-1, parc-2, and parc-3 (Fig. 1).

83	85	5
71	46	45
64	43	24

Fig. 1. Matrix of average grades in the tests 1 to 3.

The number of chosen clusters Nc is 5 [Very High (VH), High (H), Average (A), Low (L), Very Low (VL)] with no other option, so the values of default are taken (2, 100, 1e−5, 1). As output, the matrix centers is obtained (dimension 5 × 3, Fig. 2) with the coordinates of the centers of clusters and the UT matrix (dimension n × 5, Fig. 3) where the elements of each column belong to each of the five groups:

67.2672	83.5736	52.7654
63.7560	35.5156	41.5214
82.0763	89.7203	78.8954
58.0252	38.7127	23.6533
81.7372	62.5113	76.8585

Fig. 2. Matrix centers for data under study.

0.2895	0.1948	0.1365	0.2658	0.1321
0.0298	0.8384	0.0165	0.0838	0.0354
0.0294	0.1969	0.0122	0.7446	0.0188
0.1042	0.0383	0.6323	0.0284	0.1960
0.0891	0.0891	0.1185	0.0475	0.6525

Fig. 3. UT matrix for data under study.

From this point, the position of the maximum of elements of each row is determined assigning it (position index) to an element in a sixth column, as shown in Fig. 4. Subsequently, each student is grouped to one of the five categories (from left to right, VH: Very High, H: High, A: Average, L: Low; and VL: Very Low) [25].

0.1374	0.1327	0.2935	0.1947	0.2445
0.0175	0.0354	0.0363	0.8324	0.0833
0.0133	0.0188	0.0286	0.1959	0.7445
0.6324	0.1960	0.1042	0.0393	0.0283
0.1178	0.6525	0.0884	0.0941	0.0469

Fig. 4. Matrix with position index for the studied data.

After the simple observation of the maximum per row, the first student is in the category A, the second one in the L, the third one in the VL, the fourth one in the VH, etc. In percentage terms, in accordance with this technique, the clustering in the five groups shows the following format, shown in Fig. 5:

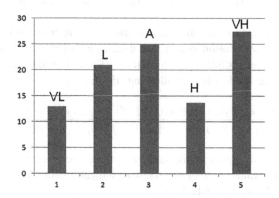

Fig. 5. Clustering of the studied data.

When the inference has ended, it is necessary to calculate a single value to represent the result. This process is called defuzzification and can be achieved by different methods. A common method is the data defuzzification in a crisp output achieved through the combination of results from the inference process and then calculating the "fuzzy centroid" of the area. The weighted strengths of each membership function of output is multiplied by their respective central points of the membership function and are added together. Finally, this area is divided by the sum of the strengths of the weighted membership function and the result is taken as the crisp output [22, 23].

Fuzzy Rule Base (Knowledge Base): The Fuzzy if-then rules and fuzzy reasoning are the backbone of the fuzzy expert systems, which are the most important modeling tools based on the fuzzy sets theory. The rule base is characterized in the form of if-then rules where the antecedents and consequents involve linguistic variables. This research considers very high, high, medium, low, and very low as linguistic variables. The collection of these rules constitutes the basis of rules for the fuzzy logic system. In this fuzzy expert system based on rules, the following rules have been used to find the knowledge base [22]:

1. If the student belongs to very high, then $(Z_1 = a_1 + b_1X_1 + c_1Y_1)$
2. If the student belongs to high, then $(Z_2 = a_2 + b_2X_2 + c_2Y_2)$
3. If the student belongs to the middle, then $(Z_3 = a_3 + b_3X_3 + c_3Y_3)$
4. If the student belongs to low, then $(Z_4 = a_4 + b_4X_4 + c_4Y_5)$
5. If the student belongs to very low, then $(Z_5 = a_5 + b_5X_5 + c_5Y_5)$

Where Xi and Yi are the grades that the students grouped in cluster i obtained in the parc-1 and parc-2 tests, respectively, Zi corresponding to parc-1. a1,...,a5, b1,...b5 and c1, c5... are constants to be determined by the method of regression analysis model.

Inference engine (Decision-Making Logic): Using the proper procedure, the true value for the antecedent of each rule is calculated and applied to the consequent part of each rule. In this case, the model of linear regression analysis for decision-making has been used. The result is a fuzzy subset to be assigned to each output variable for each rule. Once again, using a proper composition procedure, all fuzzy subsets to be assigned to each output variable are combined to form a single fuzzy subset for each output variable.

Defuzzification Interface: Defuzzification means converting the fuzzy output into crisp output. The defuzzification height was used as a technique to convert the fuzzy output into crisp output (students performance value). The defuzzification formula (Takagi-Sugeno-Kang Model) is the following, Eq. 6 [25]:

$$Y = \frac{\mu_{VH}(x,y) \cdot Z_1 + \mu_H \cdot Z_2 + \mu_A(x,y) \cdot Z_3 + \mu_L(x,y) \cdot Z_4 + \mu_{VL}(x,y) \cdot Z_5}{\mu_{VH}(x,y) + \mu_H(x,y) + \mu_A(x,y) + \mu_L(x,y) + \mu_{VL}(x,y)} \quad (6)$$

With the help of this equation, the fuzzy output can be converted into crisp output (student performance value) getting the standard indicator called Grade. The result for the first records is shown in Table 1:

Table 1. Grade indicator for the first records.

p1	p2	p3	VH	H	A	L	VL	Grade
84	84	5	0.2935	0.1947	0.1365	0.2432	0.1320	71.0740
70	44	45	0.0363	0.8284	0.0165	0.0833	0.0354	50.7925
65	44	25	0.0286	0.1959	0.0122	0.7445	0.0188	34.1655
90	93	96	0.1041	0.0393	0.6323	0.0283	0.1960	69.9056

5   Conclusions

According to the results, the research presents a new way of grouping students in terms of their academic behavior, which is more significant than using a traditional average. So, this fully developed pedagogical tool for planning according to a classification from an internationally recognized indicator is made available to teachers and administrators.

References

1. Van Dyke, T.P., Prybutok, V.R., Kappelman, L.A.: Cautions on the use of the SERVQUAL measure to ASSESS the quality of information systems services. Decis. Sci. **30**(3), 877–891 (1999)
2. Bonerge Pineda Lezama, O., Varela Izquierdo, N., Pérez Fernández, D., Gómez Dorta, R.L., Viloria, A., Romero Marín, L.: Models of multivariate regression for labor accidents in different production sectors: comparative study. In: Tan, Y., Shi, Y., Tang, Q. (eds.) DMBD 2018. LNCS, vol. 10943, pp. 43–52. Springer, Cham (2018). https://doi.org/10.1007/978-3-319-93803-5_5

3. Izquierdo, N.V., Lezama, O.B.P., Dorta, R.G., Viloria, A., Deras, I., Hernández-Fernández, L.: Fuzzy logic applied to the performance evaluation. Honduran coffee sector case. In: Tan, Y., Shi, Y., Tang, Q. (eds.) ICSI 2018. LNCS, vol. 10942, pp. 164–173. Springer, Cham (2018). https://doi.org/10.1007/978-3-319-93818-9_16

4. Pineda Lezama, O., Gómez Dorta, R.: Techniques of multivariate statistical analysis: an application for the Honduran banking sector. Innovare: J. Sci. Technol. 5(2), 61–75 (2017)

5. Viloria, A., Lis-Gutiérrez, J.P., Gaitán-Angulo, M., Godoy, A.R.M., Moreno, G.C., Kamatkar, S.J.: Methodology for the design of a student pattern recognition tool to facilitate the teaching - learning process through knowledge data discovery (big data). In: Tan, Y., Shi, Y., Tang, Q. (eds.) DMBD 2018. LNCS, vol. 10943, pp. 670–679. Springer, Cham (2018). https://doi.org/10.1007/978-3-319-93803-5_63

6. Duque Oliva, E., Finch Chaparro, C.: Measuring the perception of service quality education by students AAUCTU Duitama. Free Criterion Magazine, vol. 10, no. 16, January–July 2012

7. Yao, L.: The present situation and development tendency of higher education quality evaluation in Western Countries. Priv. Educ. Res. 3, 12–45 (2006)

8. Bertolin, J., Leite, D.: Quality evaluation of the Brazilian higher education system: relevance, diversity, equity and effectiveness. Qual. High. Educ. 14, 121–133 (2008)

9. Cronin, J., Taylor, S.: Measuring service quality: a reexamination and extension. J. Mark. 56 (3), 55–68 (1992)

10. Ballesteros Román, A.: Minería de Datos Educativa Aplicada a la Investigación de Patrones de Aprendizaje en Estudiante en Ciencias. Centro de Investigación en Ciencia Aplicada y Tecnología Avanzada, Instituto Politécnico Nacional, México City (2012)

11. Ben Salem, S., Naouali, S., Chtourou, Z.: A fast and effective partitional clustering algorithm for large categorical datasets using a k-means based approach. Comput. Electr. Eng. 68, 463–483 (2018). https://doi.org/10.1016/j.compeleceng.2018.04.023

12. Chakraborty, S., Das, S.: Simultaneous variable weighting and determining the number of clusters—a weighted Gaussian means algorithm. Stat. Probab. Lett. 137, 148–156 (2018). https://doi.org/10.1016/j.spl.2018.01.015

13. Abdul Masud, M., Zhexue Huang, J., Wei, C., Wang, J., Khan, I., Zhong, M.: Inice: a new approach for identifying the number of clusters and initial cluster centres. Inf. Sci. (2018). https://doi.org/10.1016/j.ins.2018.07.034

14. Rahman, M.A., Islam, M.Z., Bossomaier, T.: ModEx and seed-detective: two novel techniques for high quality clustering by using good initial seeds in K-Means. J. King Saud Univ. - Comput. Inf. Sci. 27, 113–128 (2015). https://doi.org/10.1016/j.jksuci.2014.04.002

15. Rahman, M.A., Islam, M.Z.: A hybrid clustering technique combining a novel genetic algorithm with K-Means. Knowl.-Based Syst. 71, 345–365 (2014). https://doi.org/10.1016/j.knosys.2014.08.011

16. Ramadas, M., Abraham, A., Kumar, S.: FSDE-forced strategy differential evolution used for data clustering. J. King Saud Univ. - Comput. Inf. Sci (2016). https://doi.org/10.1016/j.jksuci.2016.12.005

17. Vásquez, C., Torres, M., Viloria, A.: Public policies in science and technology in Latin American countries with universities in the top 100 of web ranking. J. Eng. Appl. Sci. 12 (11), 2963–2965 (2017)

18. Torres-Samuel, M., et al.: Efficiency analysis of the visibility of Latin American Universities and their impact on the ranking web. In: Tan, Y., Shi, Y., Tang, Q. (eds.) DMBD 2018. LNCS, vol. 10943, pp. 235–243. Springer, Cham (2018). https://doi.org/10.1007/978-3-319-93803-5_22

19. Bandyopadhyay, S., Maulik, U.: Genetic clustering for automatic evolution of clusters and application to image classification. Pattern Recognit. 35, 1197–1208 (2002)

20. Tam, H., Ng, S., Lui, A.K., Leung, M.: Improved activation schema on automatic clustering using differential evolution algorithm. In: IEEE Congress on Evolutionary Computing, pp. 1749–1756 (2017). https://doi.org/10.1109/CEC.2017.7969513

21. Kuo, R., Suryani Erma, E., Kuo, R.: Automatic clustering combining differential evolution algorithm and k-means algorithm. In: Lin, Y.K., Tsao, Y.C., Lin, S.W. (eds.) Proceedings of the Institute of Industrial Engineers Asian Conference 2013, pp. 1207–1215. Springer, Singapore (2013). https://doi.org/10.1007/978-981-4451-98-7_143

22. Piotrowski, A.P.: Review of differential evolution population size. Swarm Evol. Comput. **32**, 1–24 (2017). https://doi.org/10.1016/j.swevo.2016.05.003

23. Kaya, I.: A genetic algorithm approach to determine the sample size for attribute control charts. Inf. Sci. (NY) **179**, 1552–1566 (2019). https://doi.org/10.1016/j.ins.2008.09.024

24. Dobbie, G., Sing, Y., Riddle, P., Ur, S.: Research on particle swarm optimization based clustering: a systematic review of literature and techniques. Swarm Evol. Comput. **17**, 1–13 (2014). https://doi.org/10.1016/j.swevo.2014.02.001

25. Omran, M.G.H., Engelbrecht, A.P., Salman, A.: Dynamic clustering using particle swarm optimization with application in image segmentation. Pattern Anal. Appl., 332–344 (2016). https://doi.org/10.1007/s10044-005-0015-5

Classification

Classification of Radio Galaxy Images with Semi-supervised Learning

Zhixian Ma[1], Jie Zhu[1($\boxtimes$)], Yongkai Zhu[2], and Haiguang Xu[2]

[1] Department of Electronic Engineering, Shanghai Jiao Tong University,
Shanghai 200240, China
{mazhixian,zhujie}@sjtu.edu.cn
[2] Department of Astronomy, Shanghai Jiao Tong University, Shanghai 200240, China

Abstract. The physical mechanism of galaxies lead to their complicated appearances, which could be categorized and require thorough study. Though millions of radio components have been detected by the telescopes, the number of radio galaxies, whose morphologies are well-labeled and categorized, is very few. In this work, we try to mind the features of radio galaxies and classify them with a semi-supervised learning strategy. An autoencoder based on the VGG-16 net is constructed first and pre-trained with unlabeled large-scale dataset to extract the general features of the radio galaxies, and then fine-tuned with labeled small-scale dataset to obtain a morphology classifier. Experiments are designed and demonstrated based on the observations from the Faint Images of the Radio Sky at Twenty-Centimeters Survey (FIRST), where we focus on the classification of three typical morphology types namely Fanaroff-Riley Type I/II (FRI/II), and the bent tailed (C-shape) galaxies. Compared to transfer learning on the same VGG-16 network, which was not trained with enough astronomical images and may suffer from a data-unseen problem, our semi-supervised approach achieves better performance at both high and balanced precision and recall.

Keywords: Radio galaxy image · Classification ·
Semi-supervised learning · Unlabeled data

1 Introduction

There are two motivations that spur us to study and classify radio galaxies (RGs) on their morphologies. One is to reveal the relationship between morphology and intrinsic physical properties of the radio sources, e.g., active galactic nucleus (AGN) and their host super massive black holes [1]. The other is to support the design of radio telescopes (e.g., the Square Kilometer Array, SKA [2]) and their related pre-studies, e.g., simulation of the radio-frequency sky [3], point source detection algorithms [4,5], and foreground removal algorithms [6,7].

Thanks for the high sensitivity (~5 mJy) and resolutions (1.8" pixel resolution and 5.4" telescope resolution) of the very large array (VLA [11]), the FIRST

© Springer Nature Singapore Pte Ltd. 2019
Y. Tan and Y. Shi (Eds.): DMBD 2019, CCIS 1071, pp. 191–200, 2019.
https://doi.org/10.1007/978-981-32-9563-6_20

(faint images of the radio sky at twenty centimeters) survey has been completed, by which millions of radio components were observed and have been archived in data releases (DRs; e.g., DR7 [12]) for the astronomers to dig. However, only thousands of the archived radio galaxies or components have been well labeled on their morphologies [10,13–15], which is unable to drive deep-learning based classification models that have achieved outstanding performance in image classification tasks [16,17]. Such problem could be solved by semi-supervised learning or transfer learning [18,19]. By means of the big data, a semi-supervised training strategy could be applied to learn the features of radio galaxy morphologies with massive unlabeled data from the FIRST DRs and will be able to make classifications after fine-tuned with a labeled small-scale dataset. As for the transfer-learning strategy, Aniyan and Thorat [13] have utilized the transferred AlexNet to classify FRI, FRII and C-shape radio galaxy images and obtained high precision for FRIs ($\sim$91%) but low recall rate for C-shape galaxies ($\sim$75%). In our view, this could be caused by a data-unseen problem (also called dissimilar risk by Rosenstein et el. [20]), since the transferred network has not been fed with enough astronomical images though trained on large-scale daily images in the ImageNet [21].

In this work, we propose a semi-supervised learning approach to classify three radio galaxy morphology types (i.e., FRI, FRII, and C-shape), where the network is constructed based on the VGG-16 [17]. An autoencoder is trained first with an unlabeled sample including around 18,000 images from the FIRST survey [12]. Then the encoder part of the autoencoder is extracted and concatenated with two fully-connected layers to form an radio galaxy morphology classification network, which is trained with well-labeled FRI, FRII and C-shape images in literature [10,14,15].

The rest of this paper is organized as follows. In Sect. 2 we briefly introduce the definition and classification of radio galaxy morphology, and the preprocessing process of the images. In Sect. 3, the construction and semi-supervised training strategy of the network are explained. Experiments and results are displayed in Sect. 4, and we conclude in Sect. 5 with discussion and outlook.

2   Radio Galaxy Image Data

To classify the radio galaxy images, we describe the morphological properties and classification rule of them in this section, and the data preparation process as well.

2.1   Radio Galaxy Morphology

For the radio galaxy morphology, there are two general types namely Fanaroff–Riley Type I and II (FRI and FRII) through comparing the localization of the brightest region on the source and the shape of the lobes [8]. An FRI usually appears a decreasing radio emission as the distance from the core increases, and an FRII will exhibit an inverse tendency and often has bright hotspots

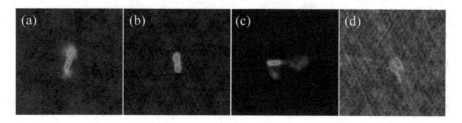

Fig. 1. Radio galaxy image examples. From (a)–(d) are FRI, FRII, WAT and NAT galaxies. (c) and (d) are combined as C-shape in this work.

at the ends of the edges called lobes. In addition to typical FR sources, some atypical morphological subtypes have also been defined in literature, including the narrow-angle tail sources (NATs) and the wide-angle tail sources (WATs; [9]), which usually appear as FRI or FRII but with bent jets or lobes that could be caused with different environment around them [10]. Since that, not only the typical FRIs and FRIIs should be distinguished, but also the WAT and NAT morphologies require consideration. In this work, we combine the WAT and NAT sources as one morphology type namely C-shape galaxy, and conduct a three-type classification task. Example images of an FRI, an FRII, a WAT and an NAT are displayed in Fig. 1.

2.2 Data Preparation

The data are retrieved from the FIRST cutout[1], by which we obtain raw radio galaxy images of size 150×150 pixels ($4'.5 \times 4'.5$) under a maximum intensity of 100 mJy [22]. Before fed into the classification model, the images should be preprocessed including denoising, multi-source separation and resizing. We illustrate the preprocessing process in Fig. 2 with an example.

Since radio galaxies leave us with different distances and usually appear variant extending scales on the images, there might be multiple sources in one image (150×150 pixels). In this work, only the galaxy centering in the image is classified, whose potential neighbors should be removed in advance to avoid confusion labeling. We propose an multi-source separation algorithm as follows,

Step-1 Detect individual components in the images with the Aegean source detection method [4];

Step-2 Locate the centered component and find the nearest neighbor of it within the rest components to calculate the center component margin;

Step-3 Cutout the center region to keep only one radio galaxy;

Step-4 Resize the image to the required shape of the classifier.

In this work, the shape of input image to the classifier is 80×80 pixels, which is decided according to the averaged extends of the radio galaxy morphologies in our data.

---

[1] FIRSTcutout: https://third.ucllnl.org/cgi-bin/firstcutout.

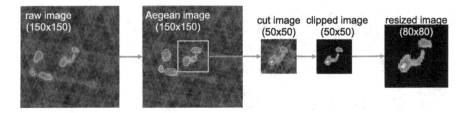

Fig. 2. An example of the data preparation process.

Besides, a radio galaxy is usually with bright luminosity and relatively sharp boundary in the image, we utilize the σ-clipping algorithm to remove the background noise and grid-like interference texture [13,23]. There are two parameters to be set, i.e., the standard deviation σ and iteration times. As suggested in paper [13], we fix the iteration times as 50 and σ as 3. Given that the parameter σ could introduce aleatoric uncertainty [24] to the classifier model and affect the classification result, we will make a simple discussion on this issue in Sect. 4 with experiments.

3   The Classification Model

3.1   Classifier Construction

We construct the radio galaxy morphology classifier based on the VGG-16 network [17]. To fit the input image shape (i.e., 80×80 pixels), we made some modifications of the original VGG-16 by removing the last three convolutional layers and fully-connected layers, and appending two new 256-dimension fully-connected layers to the output of the third max-pooling layer (i.e., the 15th layer) and a three-output softmax layer. We call the new network VGG-X and illustrate it in Fig. 3, where X will be replaced with D,T, and S (please see Sect. 4.2 for details).

Since we are training the network with a semi-supervised training strategy, an autoencoder is formed first at the unsupervised training stage (see the left and top-right subnets in Fig. 3). The decoder subnet is constructed as the same as the mirror-symmetry structure of the encoder subnet (i.e., the VGG-X network). As for the supervised training with small-scale data, we cut the connection between the encoder and decoder, and append the fully-connected and softmax layers to the end of the encoder subnet.

3.2   Training Strategy

Composed of millions of parameters, the VGG-X network requires a large-scale labeled sample to train. Given that we have only several hundred of labeled images, an optional but effective training strategy is required to figure out this issue. We propose a semi-supervised learning strategy, which is,

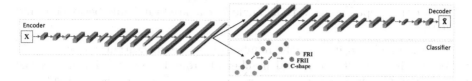

Fig. 3. The framework of the proposed VGG-X radio galaxy morphology classification network. The blue, orange and green cuboids represent the convolutional, max-pooling and up-sampling layers, respectively. Parameter configuration of each layer is kept as the original VGG-16 [17] except for the appended fully-connected layers and softmax layer, which are 256 and 3 dimensions, respectively. (Color figure online)

- Pre-training—apply the self-taught autoencoder to a large-scale unlabeled sample ($\sim$18,000 unlabeled radio galaxies selected from paper [12]) to pre-train the large number of weight and bias parameters of the VGG-X model, so as to learn the representations of the radio galaxy images and to avoid the overfitting risk when learning on a small-scale sample.
- Fine-tuning—append fully-connected layers and a softmax layer to the end of the encoder that has been pre-trained in the first step, and use it as classifier to categorize radio galaxy morphologies with a small-scale labeled sample ($\sim$600).

For the pre-training stage a large learning rate is set and we apply the mean square error (MSE) loss as the training objective. As for the fine-tuning step only the appended dense layers will be trained constrained by a small learning rate as same as the transfer learning [19], and the cross entropy objective is taken.

Some popular techniques and tricks are also applied for training the VGG-X network. A rectified linear unit (ReLU [25]) layer is appended to each convolutional or fully-connected layer as the activation function and the adaptive moment optimization function (ADAM; [26]) is applied to adjust the parameters with exponentially decaying learning rates.

4   Experiment and Result

We design and demonstrate experiments in this section to evaluate performance of our proposed approach and make comparisons to the other methods.

4.1   Data Preparation and Evaluation Indices

To avoid over-fitting problem, we perform the augmentation operation to increase the number of images for both the pre-training and fine-tuning stages. On the raw images, σ-clipping, multi-source separation, and resizing are executed to obtain the preprocessed images as introduced in Sect. 2.2. Then we deploy flipping and rotation to obtain augmented images.

1. Randomly flip the preprocessed image. The flipping mode is either left-to-right, or up-to-bottom, or diagonal;

2. Rotate the flipped image with an angle θ generated randomly between $0°$ and $360°$.

For the unlabeled images, the number of the augmentation operation is 10 for each, which results in a total number of around 1.8×10^5 images for the pre-training. For the labeled images, we select 198 FRIs from the FRICAT [14], 108 FRIIs from the FRIICAT [15] and 285 C-shape galaxies from Proctor's sample [10]. In order to obtain a balanced dataset, the number of augmentations varies in FRI, FRII and C-shape images according to their numbers, which are 100, 200 and 70, respectively.

After data augmentation, the unlabeled and labeled datasets are separated into training and validation subsets. A validation rate is set as 0.2 for pre-training where we obtain 1.4×10^5 and 3.6×10^4 images for training and validation, respectively. For the fine-tuning and classification stage, the 591 radio galaxies are firstly split into training (473 RGs) and test (118 RGs) subsets with a ratio of 4:1. The 473 radio galaxies are than augmented and separated with the same 0.2 validation rate to obtain the training (3.9×10^4 images) and validation (9×10^3 images) subsets for training.

As for the evaluation indices of the proposed radio galaxy classification approach, we apply the precision, recall (i.e., sensitivity) and F1-score measurements, where F1-score evaluates the balance of precision and recall and deserves significant consideration [13,18]. For all of the three indices, the higher of each means the better performance of the classification approach.

4.2   Experimental Result and Discussion

To show the advantages of the proposed radio galaxy morphology classification approach, we compare the proposed semi-supervised learning with two other strategies, i.e., (1) directly supervised training on the same VGG-X network with the small-scale labeled sample and (2) transfer-learning with the trained VGG-16 net [17]. To distinguish the three training strategies, we name our approach and the other two approaches as VGG-S (semi-supervised), VGG-D (directly learning), and VGG-T (transfer-learning), respectively. In addition to that, the classification results of Aniyan and Thorat [13] by transfer learning with the AlexNet is also compared, which is marked as AlexNet-T by us.

We construct the autoencoder for unsupervised pre-training and the VGG-S classifier for supervised fine-tuning as Fig. 3 illustrated. The VGG-D and VGG-T networks are with the same structure as VGG-S. For all of the networks, the training settings are fixed as the same. A hundred epochs are conducted with batch training, where the batch size is 100. Exponentially decaying learning rates for parameters optimization are initialized as 1×10^{-4} for pre-training and 5×10^{-5} for fine-tuning, and varies by a decaying rate of 0.95. For training the VGG-D and VGG-T, the initialized learning rates are set as 1×10^{-4} and 5×10^{-5} after several trials.

Figure 4 illustrates the learning curves of the VGG-S, VGG-D and VGG-T to compare the training strategies with and without pre-training on unlabeled large-scale sample and transfer-learning, respectively. In Table 1, we list the precision,

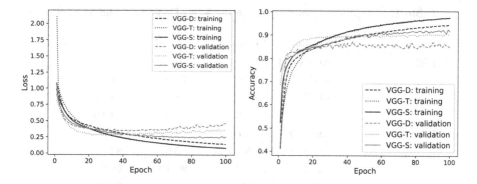

Fig. 4. Comparisons on the learning curves among the VGG-D, VDD-T and VGG-S. **Left:** classification and validation loss curves calculated with the cross entropy function. **Right:** classification and validation accuracy curves.

recall and F1-score measurements evaluated on the three morphology types by the VGG-D, VGG-S and VGG-T, respectively.

As discussed in Sect. 2.2, selection of parameter σ at σ-clipping could affect the training and classification result. Two training datasets are also processed with $\sigma = 2$ and $\sigma = 4$ to make comparisons. We find that when σ is 2, there could remain noise texture in the image, and if σ is 4, the galaxy morphology would be over-clipped to lose information. The same conclusion has also been summarized in Aniyan and Thorat's work [13].

From the experimental results, we summarize that,

(1) In general, the proposed semi-supervised training strategy that applies both unlabeled large-scale and labeled small-scale sample achieves high and balanced precision and recall rates for all of the FRI, FRII, and C-shape radio galaxy morphologies, which suggests our approach be suitable for radio galaxy image classification task.

(2) Under the same VGG-X network structure, our proposed VGG-S outperforms the VGG-D and VGG-T. In our view, the VGG-D could suffer from the overfitting problem when trained with a small-scale sample (see Fig. 4), and the VGG-T may be biased by a data-unseen problem on account that the transferred VGG-16 network has not seen enough astronomical images before.

(3) Compared to the radio galaxy classification approach AlexNet-T, we find that our VGG-X net possess the more balanced precision and recall rate for the FRII and C-shape galaxies. Though the precisions of FRIs and C-shapes by our approach are found lower than the AlexNet-T, it could be relieved by applying the voting strategy with multiple binary training instead of directly three-type classification as the AlexNet-T approach used. We have also made such test and found the precision of FRII and C-shape galaxies increasing to 99.89% and 98.36%, respectively, while the corresponding recall rates are also kept high as 98.68% and 96.77%, respectively.

Table 1. Comparisons among AlexNet-T, VGG-D, VGG-T and VGG-S.

Radio galaxy type	Approach	Precision (%)	Recall (%)	F1-score (%)
FRI	AlexNet-T	91.00	**91.00**	91.00
	VGG-D	83.78	79.48	81.58
	VGG-T	89.19	78.57	83.54
	VGG-S	**94.59**	89.38	**91.91**
FRII	AlexNet-T	75.00	**91.00**	83.00
	VGG-D	85.37	90.91	88.05
	VGG-T	87.80	90.12	88.94
	VGG-S	**90.24**	89.71	**89.05**
C-shape	AlexNet-T	**95.00**	79.00	87.00
	VGG-D	89.86	86.11	87.94
	VGG-S	88.41	89.71	89.05
	VGG-T	92.75	**90.14**	**91.43**

(4) The setting of σ in the data preprocessing, which may introduce aleatoric uncertainty [24] to the model, affects the training and classification result of the model. This should be taken into consideration in similar classification tasks.

5   Conclusion

In this work a radio galaxy morphology classification approach is proposed, which is constructed based on the VGG-16 network and trained with a semi-supervised strategy on a large-scale unlabeled sample and a small-scale well-labeled sample. Three morphology types namely Fanaroff-Type I and II, and C-shape are able to be categorized, where our proposed approach achieves ∼91% average precision, ∼90% recall rate and a balanced F1-score of ∼90%.

Compared to the transfer-learning approaches by Aniyan and Thorat [13], and the same VGG-16 based network by us, we find that the directly transfer of the deep neural networks on the astronomical images could not attain satisfied performance, especially on the cross-misclassification of the FRII and C-shape images. Meanwhile, the proposed semi-supervised learning strategy achieves relatively high precisions for all of the radio galaxy types and balanced recall rates as well, which suggests our approach be suitable for the radio galaxy morphology classification task.

In the future, we plan to include more radio galaxy morphology types to obtain a more general and robust classification model and to quantify the potential aleatoric uncertainties introduced during the data preparation process.

Acknowledgments. This work was supported by the National Natural Science Foundation of China (grant No. 11433002) and the National Key Research and Discovery Plan Nos. 2017YFF0210903 and 2018YFA0404601.

References

1. Padovani, P., et al.: Active galactic nuclei: what's in a name? Annu. Rev. Astron. Astr. **25**, 2 (2017)
2. Koopmans, L., et al.: The cosmic dawn and epoch of reionisation with SKA. In: Advancing Astrophysics with the Square Kilometre Array (AASKA14), vol. 1 (2015)
3. Wang, J., et al.: How to identify and separate bright galaxy clusters from the low-frequency radio sky. Astronphy. J. **723**, 620–633 (2010)
4. Hancock, P.J., Murphy, T., Gaensler, B.M., Hopkins, A., Curran, J.R.: Compact continuum source finding for next generation radio surveys. Mon. Not. R. Astron. Soc. **422**, 1812–1824 (2012)
5. Mohan, N., Rafferty, D.: PyBDSF: Python Blob Detection And Source Finder. Astrophysics Source Code Library (2015)
6. Chapman, E., et al.: Foreground removal using fastica: a showcase of LOFAR-EoR. Mon. Not. R. Astron. Soc. **423**(3), 2518–2532 (2012)
7. Chapman, E., Zaroubi, S., Abdalla, F.B., Dulwich, F., Jelić, V., Mort, B.: The effect of foreground mitigation strategy on EoR window recovery. Mon. Not. R. Astron. Soc. **458**, 2928–2939 (2016)
8. Fanaroff, B.L., Riley, J.M.: The morphology of extragalactic radio sources of high and low luminosity. Mon. Not. R. Astron. Soc. **167**, 31P–36P (1974)
9. Rudnick, L., Owen, F.N.: Head-tail radio sources in clusters of galaxies. Astrophys. J. L. **203**, L107–L111 (1976)
10. Proctor, D.D.: Morphological annotations for groups in the first database. Astrophys. J. Suppl. Ser. **194**, 31 (2011)
11. Becker, R.H., White, R.L., Helfand, D.J.: The FIRST survey: faint images of the radio sky at twenty centimeters. Astrophys. J. **450**, 559 (1995)
12. Best, P.N., Heckman, T.M.: On the fundamental dichotomy in the local radio-AGN population: accretion, evolution and host galaxy properties. Mon. Not. R. Astron. Soc. **421**, 1569–1582 (2012)
13. Aniyan, A.K., Thorat, K.: Classifying radio galaxies with the convolutional neural network. Astrophys. J. Suppl. Ser. **230**, 20 (2017)
14. Capetti, A., Massaro, F., Baldi, R.D.: FRICAT: a FIRST catalog of FR I radio galaxies. Astron. Astrophys. **598**, A49 (2017)
15. Capetti, A., Massaro, F., Baldi, R.D.: FRIICAT: a FIRST catalog of FR II radio galaxies. Astron. Astrophys. **601**, A81 (2017)
16. Krizhevsky, A., Sutskever, I., Hinton, G.E.: ImageNet classification with deep convolutional neural networks. Commun. ACM **60**(6), 84–90 (2017)
17. Simonyan, K., Zisserman, A.: Very deep convolutional networks for large-scale image recognition. ArXiv e-prints (2014)
18. Goodfellow, I., Bengio, Y., Courville, A.: Deep Learning. MIT Press (2016). http://www.deeplearningbook.org
19. Pan, S.J., Yang, Q.: A survey on transfer learning. IEEE Trans. Knowl. Data Eng. **22**(10), 1345–1359 (2010)
20. Rosenstein, M., Marx, Z., Kaelbling, L., Dietterich, T.: To transfer or not transfer. In: NIPS 2005 Workshop on Transfer Learning, Vancouver, Canada, vol. 1 (2005)

21. Deng, J., Dong, W., Socher, R., Li, L., Li, K., Li, F.F.: Imagenet: a large-scale hierarchical image database. In: IEEE Conference on Computer Vision and Pattern Recognition, CVPR 2009, pp. 248–255. IEEE, Miami (2009)
22. Ma, Z., Xu, H., Zhu, J., et al.: A machine learning based morphological classification of 14,251 radio AGNs selected from the best-Heckman sample. Astron. J. Suppl. S. **240**(34), 1–21 (2019)
23. Hogg, D.W.: Confusion Errors in Astrometry and Counterpart Association. Astron. J. **121**, 1207–1213 (2001)
24. Kendall, A., Gal, Y.: What uncertainties do we need in Bayesian deep learning for computer vision? ArXiv e-print arXiv:1703.04977 (2017)
25. Glorot, X., Bordes, A., Bengio, Y.: Deep sparse rectifier neural networks. J. Mach. Learn. Res. **15**(106), 275 (2011)
26. Kingma, D., Ba, J.: Adam: a method for stochastic optimization. ArXiv e-print arXiv:1412.6980 (2014)

Multi-label Text Classification Based on Sequence Model

Wenshi Chen[1], Xinhui Liu[1], Dongyu Guo[2], and Mingyu Lu[1(✉)]

[1] Information Science and Technology College, Dalian Maritime University,
Dalian, China
lnjzcws@sohu.com, lumingyu@dlmu.edu.cn
[2] Wuhan University of Technology, Wuhan, China

Abstract. In the multi-label text classification problem, the category labels are frequently related in the semantic space. In order to enhance the classification performance, using the correlation between labels and using the Encoder in the seq2seq model and the Decoder model with the attention mechanism, a multi-label text classification method based on sequence generation is proposed. First, the Encoder encodes the word vector in the text to form a semantic coding vector. Then, the LSTM neural network in the Decoder stage is utilized to process the dependency of the category label sequence to consider the correlation between the category labels, and the attention mechanism is added to calculate the probability of attention distribution. Highlight the effect of key input on the output, and improve the missing semantic problem caused by the input too long, and finally output the predicted label category. The experimental results show that our model is better than the existing model after considering the label correlation.

Keywords: Multi-label text classification · Label correlation · Seq2seq model

1 Introduction

The traditional classification is mainly the classification of single labels. Each sample can only be represented by one category. However, in real life, one sample can be characterized by multiple labels at the same time. In Zhihu, each topic gets one or more labels. In the blog, you can add one or more labels to an article. In the image classification [1], each image can contain different scenes at the same time. In text categorization [2], each text can belong to more than one category at the same time. In the bioinformatics problem [3], each gene can have multiple functional attributes at the same time. This paper mainly studies the application of multi-label classification problem in text classification. Text classification problems can be divided into two categories: single-label text questions and multi-label text problems. Single-label text classification assumes that labels are independent of each other, each text can only belong to one category label, multi-label text classification considers that category labels are related, and one text can be divided into several different categories simultaneously [4]. Therefore, for a sample containing multiple categories of multi-label data, the traditional single-label classification method is no longer applicable.

© Springer Nature Singapore Pte Ltd. 2019
Y. Tan and Y. Shi (Eds.): DMBD 2019, CCIS 1071, pp. 201–210, 2019.
https://doi.org/10.1007/978-981-32-9563-6_21

In this paper, the multi-label text classification task is regarded as the sequence generation problem, the seq2seq model (Encoder-Decoder) is used for multi-label text classification which effectively learns both the semantic redundancy and the co-occurrence dependency in an end-to-end way. The Attention mechanism in Decoder network is to give different weights to the input sequence by considering the contribution of different text content to the category label, which improves the semantic missing problem caused by the long input sequence.

Suppose $X = R^d$ indicates that there is a D-dimensional sample space in the real number field R, $Y = \{y_1, y_2, \ldots, y_q\}$ represents a category label set containing q label spaces, and multi-label classification refers to passing the training data set. $D = \{(x_i, Y_i) | 1 \leq i \leq m\}$ obtains a mapping function $f : X \rightarrow 2^Y$, where $x_i \in X$, i.e. x_i is a training data in the input space X, and $Y_i \in Y$, i.e. Y_i is a sample. The collection of labels for x_i [5]. When a sample to be classified $x \in X$ is input, the prediction label set $P_x \subset Y$ of x is obtained by the mapping function f such that P_x is closest to the real label set Y_x of the sample x. The single-label classification problem is a special case of the multi-label classification problem. When $|Y_i| = q = 1$, the multi-label classification is converted into a single-label classification. As the number of category labels increases in multi-label classification problem, the size of the output space will increase exponentially. Since the labels and labels in the multi-label classification are not independent of each other [6], they are under a certain relationship in the semantic space. The classification label correlation is considered in the classification to improve the classification precision.

2   Related Research

There are two types of multi-label classification algorithms. One method converts the multi-label classification problem into a traditional classification problem. The problem conversion method which including Binary Relevance algorithm [7], LabelPowerset algorithm [8]; Random k-Labelsets algorithm [9], Classifier Chains algorithm [10]. Another method is to adapt the existing algorithm to multi-label classification, namely algorithm adaptation method, including ML-DT algorithm, RankSVM algorithm [11], etc. These two algorithms respectively perform decision tree algorithm and support vector machine algorithm. Improved to adapt it to multi-label classification algorithms. The labels in the multi-label classification are not independent of each other, but have a certain relationship. If the label correlation is considered in the classification, the classification accuracy is improved. The above-mentioned Classifier Chains algorithm uses the label. The correlation between the categories is classified.

Deep neural networks have achieved good performance in Natural Language Processing (NLP). The BP-MLL (Back-Propagation for Multi-label Learning) method proposed by Zhou et al. [12] used neural networks for multi-label text classification at the first time. The Sequence to Sequence (Seq2seq) was proposed by Sutskever et al. in 2014 [13]. The model uses two Recurrent Neural Networks (RNN) to construct an encoder and a decoder respectively which can map variable length input sequences to variable length output sequences. Cho et al. proposed the Encoder-Decoder model for machine translation [14].

Google proposed a dialogue model based on seq2seq [15], which is also used in other areas of NLP, such as text abstracts [16], multi-task learning [17].

The Attention mechanism was first applied in the field of image processing for image subject generation [18], character recognition [19] and so on. Bahdanau et al. [20] added the attention mechanism to the Encoder-Decoder, which is the first to use the attention mechanism in the NLP field, effectively solving the problem of semantic information loss caused by the input statement being too long. The effect of machine translation has improved significantly. In this paper, the attention mechanism is added in the Decoder stage of the seq2seq model to consider the contribution of different text content to the category label, and assign different weights to the input sequence, which improves the semantic missing problem caused by the input sequence being too long.

3   Multi-label Text Classification

The multi-label text classification model based on sequence generation proposed in this paper consists of seq2seq model and attention mechanism. The seq2seq model can capture the correlation between category labels through the LSTM neural network in the Decoder stage. The seq2seq model with the attention mechanism can not only consider the correlation between category labels, but also automatically select a word with a large contribution to give a larger weight when predicting different category labels. The seq2seq model solves the problem of both input and output being indefinitely long. The seq2seq model consists of three parts, an encoder, a decoder, and an intermediate semantic vector C that connects to the Encoder-Decoder. The Encoder and Decoder respectively correspond to two neural networks of the input sequence and the output sequence. The input sequence is encoded by the Encoder to form an intermediate semantic vector, which is passed to the Decoder output target vector. The Encoder-Decoder consists of two neural network units: the Encoder network takes the information from the entire input sequence to generate an intermediate semantic vector, which is then passed to the Decoder network and decoded into a series of category labels.

Since in the seq2seq model, the input of the first step of the Decoder is the encoding vector output by the last step of the Encoder, the information used by the final output of the Decoder is all from the last output vector of the Encoder. When the input is long, the semantic vector obtained by Encoder can not completely contain all the information of the sentence, which leads to the loss of some information. Bahdanau proposed the attention mechanism based on the seq2seq model [20], which improved the problem of semantic missing due to excessive input data. Combining the characteristics of text classification tasks, this paper uses the attention mechanism in the Decoder stage to emphasize the effect of key inputs on the output by calculating the attention probability distribution. Each hidden layer of the Encoder section is weighted so that Decoder has multiple semantic information to get the semantic representation of the input.

Encoder: Encodes the input sequence into an intermediate semantic vector C that contains the information of the input sequence. The input sequence is represented as a sequence of word vectors $x = (x_1, \cdots, x_{T_x})$ for each document, and the encoding process is:

$$h_t = f(x_t, h_{t-1}) \tag{1}$$

$$c = q(\{h_1, h_2, \cdots, h_{T_x}\}) \tag{2}$$

Where h_t is the hidden layer state at time t, and f is the neural network unit. RNN (Recurrent Neural Network) variant LSTM (Long Short-Term Memory) or GRU (Gated Recurrent Unit) is usually used instead of RNN to avoid RNN gradient. For disappearance and other issues, this paper uses LSTM neural network, q is a linear function. Enter the encoding vector of the sentence as the last state vector of the encoder, therefore, there is

$$q(\{h_1, h_2, \cdots, h_{T_x}\}) = h_{T_x} \tag{3}$$

Decoder: Assume that the output sequence is the category label $y = (y_1, \cdots, y_{T_y})$, the output of the time t is determined by the previous output and the semantic vector C, and the probability value of the target word is calculated for the vector C.

$$p(y) = \prod_{t=1}^{T} p(y_t | \{y_1, \cdots, y_{t-1}\}, c) \tag{4}$$

$$p(y_t | \{y_1, \cdots, y_{t-1}\}, c) = g(y_{t-1}, s_t, c) \tag{5}$$

$$s_t = f(s_{t-1}, y_{t-1}, s_t, c) \tag{6}$$

Where g represents the LSTM neural network unit and s_t represents the state of the LSTM hidden layer at time t.

LSTM: LSTM can solve the problem of RNN gradient disappearing. The standard LSTM network consists of three layers: input layer, hidden layer, and output layer. The hidden layer contains Input Gate, Output Gate, Forget Gate, and Memory Cell which determines the information in the reserved or forgotten memory unit through three gate structures.

$$i_t = \delta(W_i \cdot |h_{t-1}, x_t| + b_i) \tag{7}$$

$$f_t = \delta(W_f \cdot |h_{t-1}, x_t| + b_f) \tag{8}$$

$$o_t = \delta(W_o \cdot |h_{t-1}, x_t| + b_o) \tag{9}$$

$$\overline{C_t} = \tanh(W_c \cdot |h_{t-1}, x_t| + b_c) \tag{10}$$

$$C_t = f_t * C_{t-1} + i_t * C_{t-1} \tag{11}$$

$$h_t = o_t * \tanh(C_t) \tag{12}$$

Where i, f, o represent input, forgetting, and output gates respectively; x, h, c represent input layer, hidden layer, memory unit; W, b represents weight matrix and offset.

Attention mechanism: Combined with the characteristics of the text classification task, the attention mechanism is added in the Decoder stage, and the effect of the key input on the output can be highlighted by calculating the attention probability distribution. Each hidden layer of the Encoder section is weighted so that Decoder has multiple semantic information to get the semantic representation of the input.

In the Decoder stage,

$$p(y_t | \{y_1, \cdots, y_{t-1}\}, c) = g(y_{t-1}, s_t, c_t) \tag{13}$$

$$s_t = f(s_{t-1}, y_{t-1}, s_t, c_t) \tag{14}$$

Where s_t represents the hidden layer state of the Decoder at time t. Unlike the basic seq2seq model, the semantic encoding vector C_t is not only the output of the last step of the Encoder stage, but all the hidden layers of the Encoder stage $(h_1, h_2, \cdots, h_{T_x})$ Weighted summation, which gives the input of the new state. The coding vector C_t in time t can be expressed as:

$$c_t = \sum_{i=1}^{T_x} \alpha_{ti} h_i \tag{15}$$

Where h_i represents the Decoder hidden layer state at time t, and α_{ti} represents the weight. For each hidden layer state h_i, the weight α_{ti} is calculated as:

$$\alpha_{ti} = \frac{\exp(e_{ti})}{\sum_{k=1}^{T_x} \exp(e_{tk})} \tag{16}$$

$$e_{ti} = \alpha(s_{t-1}, h_i) \tag{17}$$

e_{ti} is the Decoder obtained from the hidden layer state at time t $-$ 1 and the hidden layer state h_i at time i.

The sequence generation model of this paper consists of an Encoder and a Decoder with an attention mechanism. The Encoder part is shown on the left side of Fig. 1. The input layer unifies the text to be classified into the same length sequence and passes it to the Embedding layer. The Embedding layer uses the word embedding method. Convert each word into a vector, enter it into the LSTM neural network, and encode the input vector. When the model predicts different category labels, different input words will make different contributions. The attention mechanism generates intermediate semantic vectors by assigning different weights to different words in the input text, and gives greater weight to words that contribute to the output category label, otherwise it gives less weight. The weight calculation is as in Eq. 16, so that when the model predicts different categories of labels. The attention mechanism can consider the difference in word contributions of the input text.

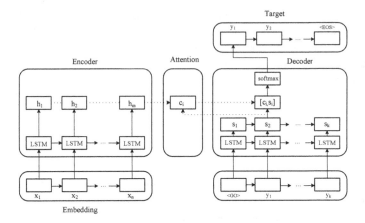

Fig. 1. Model architecture

The Decoder part of the model is shown on the right side of Fig. 1. The input sequence of Decoder uses <GO> as the head of the category label sequence, and the output sequence uses <EOS> as the tail of the prediction category label. The dependency of the tag labels is considered by the LSTM processing label sequence dependency, the labels are generated sequentially, and the next label is predicted based on the previously predicted label, so that the softmax activated Dense layer converts it into a category label output.

4 Experiments

4.1 Experimental Data

Reuters-21578 dataset: Distributed in 22 files, from reut2-000.sgm to reut2-020.sgm, each file contains 1000 documents, reut2-021.sgm contains 578 documents, a total of 21578 documents, divided into 672 categories, and this experiment selects the 20 category labels with the largest number of articles.

IMDB dataset: Contains 117,352 movie profiles, including 117,190 introductory films in English, each of which can be associated with 27 categories.

4.2 Experimental Evaluation

1. Precision: The proportional relationship between the correctly predicted label and the sum of the predicted label and the actual label. The larger the value, the more the number of labels that are correctly predicted.
2. Recall: The ratio of the correctly predicted label to the total number of actual labels. The larger the value, the better the system performance.
3. F1 Score: This evaluation indicator is a comprehensive indicator that combines accuracy and recall. The larger the value, the better the system performance.

4. Hamming Loss: It is the XOR between the actual result and the predicted result in the sample, indicating the number of times the sample label pair is misclassified. The smaller the value, the better the performance of the model.
5. Subset Accuracy: Calculate the correct proportion of all predictions in all samples. Full prediction of the correct sample requires that the predicted label set be identical to the correct label set. The larger the value of this indicator, the better the system performance.

4.3   Model Parameter

Because there are irregular characters (HTML mark, xml mark, etc.), extra punctuation marks, meaningless stop words, etc. in the original text, the noise data in the text is removed to form a pure English text corpus. The length of the words contained in each text is different. If the number of words is too short, the text can not be accurately judged. If the number of words is too long, the space will be wasted. Therefore, remove the text with less than 20 words and select 70% of the text. The maximum length is used as the input length, and the text less than the number of words is filled with <pad>, and more than the length is used to truncate the text. To evaluate the model, we divided the data set into two parts of 2/3 of the data for training and 1/3 of the data to verify the performance of the classification model. Five evaluation indicators precision (denoted as P), recall rate (denoted as R), comprehensive index F1-Score (denoted as F1), Hamming Loss (denoted as HL), Subset Accuracy (denoted as SA). Use these five indicators to compare the classification effects of different classification methods.

Table 1 shows an overview of each dataset, T represents the total number of processed datasets, L represents the number of label categories, V represents the number of words in the dictionary of each data set, and A represents the average number of label categories included in each sample.

Table 1. Overview of the dataset.

Datasets	T	L	V	A
Reuters-21578	13306	20	36035	1.34
IMDB	93458	27	141045	2.41

In this paper, Adam (Adaptive Moment Estimation) is used as the optimizer function to optimize the neural network. The loss function uses binary_crossentropy. To prevent over-fitting of the neural network, we use dropout in the encoder with a value of 0.3 and epoch is 10. The Encoder and Decoder have an LSTM layer of 1, an Encoder hidden layer size of 128, and a Decoder hidden layer size of 64.

Reuters-21578: The maximum text input length is 915, the maximum length of 70% text is 83, the maximum number of labels for text is 8, and the number of labels for 70% text is 2. The training set text number is 8870, the verification set text number is 2218, the test set text number is 2218, and the batch size is 16.

IMDB: The maximum text input length is 1106, the maximum length of 70% text is 65, the maximum number of labels for text is 12, and the number of labels for 70% text is 3. The training set text number is 62304, the verification set text number is 15577, and the test set text number is 15577.

4.4   Experimental Results

In this paper, a multi-label classification algorithm based on sequence generation is implemented. The model based on sequence generated named seq2seq+attention proposed in this paper compares with the machine learning methods includes Binary Relevance, Label Powerset, Classifier Chains and seq2seq model in five indicators as P, R, F1, HL and SA. The results on each dataset are shown in Tables 2 and 3. The "+" indicates that the larger the value, the better the performance of the model. The "−" indicates that the smaller the value, the better the model performance.

Table 2. The results on Reuters-21578

Models	P(+)	R(+)	F1(+)	HL(−)	SA(+)
Binary Relevance	73.44	65.28	69.12	0.2545	24.24
Label Powerset	77.55	50.00	60.80	0.2969	27.27
Classifier Chains	79.24	53.85	64.12	0.2848	24.24
seq2seq	86.42	84.47	85.42	0.0220	74.68
seq2seq+attention	88.13	84.71	86.36	0.0203	76.06

From Table 2 that on the Reuters-21578 dataset, the proposed model is superior to the traditional machine learning model and the seq2seq model. Binary Relevance and Label Powerset do not consider label correlation. Classifier Chains Considering the label correlation, the results show that the indicators proposed in this paper have improved, the F1 value increased by 22.24%, and the SA increased by 51.82%.

Table 3. The results on IMDB

Models	P(+)	R(+)	F1(+)	HL(−)	SA(+)
Binary Relevance	65.57	50.63	57.14	0.3636	9.09
Label Powerset	68.57	70.59	69.57	0.2445	18.18
Classifier Chains	76.19	79.01	77.58	0.2242	30.30
seq2seq	80.12	76.62	78.32	0.0326	59.67
seq2seq+attention	80.12	76.67	78.35	0.0325	60.12

From Table 3 that on the IMDB dataset, the model proposed in this paper is superior to the machine learning model and the seq2seq model in all indicators, and the indicators are improved, the F1 value is increased by 0.77%, and the SA is improved by 29.82%.

From Tables 2 and 3, the model Classifier Chains, seq2seq and seq2seq+attention which consider the label correlation, are superior to the algorithms that do not consider the label correlation in each index, and the experimental results of the seq2seq model are superior to the traditional machine learning. The algorithm has improved on various indicators. Therefore, when conducting multi-label text classification research, the influence of label correlation on classification results should be considered. Since different parts of the input sequence may have different contributions to predicting different category labels, the classification performance of the seq2seq model with the attention mechanism is better than the seq2seq model without the attention mechanism. The attention mechanism shows good results in multi-label text classification.

5   Conclusion

The main work of this paper is to regard the multi-label classification problem as the sequence generation problem. The LSTM sequence of the Decoder part of the seq2seq model generates labels to solve the correlation problem between labels. The seq2seq model added to the attention mechanism automatically finds the correspondence between the input text and the predicted category label, and gives greater weight to the words that contribute greatly. The experimental results show that the proposed model achieves good results in terms of accuracy, recall, F1, hamming loss and subset accuracy. Compared with the baseline model, the effect has been significantly improved.

References

1. Wang, J., Yang, Y., Mao, J., Huang, Z., Huang, C., Xu, W.: CNN-RNN: a unified framework for multi-label image classification. In: Proceedings of the IEEE Conference on Computer Vision and Pattern Recognition, pp. 2285–2294 (2016)
2. Chen, G., Ye, D., Xing, Z., Chen, J., Cambria, E.: Ensemble application of convolutional and recurrent neural networks for multi-label text categorization. In: 2017 International Joint Conference on Neural Networks (IJCNN), pp. 2377–2383. IEEE (2017)
3. Feng, S., Fu, P., Zheng, W.: A hierarchical multi-label classification algorithm for gene function prediction. Algorithms **10**, 138 (2017)
4. Tsoumakas, G., Katakis, I.: Multi-label classification: an overview. Int. J. Data Warehous. Min. (IJDWM) **3**, 1–13 (2007)
5. Zhang, M.-L., Zhou, Z.-H.: A review on multi-label learning algorithms. IEEE Trans. Knowl. Data Eng. **26**, 1819–1837 (2014)
6. Xu, C., Tao, D., Xu, C.: A survey on multi-view learning. arXiv preprint arXiv:1304.5634 (2013)
7. Luaces Rodríguez, Ó., Díez Peláez, J., Barranquero Tolosa, J., Coz Velasco, J.J.D., Bahamonde Rionda, A.: Binary relevance efficacy for multilabel classification. Prog. Artif. Intell. **1**(4) (2012)
8. Cherman, E.A., Monard, M.C., Metz, J.: Multi-label problem transformation methods: a case study. CLEI Electron. J. **14**, 4 (2011)

9. Tsoumakas, G., Vlahavas, I.: Random *k*-labelsets: an ensemble method for multilabel classification. In: Kok, J.N., Koronacki, J., Mantaras, R.L., Matwin, S., Mladenič, D., Skowron, A. (eds.) ECML 2007. LNCS (LNAI), vol. 4701, pp. 406–417. Springer, Heidelberg (2007). https://doi.org/10.1007/978-3-540-74958-5_38
10. Read, J., Pfahringer, B., Holmes, G., Frank, E.: Classifier chains for multi-label classification. Mach. Learn. **85**, 333 (2011)
11. Elisseeff, A., Weston, J.: A kernel method for multi-labelled classification. In: Advances in Neural Information Processing Systems, pp. 681–687 (2002)
12. Zhang, M.-L., Zhou, Z.-H.: Multilabel neural networks with applications to functional genomics and text categorization. IEEE Trans. Knowl. Data Eng. **18**, 1338–1351 (2006)
13. Sutskever, I., Vinyals, O., Le, Q.V.: Sequence to sequence learning with neural networks. In: Advances in Neural Information Processing Systems, pp. 3104–3112 (2014)
14. Cho, K., et al.: Learning phrase representations using RNN encoder-decoder for statistical machine translation. arXiv preprint arXiv:1406.1078 (2014)
15. Vinyals, O., Le, Q.: A neural conversational model. arXiv preprint arXiv:1506.05869 (2015)
16. Rush, A.M., Chopra, S., Weston, J.: A neural attention model for abstractive sentence summarization. arXiv preprint arXiv:1509.00685 (2015)
17. Luong, M.-T., Le, Q.V., Sutskever, I., Vinyals, O., Kaiser, L.: Multi-task sequence to sequence learning. arXiv preprint arXiv:1511.06114 (2015)
18. Xu, K., et al.: Show, attend and tell: Neural image caption generation with visual attention. In: International Conference on Machine Learning, pp. 2048–2057 (2015)
19. Lee, C.-Y., Osindero, S.: Recursive recurrent nets with attention modeling for OCR in the wild. In: Proceedings of the IEEE Conference on Computer Vision and Pattern Recognition, pp. 2231–2239 (2016)
20. Bahdanau, D., Cho, K., Bengio, Y.: Neural machine translation by jointly learning to align and translate. arXiv preprint arXiv:1409.0473 (2014)

A Survey of State-of-the-Art Short Text Matching Algorithms

Weiwei Hu[1(✉)], Anhong Dang[1], and Ying Tan[2]

[1] State Key Laboratory of Advanced Optical Communication Systems and Networks, School of Electronics Engineering and Computer Science, Peking University, Beijing 100871, China
{weiwei.hu,ahdang}@pku.edu.cn
[2] Key Laboratory of Machine Perception (MOE), and Department of Machine Intelligence, School of Electronics Engineering and Computer Science, Peking University, Beijing 100871, China
ytan@pku.edu.cn

Abstract. The short text matching task uses an NLP model to predict the semantic relevance of two texts. It has been used in many fields such as information retrieval, question answering and dialogue systems. This paper will review several state-of-the-art neural network based text matching algorithms in recent years. We aim to provide a quick start guide to beginners on short text matching. The representation based model DSSM is first introduced, which uses a neural network model to represent texts as feature vectors, and the cosine similarity between vectors is regarded as the matching score of texts. Word interaction based models such as DRMM, MatchPyramid and BERT are then introduced, which extract semantic matching features from the similarities of word pairs in two texts to capture more detailed interaction information between texts. We analyze the applicable scenes of each algorithm based on the effectiveness and time complexity, which will help beginners to choose appropriate models for their short text matching applications.

Keywords: Short text matching · Deep learning · Representation learning · Neural networks

1 Introduction

Short text matching is a widely used NLP technology which aims to model the semantic relationship between two texts. Information retrieval, question answering and dialogue systems are the main application areas of short text matching.

In information retrieval, users want to find relevant documents for a given query. How to match the query with appropriate documents is crucial to search engines. Text matching can also be used to match question with appropriate answers, which is very helpful in automatic custom service and will significantly reduce the labor cost.

Y. Tan and Y. Shi (Eds.): DMBD 2019, CCIS 1071, pp. 211–219, 2019.
https://doi.org/10.1007/978-981-32-9563-6_22

Recent researches show that neural network based text matching algorithms outperform traditional text matching algorithms such as TFIDF, latent semantic analysis (LSA) [3] and latent Dirichlet allocation (LDA) [2]. Using neural networks to represent text and learn the interaction pattern between texts will make the model able to mine the complex semantic relationship of texts.

Many neural network based text matching algorithms have been proposed in recent years. This paper only focuses several state-of-the-art algorithms among them. Considering the effectiveness and time efficiency, DSSM, DRMM, Match-Pyramid and BERT are chosen to present in this paper. There are many other famous text matching algorithms, such as ARC [9] and DURT [13]. Due to the space limit, we will not introduce them in this paper. An experimental evaluation of different text matching algorithms can be found in [5,15].

2   Deep Structured Semantic Models (DSSM)

DSSM is a well-known short text matching algorithm, which is the abbreviation for deep structured semantic models [10]. It is first proposed to match query and documents in web search applications. DSSM uses neural networks to represent queries and documents as vectors. The vector distance between a query and a document is regarded as the matching score of them. There are mainly three kinds of neural networks to represent text for DSSM, and we will introduce them in the following three sub-sections.

2.1   Feed-Forward Network Based DSSM

The basic DSSM algorithm uses feed-forward networks to represent queries and documents. The architecture of basic DSSM is shown in Fig. 1.

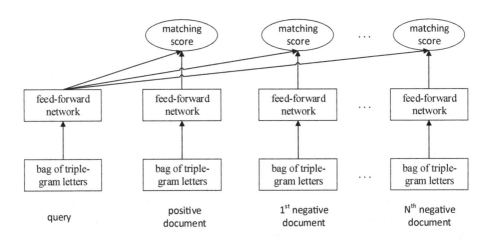

Fig. 1. The architecture of DSSM

The input query and documents are firstly represented as bag of triple-gram letters by word hashing. Word hashing by triple-gram of letters has two advantages. The first one is that it is able to significantly reduce the size of the vocabulary. A small vocabulary will make calculation of the bottom layer of neural network very fast. The second advantage is that triple-gram representation of query letters is more robust to misspelling than word based representation. The search engine usually faces a large number of users, and many users may misspell the queries. For misspelled words, the correct parts will be preserved by triple-gram letters.

The bag of triple-gram letters is then represented as a boolean vector. If a triple-gram is present in a query or a document, the corresponding dimension of the boolean vector is set to one. Otherwise, it is set to zero. This vector is fed into a feed-forward network. The network will output a vector which is the semantic representation of the query or the document.

The cosine similarity between a query and a document is regarded as the matching score of their similarity. Let the cosine similarity between a query and its positive document vector be s^+, and the cosine similarity with its i-th negative document be s_i^-, assuming there are N negative documents in total. The loss function for training DSSM is shown in Formula 1.

$$loss = -log \frac{e^{\gamma s^+}}{e^{\gamma s^+} + \sum_{i=1}^{N} e^{\gamma s_i^-}}. \tag{1}$$

γ in Formula 1 is the hyper-parameter to control the smoothness of the softmax function. Minimizing this loss function will push the matching score of positive document become larger and push the matching scores of negative documents become smaller. Such process will make DSSM able to discriminate relevant documents from irrelevant ones for a given query.

2.2 Convolutional DSSM (CDSSM)

The basic DSSM only uses feed-forward networks to represent the query and documents. However, feed-forward networks are not originally designed for processing sequential data, while query and documents are sequence of words. To better capture the sequential nature of query and document text, the DSSM with convolutional-pooling structure (abbr. CDSSM) is proposed [19].

CDSSM uses a sliding window to split the input text (i.e. a query or a document) into word n-grams. In each sliding window, the triple-gram letters are extracted for each word, and each word is represented as a boolean vector of triple-gram letters. The boolean vectors of all words in the sliding window are concatenated as the feature vector of the sliding window.

The convolutional operation is applied on each sliding window. The feature vector of each sliding window is fed into a feed-forward network, in order to extract higher level semantic feature of the sliding window. The max-pooling layer takes the element-wise maximum of the network's output vectors of all sliding windows. The variable-length input text will become a feature vector with fixed length. At last another feed-forward network is applied on top of the max-pooling layer, and output the final semantic feature vector of the input text.

2.3 LSTM-DSSM

For many sequential applications, long short-term memory (LSTM) [8] works better than convolutional networks. LSTM-DSSM [16] is a variant of DSSM which uses LSTM to extract the feature vector of input text.

LSTM is a special kind of recurrent neural networks [6]. Recurrent neural networks process the input text sequentially, and use recurrent layers to remember previous states. For long sequences, the recurrent neural networks may encounter the problem of gradient vanishing during training. LSTM introduces input gate, output gate and forget gate to the recurrent layers. The opening and closing of the gates make LSTM able to reserve long-term information for long sequences.

3   Deep Relevance Matching Model (DRMM)

DRMM [7] matches two pieces of text in word level. Word interaction matrix is first calculated, and then the high-level feature is extracted from the word interaction matrix.

Supposing the embeddings of query words are represented as $q_1, q_2, ..., q_Q$, and the embeddings of the document words are represented as $d_1, d_2, ..., d_D$, where Q is the length of the query and D is the length of the document.

For the i-th query word q_i, a term gate is first applied to get its weight among all the query words. The term gate g_i is calculated as Formula 2.

$$g_i = \frac{e^{wq_i}}{\sum_{j=1}^{Q} e^{wq_j}}. \tag{2}$$

w in Formula 2 is a learnable parameter. It can be seen that g_i is a normalized weight of the i-th query word, and the term gates of all query words sum up to 1.

To match the document for the i-th query word, the cosine distance of q_i with all the words in the document is calculated. That is:

$$s_{ij} = cosine(q_i, d_j) \tag{3}$$

For the i-th query word, the histogram of s_{i1} to s_{iD} is calculated. The range of cosine distance is between -1 and 1. This range is split into several intervals and the number of cosine values fallen in each interval is calculated. The counts of words in all intervals forms a feature vector, which is able to reflect the matching status of the i-th query word with the document. A feed-forward network with one output neuron is applied on this feature vector. Let the output of the network be m_i, which can be regarded as the matching score of the i-th query word with the document.

The final matching score of the query and the document is calculated as follows:

$$score = \sum_{i=1}^{Q} g_i m_i. \tag{4}$$

That is, the matching score is the weighted sum of all the query words' matching scores with the document.

The original DRMM paper uses histogram of bins to represent the interaction feature. Besides, the maximum k elements of s_{i1} to s_{iD} can also be used as an interaction feature, where k is a hyper parameter which controls the feature size. Using the top k elements from the word matching scores will make the interaction between query and document become more finer-grained. Such feature usually performs better than the original DRMM with the histogram feature.

4   MatchPyramid

DRMM uses human-defined features on top of the matching matrix to calculate the matching score. Instead of using human-defined feature, MatchPyramid [17] uses convolutional neural networks to learn feature representations from the matching matrix.

The architecture of MatchPyramid is shown in Fig. 2.

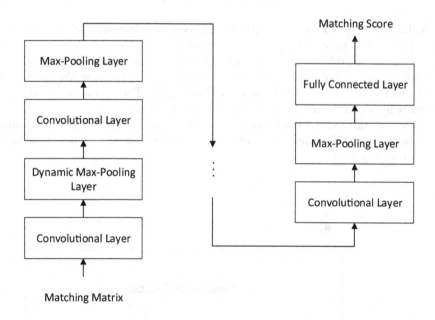

Fig. 2. The architecture of MatchPyramid.

The matching matrix between the query word sequence and the document word sequence is firstly calculated. The matching matrix is then regarded as an 2D image and convolutional neural networks is used to predict the matching score. A serial of alternating convolutional and max-pooling layers are applied on the image of matching matrix. The first max pooling layer uses dynamic pooling size to convert text with variable length to fixed size. Full connected layer is used to output the matching score of MatchPyramid.

Due to the representation learning nature of convolutional neural networks, MatchPyramid is able to learn the hierarchical interaction between query words and document words, and mine complex matching pattern between texts.

5   Bidirectional Encoder Representations from Transformers (BERT) for Short Text Matching

BERT is a pre-trained text encoder model which is based on the transformer architecture [4]. It has achieved state-of-the-art performance on many tasks, such as GLUE benchmark [21] and SQUAD [18].

BERT uses transformer to encode a word sequence of text into an output vector sequence. Transformer is firstly proposed for machine translation [20]. The encoder of transformer consists of a serial of layers. Each layer can be further divided into two sublayers, i.e. the multi-head self-attention layer and the feed-forward layer. Due to the space limit, we will not describe the detailed structure of transformer here.

BERT firstly pre-trains the transformer encoder on unsupervised text corpus. Traditional NLP pre-training algorithms use language model to predict next words. BERT introduces two new pre-training methods. The first method is to randomly mask out some words in a sentence, and then use the output of transformer encoder to predict these masked words. The second method is to predict whether two sentences are adjacent in the corpus.

After pre-training the transformer encoder, many specific tasks can be fine-tuned on the trained model. The architecture for fine-tuning text matching task in shown in Fig. 3.

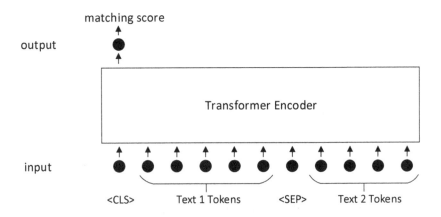

Fig. 3. BERT for text matching.

In the input layer, the two texts are concatenated as one sequence. A special $<SEP>$ token is added between the two sequences. A beginning token $<CLS>$ is inserted at the beginning of BERT input. The output corresponding to the $<CLS>$ is used to predict the matching score.

Using BERT to match short text will take full advantage the interaction feature between two texts because of the self-attention mechanism of transformer encoder. In many text matching task such question answering, BERT has outperformed existing algorithms by a large margin.

6   Model Selection in Real-World Applications

DSSM uses cosine similarity between feature vectors to measure the matching score between two texts. The calculation of cosine similarity is usually very fast. This makes the DSSM very suitable for online services which have real-time requirements on inferring the text similarity, on condition that the feature vectors have been pre-calculated offline.

For example, in search engines, the feature vector of frequent queries and all documents can be calculated in offline machines. The online servers only need to calculate infrequent queries. The matching scores of a query and candidate documents can be simply calculated by the cosine distances. This process is able to return search results quickly and will significantly reduce the load of online servers.

Another applicable scene of DSSM is when we need to find semantic similar texts from a huge collection of texts. It is inefficient to calculate matching scores between the query text and all texts in the huge collection, since the time complexity is $O(N)$ where N is the size of the huge collection. For cosine similarity, the approximate nearest neighbor (ANN) search [1,14] can be used to accelerate the process of searching for nearest neighbors. ANN uses some index structures to avoid the comparison with all samples in the huge collection, and is able to control the time of searching for similarity vectors from a huge collection in an acceptable range. State-of-the-art ANN algorithms includes HNSW [12] and Faiss [11], etc.

DSSM does not uses direct word level interaction feature to calculate the matching score, which may limit its effectiveness on matching short texts. DRMM, MatchPyramid and BERT are able to result in better matching performance, because they take advantages of fine-grained matching information based on word interaction feature. However, such word interaction based algorithms cannot be pre-calculated offline and are unable to be accelerated in online services like the way of DSSM.

If the online efficiency is crucial, DRMM is a good choice for online services. The main components of DRMM are feed-forward networks, which are rather faster than other kinds of neural networks. BERT usually performs better than other algorithms, but its architecture is too complex and its time complexity is much higher than other algorithms. It is suitable for the applications that the time efficiency is not the major constraint and the state-of-the-art matching performance has much more positive influences. MatchPyramid can be a trade-off between the matching performance and the time efficiency.

7 Conclusion

This paper reviews some state-of-the-art short text matching algorithms. DSSM uses a feature vector to represent each text and the vector distances is regarded as the matching score. Convolutional networks and LSTM are able to further enhance the representation ability of DSSM. DRMM extracts features from word interaction matrix between words of two text, which is able to better capture the interaction information between text. MatchPyramid regards the matching matrix as an image, and uses convolutional neural networks to predict the matching score. BERT utilizes the unsupervised pre-trained model to enhance the feature learning of text. It uses the self-attention mechanism of transformer encoder to learn the interaction between two texts, which is able to reflect higher level semantic interaction.

Acknowledgments. This work was supported by National Key Research and Development Program of China under grant no. 2016QY02D0304.

References

1. Aumüller, M., Bernhardsson, E., Faithfull, A.: ANN-benchmarks: a benchmarking tool for approximate nearest neighbor algorithms. Inf. Syst. (2019)
2. Blei, D.M., Ng, A.Y., Jordan, M.I.: Latent Dirichlet allocation. J. Mach. Learn. Res. **3**(Jan), 993–1022 (2003)
3. Deerwester, S., Dumais, S.T., Furnas, G.W., Landauer, T.K., Harshman, R.: Indexing by latent semantic analysis. J. Am. Soc. Inf. Sci. **41**(6), 391–407 (1990)
4. Devlin, J., Chang, M.W., Lee, K., Toutanova, K.: BERT: pre-training of deep bidirectional transformers for language understanding. arXiv preprint arXiv:1810.04805 (2018)
5. Fan, Y., Pang, L., Hou, J., Guo, J., Lan, Y., Cheng, X.: MatchZoo: a toolkit for deep text matching. arXiv preprint arXiv:1707.07270 (2017)
6. Graves, A.: Supervised sequence labelling. In: Graves, A. (ed.) Supervised Sequence Labelling with Recurrent Neural Networks. SCI, vol. 385, pp. 5–13. Springer, Berlin (2012). https://doi.org/10.1007/978-3-642-24797-2_2
7. Guo, J., Fan, Y., Ai, Q., Croft, W.B.: A deep relevance matching model for ad-hoc retrieval. In: Proceedings of the 25th ACM International on Conference on Information and Knowledge Management, pp. 55–64. ACM (2016)
8. Hochreiter, S., Schmidhuber, J.: Long short-term memory. Neural Comput. **9**(8), 1735–1780 (1997)
9. Hu, B., Lu, Z., Li, H., Chen, Q.: Convolutional neural network architectures for matching natural language sentences. In: Advances in Neural Information Processing Systems, pp. 2042–2050 (2014)
10. Huang, P.S., He, X., Gao, J., Deng, L., Acero, A., Heck, L.: Learning deep structured semantic models for web search using clickthrough data. In: Proceedings of the 22nd ACM International Conference on Information & Knowledge Management, pp. 2333–2338. ACM (2013)

11. Johnson, J., Douze, M., Jégou, H.: Billion-scale similarity search with GPUs. arXiv preprint arXiv:1702.08734 (2017)
12. Malkov, Y.A., Yashunin, D.A.: Efficient and robust approximate nearest neighbor search using hierarchical navigable small world graphs. IEEE Trans. Pattern Anal. Mach. Intell. (2018)
13. Mitra, B., Diaz, F., Craswell, N.: Learning to match using local and distributed representations of text for web search. In: Proceedings of the 26th International Conference on World Wide Web, pp. 1291–1299. International World Wide Web Conferences Steering Committee (2017)
14. Naidan, B., Boytsov, L., Nyberg, E.: Permutation search methods are efficient, yet faster search is possible. Proc. VLDB Endow. **8**(12), 1618–1629 (2015)
15. NTMC-Community: Matchzoo (2017). https://github.com/NTMC-Community/MatchZoo/tree/1.0
16. Palangi, H., et al.: Semantic modelling with long-short-term memory for information retrieval. arXiv preprint arXiv:1412.6629 (2014)
17. Pang, L., Lan, Y., Guo, J., Xu, J., Wan, S., Cheng, X.: Text matching as image recognition. In: Thirtieth AAAI Conference on Artificial Intelligence (2016)
18. Rajpurkar, P., Zhang, J., Lopyrev, K., Liang, P.: SQuAd: 100,000+ questions for machine comprehension of text. arXiv preprint arXiv:1606.05250 (2016)
19. Shen, Y., He, X., Gao, J., Deng, L., Mesnil, G.: A latent semantic model with convolutional-pooling structure for information retrieval. In: Proceedings of the 23rd ACM International Conference on Conference on Information and Knowledge Management, pp. 101–110. ACM (2014)
20. Vaswani, A., et al.: Attention is all you need. In: Advances in Neural Information Processing Systems, pp. 5998–6008 (2017)
21. Wang, A., Singh, A., Michael, J., Hill, F., Levy, O., Bowman, S.R.: GLUE: a multi-task benchmark and analysis platform for natural language understanding. arXiv preprint arXiv:1804.07461 (2018)

Neuronal Environmental Pattern Recognizer: Optical-by-Distance LSTM Model for Recognition of Navigation Patterns in Unknown Environments

Fredy Martínez$^{(\boxtimes)}$, Edwar Jacinto , and Holman Montiel

Universidad Distrital Francisco José de Caldas, Bogotá D.C., Colombia
{fhmartinezs,fmartinezs,ejacintog}@udistrital.edu.co
http://www.udistrital.edu.co

Abstract. The level of autonomy and intelligence of a robot is a fundamental characteristic of tasks related to service robotics, particularly due to its interaction with human environments. These environments tend to change constantly over time (dynamic environments) so a mapping turns out to be inefficient in the long run. A service robot must be able to perform basic actions (displacements in the environment) autonomously with the lowest computational requirement. As for indoor navigation, it is ideal for the robot to be able to identify certain characteristics in the environment that allow it to plan its movement. This paper proposes a direct data-driven path planner for small robots capable of producing control actions directly from sensor readings. The motion decision is based on the similarity of the captured data with respect to models previously stored on the robot. These models were developed for different types of obstacles (obstacles in different positions and free space) using deep regression. The proposed scheme was successfully evaluated on a real prototype in the laboratory.

Keywords: Autonomous · Big data · Data-driven · Sensor ·
Similarity · Motion planner

1 Introduction

Robotics is part of the daily life of the human being. Its integration with people's routines is increasing, both in robotic platforms performing tasks (cleaning, monitoring, fun, etc.) and in the professional training of individuals [1,15]. This integration is growing and will end up producing a world in which many of the personal needs of the human being will be supported by a robot [20,22].

The objective of robotics research is to develop artificial systems that support human beings in certain tasks. The design of these robots focuses on the problems of sensing, actuation, mobility and control [7,11]. These are large fields of research with a large number of open problems, even greater if it is considered

© Springer Nature Singapore Pte Ltd. 2019
Y. Tan and Y. Shi (Eds.): DMBD 2019, CCIS 1071, pp. 220–227, 2019.
https://doi.org/10.1007/978-981-32-9563-6_23

that the design of each solution is highly dependent on the specific task to be developed by the robot, and the conditions under which it must be developed. Some design criteria used in the development of modern robots are: simplicity, low cost, high performance and reliability [10].

One of the biggest challenges in mobility robots, and particularly in movement planning, is the autonomous navigation. The problem of finding a navigation path for an autonomous mobile robot (or several autonomous mobile robots in the case of robot swarms) is to make the agent (a real object with physical dimensions) find and follow a path that allows it to move from a point of origin to a point of destination (desired configuration) respecting the constraints imposed by the task and/or the environment (obstacles, free space, intermediate points to visit, and maximum costs to incur in the task) [2,4,21]. This problem is very investigated given the need for autonomy and robustness of the robots [5,12].

For correct navigation in a reactive autonomous scheme, the robot must sense the pertinent information of the environment [8,14]. This includes obstacles and restrictions of the environment, communication with nearby agents, and the detection of certain elements in the environment [16]. From this information, the robot performs calculations of its estimated position with respect to the destination, traces a movement response, executes it, and verifies the results. This navigation strategy is heavily dependent on the quantity and quality of information processed [18].

There are many robot designs for this type of task. Most of these designs include robust and complex hardware with high processing capacity [9,17]. This hardware is equipped with a high-performance CPU, often accompanied by a programmable logic device (CPLD or FPGA) for dedicated processing, and a specialized communication unit. In addition, these robots often have advanced and complex sensors [6]. Due to these characteristics, these robots are expensive and with a high learning curve for an untrained user [13].

Unlike this trend, our research proposes the use of a minimalist solution, particularly in terms of computational cost. We propose a solution to the problem of motion planning that uses as little information as possible [3]. For this purpose, the robot is equipped with a set of distance sensors, and from the data captured by them, and using behavior models previously identified in the laboratory, we program movement policies that determine the final behavior of the swarm [19]. The sensor data is processed in real time by comparing them with the laboratory models, using different metrics of similarity.

The paper is organized as follows. In Sect. 2 presents a description of the problem. Section 3 describes the strategies used to analyze raw data and generate data models for specific environmental characteristics. Section 4 introduces some results obtained with the proposed strategy. Finally, conclusion and discussion are presented in Sect. 5.

2   Problem Statement

When a robot must move in an unknown environment it is vitally important to detect adequate and sufficient information about the environment, particularly when the environment is susceptible to frequent changes. In these situations, it is not possible to previously store a map of the environment in the robot, nor to program it to carry out mapping. The information detected by the sensors must be processed in real time by the robot in order to determine a suitable movement. One way to interpret this information is to compare it with behavior models previously stored in the robot.

In this sense, we propose a set of pre-recorded models in the robot that allow it to recognize directly from the raw data of the sensors the type of local interaction that it is detecting (obstacle, free space, edges of the environment or other robots).

Let $W \subset \mathbb{R}^2$ be the closure of a contractible open set in the plane that has a connected open interior with obstacles that represent inaccessible regions. Let $\mathcal{O}$ be a set of obstacles, in which each $O \subset \mathcal{O}$ is closed with a connected piecewise-analytic boundary that is finite in length. Furthermore, the obstacles in $\mathcal{O}$ are pairwise-disjoint and countably finite in number. Let $E \subset W$ be the free space in the environment, which is the open subset of W with the obstacles removed.

Let us assume an agent in this free space. This agent knows the environment W from observations, using sensors. These observations allow them to build an information space I. An information mapping is of the form:

$$q : E \longrightarrow S \tag{1}$$

where S denote an *observation space*, constructed from sensor readings over time, i.e., through an observation history of the form:

$$\tilde{o} : [0, t] \longrightarrow S \tag{2}$$

The interpretation of this information space, i.e., $I \times S \longrightarrow I$, is that which allows the agent to make decisions. The problem can be expressed as the search for a function u for specific environment conditions from a set of sensed data in the environment $y \subset S$ and a target function g.

$$f : y \times g \longrightarrow u \tag{3}$$

Each of these identified functions (models) can then be used for the robot to define motion actions in unknown environments, particularly when detecting other robots in the environment.

Each one of the models is compared with the readings of the sensors by means of a similarity metric. A high degree of similarity allows to characterize the readings with the model, and therefore to choose a movement policy. Motion policies are designed for each behavior modeled in the robot.

3   Methodology

An autonomous service robot must perform many complex tasks in parallel and in real time. One of these tasks is motion planning. In a real application, it would be ideal if the robot could perform this activity without compromising its performance and with a high degree of reliability. The motivating idea of our research is to simplify the decision-making process for the agent from the sensor data. However, the function between input data and movement policies corresponds to a complex model. This model, however, can be identified and used as a parameter to define actions.

Sensor information tends to behave similarly regardless of previous readings. The problem of model development for specific environmental conditions (obstacles and clear path) is analyzed as a regression problem. Throughout the tests with the robot in known environments, a sequence of temporary data is produced that must allow estimating the type of characteristics in the environment for any similar situation. The sequence of data for an obstacle is shown in Fig. 1.

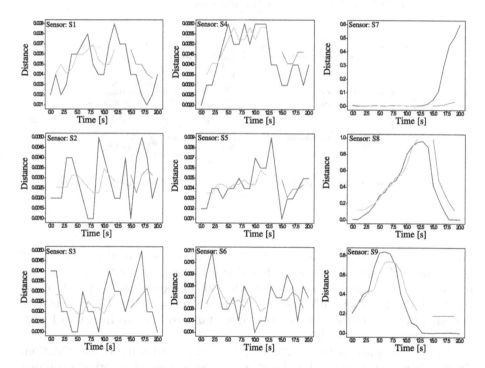

Fig. 1. Obstacle detected by sensors S7, S8 and S9. Normalized reading of infrared sensors (blue), LSTM model (red), and predictions for the training dataset (green). (Color figure online)

We have chosen to identify the behavior of the dataset through a recurrent neural network, with the intention of knowing the future state of the system from its past evolution. This particular type of data is known as time series, and are characterized by the output that is influenced by the previous events. For this, we propose the use of an LSTM (Long Short-Term Memory) network.

The models for each characteristic of the environment are used as a reference for the calculation of similarity with respect to the data captured by the robot during navigation. The selected metric corresponds to the two-dimensional representation of the data against time, i.e. by image comparison.

The comparison is made against all channels, i.e. each of the nine channels is compared with each of the nine channels of the model (81 comparisons in each iteration). The distance used as a metric for the calculation of similarity was Chi-Square due to its better performance in tests.

Control policies were simplified as a sequence of actions specific to the identified environmental characteristics (Fig. 2). For example, an obstacle to the front activates the Evasion Policy, which in this case consists of stopping, turning a random angle in a random direction, and finally moving forward. Each characteristic of the environment has a control policy, which is adjusted according to how the characteristic is detected by the sensors.

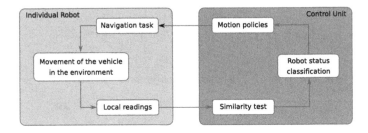

Fig. 2. Flowchart of the proposed motion planner decision making system.

4   Results

We have tested our proposed algorithm on a dataset generated by a 45 cm × 61 cm robot (Fig. 3(a)). The robot has nine uniformly distributed infrared sensors around its vertical axis (S1 to S9, with 40° of separation from each other, counterclockwise), each with a range of 0.2 to 0.8 m (Fig. 3(b)). The data matrix delivered by the sensors corresponds to nine standard columns from 0 to 1 with reading intervals between 900 ms.

The performance of the LSTM models is evaluated using cross-validation. To do this we separated each dataset in an orderly manner, creating a training set and a test set. For training, we use 70% of the data, and we use the rest to test the model. The network has a visible layer with one input, a hidden layer with eight LSTM blocks or neurons, and an output layer that makes a single value prediction. The default sigmoid activation function is used for the LSTM blocks. The network is trained for 100 epochs and a batch size of one is used.

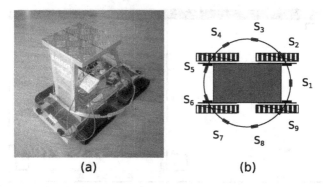

Fig. 3. (a) ARMOS TurtleBot 1 equipped with a set of nine infrared sensors and one DragonBoard 410C development board. (b) Top view of the robot and distribution of the distance sensors.

The model fit code was written in Python using Keras. We set up a SciPy work environment with Pandas support. Models were evaluated using Keras 2.0.5, TensorFlow 1.2.0 and scikit-learn 0.18.2.

The evaluation of the strategy was carried out by programming a simple navigation task. A series of obstacles are placed in the environment, and the robot is placed at a starting point. The robot is programmed to freely navigate the environment by reacting to sensor readings. The navigation task ends when it reaches the target point (in the simulation this point is located on the upper right side of the environment).

The navigation scheme has been tested in simulation (Fig. 4) and in the laboratory with different configurations on a 6 × 6.5 m environment. We have performed more than 100 tests, 92% of them completely successful, that is, the robot managed to navigate the environment and after a certain amount of time, he manages to locate the target point and stay in it.

The navigation scheme does not use any other information, only the reading of the sensors. Nor does it use communication with an external control unit, the robot is completely autonomous. Under these operating conditions, the robot is able to identify obstacles in the environment and keep moving while some other parameter of the task tells it some other action.

5   Conclusions

This paper presents a behavior model based on LSTM networks for the local interaction of autonomous robots in unknown environments. From the raw readings of a set of nine distance sensors, the robot is able to identify a reading model and apply an appropriate movement policy. The behavior models are characterized by offline using LSTM networks. These models are then recorded in the robot, which compares them continuously by similarity with the online readings

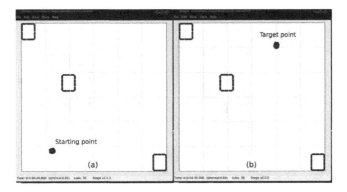

Fig. 4. Simulation of navigation task. (a) Initial random distribution of the robot. (b) Position of robot after 4:30 min. Simulation performed in Player-Stage.

of the sensors. This motion planning does not require a high computational cost and can be done in parallel with more complex tasks. The proposed scheme was tested by simulation and on a real robot, in both cases, the success rate was over 90%. Future developments will focus on integrating the algorithm into more complex tasks.

Acknowledgments. This work was supported by the Universidad Distrital Francisco José de Caldas and the Scientific Research and Development Centre (CIDC). The views expressed in this paper are not necessarily endorsed by Universidad Distrital. The authors thank the research group ARMOS for the evaluation carried out on prototypes of ideas and strategies.

References

1. Ackovska, N., Kirandziska, V.: The importance of hands-on experiences in robotics courses. In: 17th International Conference on Smart Technologies, IEEE EURO-CON 2017, pp. 56–61 (2017)
2. Almasri, M., Alajlan, A., Elleithy, K.: Trajectory planning and collision avoidance algorithm for mobile robotics system. IEEE Sens. J. **16**(12), 5021–5028 (2016)
3. Benitez, J., Parra, L., Montiel, H.: Diseño de plataformas robóticas diferenciales conectadas en topología mesh para tecnología Zigbee en entornos cooperativos. Tekhnê **13**(2), 13–18 (2016)
4. Hyun, N., Vela, P., Verriest, E.: A new framework for optimal path planning of rectangular robots using a weighted Lp norm. IEEE Robot. Autom. Lett. **2**(3), 1460–1465 (2017)
5. Jacinto, E., Giral, M., Martínez, F.: Modelo de navegación colectiva multi-agente basado en el quorum sensing bacterial. Tecnura **20**(47), 29–38 (2016)
6. Mane, S., Vhanale, S.: Real time obstacle detection for mobile robot navigation using stereo vision. In: International Conference on Computing, Analytics and Security Trends (CAST), pp. 1–6 (2016)

7. Martínez, F., Acero, D.: Robótica Autónoma: Acercamientos a algunos problemas centrales. CIDC, Distrital University Francisco José de Caldas (2015). ISBN 9789588897561
8. Melik, N., Slimane, N.: Autonomous navigation with obstacle avoidance of tricycle mobile robot based on fuzzy controller. In: 4th International Conference on Electrical Engineering, ICEE 2015, pp. 1–4 (2015)
9. Nasrinahar, A., Huang, J.: Effective route planning of a mobile robot for static and dynamic obstacles with fuzzy logic. In: 6th IEEE International Conference on Control System, Computing and Engineering, ICCSCE 2016, pp. 1–6 (2016)
10. Nattharith, P., Serdar, M.: An indoor mobile robot development: a low-cost platform for robotics research. In: International Electrical Engineering Congress, iEECON 2014, pp. 1–4 (2014)
11. Ning, T., Xiaoyi, G., Hongliang, R.: Simultaneous robot-world, sensor-tip, and kinematics calibration of an underactuated robotic hand with soft fingers. IEEE Access 6, 22705–22715 (2017)
12. Oral, T., Polat, F.: MOD* Lite: an incremental path planning algorithm taking care of multiple objectives. IEEE Trans. Cybern. 46(1), 245–257 (2016)
13. Ortiz, O., Pastor, J., Alcover, P., Herrero, R.: Innovative mobile robot method: improving the learning of programming languages in engineering degrees. IEEE Trans. Educ. 60(2), 143–148 (2016)
14. Rendón, A.: Evaluación de estrategia de navegación autónoma basada en comportamiento reactivo para plataformas robóticas móviles. Tekhnê 12(2), 75–82 (2015)
15. Rezeck, P., Azpurua, H., Chaimowicz, L.: HeRo: an open platform for robotics research and education. In: 2017 Latin American Robotics Symposium (LARS) and 2017 Brazilian Symposium on Robotics (SBR), pp. 1–6 (2017)
16. Schmitt, S., Will, H., Aschenbrenner, B., Hillebrandt, T., Kyas, M.: A reference system for indoor localization testbeds. In: International Conference on Indoor Positioning and Indoor Navigation, IPIN 2012, pp. 1–8 (2012)
17. Yang, S.-J., Kim, T.-K., Kuc, T.-Y., Park, J.-K.: Geomagnetic localization of mobile robot. In: International Conference on Mechatronics, ICM 2017, pp. 1–6 (2017)
18. Sheikh, U., Jamil, M., Ayaz, Y.: A comparison of various robotic control architectures for autonomous navigation of mobile robots. In: International Conference on Robotics and Emerging Allied Technologies in Engineering, iCREATE 2014, pp. 239–243 (2014)
19. Sztipanovits, J., et al.: Toward a science of cyber-physical system integration. Proc. IEEE 100(1), 29–44 (2012)
20. Tajti, F., Szayer, G., Kovács, B., Dáiel, B., Korondi, P.: CRM TC covering paper - robotics trends. In: 39th Annual Conference of the IEEE Industrial Electronics Society, IECON 2013, pp. 48–53 (2013)
21. Teatro, T., Eklund, M., Milman, R.: Nonlinear model predictive control for omnidirectional robot motion planning and tracking with avoidance of moving obstacles. Can. J. Electr. Comput. Eng. 37(3), 151–156 (2014)
22. Zillich, M., Vincze, M.: To explore and to serve - robotics between basic research and the actually useful. In: Advanced Robotics and Its Social Impacts, pp. 80–82 (2011)

Mining Patterns

Discovering Strategy in Navigation Problem

Nurulhidayati Haji Mohd Sani[1(✉)], Somnuk Phon-Amnuaisuk[1,2],
and Thien Wan Au[1]

[1] School of Computing and Informatics, Universiti Teknologi Brunei,
Gadong, Brunei Darussalam
p20151003@student.utb.edu.bn, twan.au@utb.edu.bn
[2] Centre for Innovative Engineering, Universiti Teknologi Brunei,
Jalan Tungku Link, Gadong BE1410, Brunei Darussalam
somnuk.phonamnuaisuk@utb.edu.bn

Abstract. This paper explores ways to discover strategy from a state-action-state-reward log recorded during a reinforcement learning session. The term strategy here implies that we are interested not only in a one-step state-action but also a fruitful sequence of state-actions. Traditional RL has proved that it can successfully learn a good sequence of actions. However, it is often observed that some of the action sequences learned could be more effective. For example, an effective five-step navigation to the north direction can be achieved in thousands of ways if there are no other constraints since an agent could move in numerous tactics to achieve the same end result. Traditional RL such as value learning or state-action value learning does not directly address this issue. In this preliminary experiment, sets of state-action (i.e., a one-step policy) are extracted from 10,446 records, grouped together and then joined together forming a directed graph. This graph summarizes the policy learned by the agent. We argue that strategy could be extracted from the analysis of this graph network.

Keywords: Navigation · Discovering strategy ·
Monkey Banana Problem

1 Introduction

This paper discusses a technique of knowledge discovery for strategies in solving the Monkey and Banana Problem (MBP). MBP, as introduced in the early work of artificial intelligence by McCarthy [12], is often used as a simple example to investigate problem-solving techniques for decision making tasks such as path planning and navigation. In it, the agent (i.e. monkey) is placed in a room containing a banana which is suspended from a ceiling and a movable crate. The goal is for the monkey to reach and get the banana. The monkey can perform actions, such as walking or climbing.

© Springer Nature Singapore Pte Ltd. 2019
Y. Tan and Y. Shi (Eds.): DMBD 2019, CCIS 1071, pp. 231–239, 2019.
https://doi.org/10.1007/978-981-32-9563-6_24

The MBP is commonly used in the AI literature due to its simple yet sufficient detail and aspect for understanding a problem [6]. Furthermore, MBP is closely and relatively similar in illustrating the interaction of human or robot with the real world. For this investigation, we modify the domain into a simple navigating problem where an agent (i.e. monkey) and a goal are placed in a labyrinth room and the agent is required to find and reach the goal.

Given a map and a goal state, a typical planning problem such as MBP involves identifying a sequence of actions that leads the agent to the goal state. This sequence of actions, also known as *sub-goals*, can be identified as a plausible strategy in solving the problem [1,4,10]. To date, contemporary techniques investigated in the literature such as [15] and [13] uses data mining to show that the learned actions can lead to the goal. Even so, the technique lacks strategy identification.

In addition, strategies in classical planning problems were commonly formulated and elicited by a knowledge engineer. As strategies were provided, planning was made easier by the use of search algorithms such as A* [5,8] to search through these formulated sub-goals for an optimal strategy. As the planning system is often explicitly formulated, the reinforcement learning (RL) technique allows the system to automatically learn to build the policy to solve problems without explicitly declaring them. Similarly, this RL technique also cannot identify strategies although the investigation found solutions [16].

Here, we investigate how a strategy evolved from repeated gameplay in a navigation problem through the application of data mining and other computing techniques without explicitly declaring them. The agent implemented in this experiment is a simple self-exploratory and self-learning reinforcement learning agent operating in a 3-dimensional environment using vision perceptions. The intuition for this investigation is to guide us to discover strategy in a more complex problem.

The rest of the paper is organized as follows. In Sect. 2, we define the problem and provide a detailed formulation of the problem as well as how the experiment is operated and then we discuss how data is recorded and how strategy is mined in Sect. 3. Section sect:results reports the discussions for our findings. Finally, we conclude our findings and propose future work in Sect. 5.

2   Learning to Navigate Using Q-Learning

2.1   Problem Description

A provision of a map or a plan allows a planning system to find the optimal path or sequence of actions that result in the best performance in the literature. However, this rule-based approach is often tedious for a large and complex problem and is also not normally flexible and reusable. Many researchers have been investigating ways to resolve this issue. Inclusively, several of these works have shown that strategies can be learned from obtained data. An instance is by extracting patterns from replays of a game and game logs to learn and predict player strategies [2,9]. One, in particular, is that traditional RL has proven that actions can be learned [15]. However, the learned strategy could be more effective.

While the strategy allows the system to successfully achieve the goal, it is often observed that there are several ways to accomplish it where one of them may be a better strategy from the other. For instance, in the first person shooter game, where a player first prioritizes in healing due to its low health and then refocus on killing an enemy although the player can kill that enemy instantly due to enemy's low health [18] or a player can instantly kill an enemy by shooting a head blow but instead, the player chooses to injure the enemy by shooting on the other parts of the enemy's body since they are easier targets [7].

The hidden strategy might not be accessible due to the horizon effect. As a result, many works of literature use methods like graph mining which allows the system to look at the problem in a bigger picture which may be useful in strategy discovery in many areas including security [3,11], and design [17].

In our experiment, the agent focuses on automatically navigating its way in the environment using RL while avoiding obstacles along the way using vision perceptions as its state-value function. Then using a directed graph, we analyse and search for a strategy that may solve the problem more effectively.

2.2 Experimental Setup

The environment, E, is a virtual 3D space populated by m game objects O_m. An object may be an agent M, a goal G, or an obstacle W. For every time step, the agent perceives the states of the environment S and performs A actions. For every action selected, a feedback R is given. An episode ends either when: (i) agent M reaches the goal; or (ii) agent reaches the maximum number of steps.

We employ reinforcement learning as the learning paradigm for the agent to perform in the environment. The general outline of the environment is a square-shaped room surrounded by 40 exterior walls populated by a single agent M and a single goal G (see Fig. 1. For every gameplay, a randomly generated number of walls between 45 and 90 is randomly positioned within the environment. An episode ends when an agent reaches a maximum number of 300 steps or when the agent reaches the goal. A total of 300 episodes are recorded for every gameplay and a total of 30 gameplay are repeated.

Fig. 1. The outline of the environment: an environment, an agent, a goal and walls

For each time step, a state is perceived and action is selected leading to the next state giving feedback for every transition. These chains of data were recorded to be analysed. The employed solutions were extracted and grouped according to their respective state values and then using network directed graphs, we display the relationships between the states.

Q-Learning Agent. Our agent is a human model configured following [16] where it can perceive its environment through sensors and react accordingly through actions [14]. The agent implements Q-learning as the learning paradigm to build its policy. In our implementation, the agent first percepts the current state and then decides between exploration and exploitation of actions using ϵ-greedy action-selection policy with decay. The feedback (either a reward or punishment) for every action selected depends on the state of the environment as illustrated in Table 1.

Table 1. Parameter settings for the experiments

Parameter settings	Values
Environment	
Grid size	10×10
Number of walls, W	
Exterior	40
Interior	45–90
Number of agent, M	1
Number of goal, G	1
Possible actions, A	$\{N, NE, E, SE, S, SW, W, NW\}$
Rewards, R	
Hits a wall	-2.0 point
Achieve a goal	50.0 point
Move towards goal	5.0 point
Every step	-0.001 point
RL parameters	
Learning rate, α	0.5
Discount rate, γ	0.99
ϵ-greedy probability	0.8

3   Mining Strategy from Gameplay

3.1   Recording Patterns

Let the elements $D = \{d_0, d_1, d_2, \cdots d_n\}$ describe the recorded data for our problem. A data $d \epsilon D$ of our experiment is given by the element:

$$d = \{s_t, a, s_{t+1}, r\}$$

where the set s_t represents a set of game state at time t. The state $s \in S$ of the environment is as perceived by the agent consisting of eight cardinal and ordinal direction of the compass surrounding the agent (i.e. eight directions of the compass) as illustrated in Fig. 2. Each direction comprises of one of the following data: $1W, 2W, 1G, 2G$ (where $1 =$ near, $2 =$ far, $W =$ wall and $G =$ goal). The distances (i.e. near and far) helps to identify and manages the feedback given to the agent for selecting an action. Action a is the agent's actions. The agent performs $a \in A$ action for every time step, which denotes the movement of one unit in any one of the eight directions $(N, NE, E, SE, S, SW, W, NW)$. For every state-action $Q(s,a)$ pair, a reward $r \in R$ is given to the agent for the decision leading to a next state s_{t+1}.

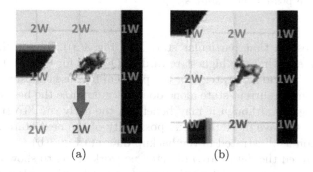

(a)   (b)

Fig. 2. The state as perceived by the agent navigating in the environment: (a) The current state as perceived by the agent before performing an action N; (b) The next state as perceived by the agent after moving one step in the N direction

The recorded data is summarized and the rewards r of data with matching state s_t, action a and next state s_{t+1} is averaged to obtain the final reward r for redundancy prevention. The states are analysed and grouped to find the relationship and links between them.

4   Results and Discussion

In our experiment, we decided to record (i) the current state; (ii) action performed by the agent based on the current state; (iii) the next state; and (iv) the feedback received. A summarized total of 10,446 reasonable policy emerged from 1,020,976 data recorded from a total of 30 different game sessions.

The recorded data consists of the transition:

$$(s_t, a_t) \rightarrow s_{t+1}$$

where s_t is the current state at time t, a_t is the action taken for the time t and s_{t+1} is the next state leading from the pair (s_t, a_t). The summarized data can be further reduced by taking the highest reward which ultimately shows the best

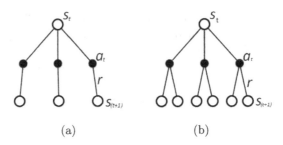

(a)   (b)

Fig. 3. A backup diagram illustrating the transition of state, action (s_t, a_t) to the next state $s_{(t+1)}$: (a) A transition of state, action (s_t, a_t) to the optimal next state $s_{(t+1)}$ which consist only one optimal state for every pair; (b) A transition of state, action (s_t, a_t) to several possible next state $s_{(t+1)}$

action to choose in that particular state (see Fig. 3(a)). This technique is no different from selecting the highest reward in Q-table for Q-learning technique which may not provide a strategy for the problem because choosing the optimal action based on the current state alone does not guarantee the best performance in the overall game although it may benefit in the early run. To show this, we allow the data to have more than one possibility of the next state for choosing the same action as illustrated in the backup diagram Fig. 3(b).

We interpreted the data into a directed network graph to show the relationship between the states and how they relate to the action selected. To avoid confusion in reading the lengthy data, the state values were first reshaped into simpler values as illustrated in Table 2 before drawn into a directed network graph. Table 3 shows some sample data recorded from the experiments used in the formation of the directed network graph.

The overall directed network graph consisting of 10,446 connectivity is illustrated in Fig. 4(a). We extract a partial network sample of this graph (see Fig. 4(b)) as an example to display the close-up of how the states relate to one another through actions selected. If we refer to this graph, more efficient strategies to solve the navigation problems may be discovered. For example, in the graph, if the agent is in state 424, a better strategy is to choose the direction N as it can reach the goal in two movements (i.e. $N \rightarrow N$) rather than choosing NE where it requires three movements (i.e. $NE \rightarrow NW \rightarrow W$) to reach the goal. However, based on the rewards, the agent is likely to choose NE as it provides a higher reward (0.45 points) as compared to N which gives a slightly lower reward (0.44 points). Using this graph, the agent is able to foresee the best strategy, overcoming the horizon effect issue in solving the navigation problem.

This scenario provides a simple strategy discovery technique for a simple navigation problem. As this paradigm allows a strategy to be extracted, the next question we would ask is would it be possible to extract a more complex gameplay strategy? We believe the current technique can be expanded further into a more complex problem by increasing the complexity of the environment such as using sub-goals for planning problems.

Table 2. A sample list of reshaped state values recorded in our experiments where the left column shows the actual recorded state and the right column shows the simplified reshaped values

State	References
1G, 2W, 2W, 2W, 2W, 1W, 1W, 1W	15
1W, 1W, 1W, 1G, 1W, 1W, 2W, 2W	30
1W, 1W, 1W, 1G, 2W, 1W, 1W, 2W	31
1W, 1W, 2W, 1G, 1W, 1W, 2W, 2W	134
1W, 1W, 2W, 2W, 2G, 1W, 2W, 2W	191
1W, 2W, 1G, 1W, 1W, 1W, 2W, 1W	245
2G, 2W, 2W, 1W, 2W, 1W, 1W, 1W	424
2W, 2W, 1W, 1W, 2W, 1G, 2W, 2W	431
1W, 1W, 2W, 2W, 2W, 2W, 1G, 2W	453
2W, 1W, 1W, 2G, 2W, 1W, 2W, 2W	482
2W, 1W, 1W, 2W, 2G, 1W, 1W, 2W	487
END	END

Table 3. Few sample data recorded from the experiments used in the formation of the directed network graph consisting of the current state, selected action, the next state and the reward

Current state	Action	Next state	Reward
15	N	END	4.94168826
30	SE	END	4.917404234
30	W	191	−0.02700002
30	W	487	−0.132000021
30	N	30	−2.07249975
31	SE	END	4.939999991
31	N	31	−2.066370222
134	SE	END	4.945
134	N	134	−2.024
191	N	191	−2.062857143
191	E	482	−0.1859997
245	N	245	−2.073333167
245	E	END	4.948437563
245	NE	31	−0.088999667
424	N	15	0.439353024
424	NE	431	0.4526924
431	NW	453	0.44133315
453	W	END	4.93187588
482	SE	424	0.4135001

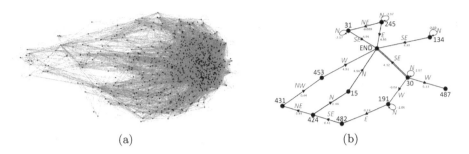

(a)   (b)

Fig. 4. A directed graph network showing the relationship between states and action: (a) the overall network consisting of 10,446 nodes (states); (b) partial network showing 10 nodes (states)

5   Conclusion and Future Work

In this paper, the state-action-state-reward log recorded from a reinforcement learning session is examined. The recorded data consists of the current state, action taken, next state; and reward received. A total of 10,446 recorded data was used in this experiment from 30 different gameplay. We illustrated states as nodes using a directed graph, links between nodes express actions and corresponding rewards for the transitions between nodes.

We argue that by transforming RL activities log into a graph, it is possible to discover an effective strategy from this graph structure using A* or other graph analysis such as graph cut e.g., to group similar activities together. This preliminary experiment only discussed a simple navigation problem and we hope to further expand the investigation into a more complex environment. Furthermore, this approach may provide a means to extract rules which will be useful in understanding the problem-solving process in RL.

References

1. Atyabi, A., Phon-Amnuaisuk, S., Ho, C.K.: Navigating a robotic swarm in an uncharted 2D landscape. Appl. Soft Comput. **10**(1), 149–169 (2010). https://doi.org/10.1016/j.asoc.2009.06.017
2. Dereszynski, E., Hostetler, J., Fern, A., Dietterich, T., Hoang, T.T., Udarbe, M.: Learning probabilistic behavior models in real-time strategy games. In: Proceedings of the Seventh AAAI Conference on Artificial Intelligence and Interactive Digital Entertainment, AIIDE 2011, pp. 20–25. AAAI Press (2011). http://dl.acm.org/citation.cfm?id=3014589.3014594
3. Dharmarajan, K., Dorairangaswamy, M.A.: Web user navigation pattern behavior prediction using nearest neighbor interchange from weblog data. Int. J. Pure Appl. Math. **116**, 761–775 (2017)
4. Dhiman, V., Banerjee, S., Griffin, B., Siskind, J.M., Corso, J.J.: A critical investigation of deep reinforcement learning for navigation. CoRR abs/1802.02274 (2018)
5. Duchoň, F., et al.: Path planning with modified a star algorithm for a mobile robot. Procedia Eng. **96**, 59–69 (2014). https://doi.org/10.1016/j.proeng.2014.12.098

6. Feldman, J.A., Sproull, R.F.: Decision theory and artificial intelligence II: the hungry monkey. Cogn. Sci. **1**(2), 158–192 (1977)

7. Glavin, F.G., Madden, M.G.: Adaptive shooting for bots in first person shooter games using reinforcement learning. IEEE Trans. Comput. Intell. AI Games **7**(2), 180–192 (2015)

8. Guruji, A.K., Agarwal, H., Parsediya, D.: Time-efficient $A*$ algorithm for robot path planning. Procedia Technol. **23**, 144–149 (2016). https://doi.org/10.1016/j.protcy.2016.03.010

9. Hsieh, J.L., Sun, C.T.: Building a player strategy model by analyzing replays of real-time strategy games. In: 2008 IEEE International Joint Conference on Neural Networks (IEEE World Congress on Computational Intelligence). IEEE, June 2008. https://doi.org/10.1109/ijcnn.2008.4634237

10. Hussein, A., Elyan, E., Gaber, M.M., Jayne, C.: Deep imitation learning for 3D navigation tasks. Neural Comput. Appl. **29**(7), 389–404 (2018)

11. Li, Y., Lin, Q., Zhong, G., Duan, D., Jin, Y., Bi, W.: A directed labeled graph frequent pattern mining algorithm based on minimum code. In: 2009 Third International Conference on Multimedia and Ubiquitous Engineering. IEEE, June 2009. https://doi.org/10.1109/mue.2009.67

12. McCarthy, J.: Situations, actions, and causal laws, p. 14, July 1963

13. Phon-Amnuaisuk, S.: Evolving and discovering Tetris gameplay strategies. Procedia Comput. Sci. **60**, 458–467 (2015). https://doi.org/10.1016/j.procs.2015.08.167

14. Russell, S., Norvig, P.: Artificial Intelligence: A Modern Approach. Pearson, London (2010)

15. Haji Mohd Sani, N., Phon-Amnuaisuk, S., Au, T.W., Tan, E.L.: Learning to navigate in a 3D environment. In: Sombattheera, C., Stolzenburg, F., Lin, F., Nayak, A. (eds.) MIWAI 2016. LNCS (LNAI), vol. 10053, pp. 271–278. Springer, Cham (2016). https://doi.org/10.1007/978-3-319-49397-8_23

16. Sani, N.H.M., Phon-Amnuaisuk, S., Au, T.W., Tan, E.L.: Learning to navigate in 3D virtual environment using Q-learning. In: Omar, S., Haji Suhaili, W.S., Phon-Amnuaisuk, S. (eds.) CIIS 2018. AISC, vol. 888, pp. 191–202. Springer, Cham (2019). https://doi.org/10.1007/978-3-030-03302-6_17

17. Tekin, U., Buzluca, F.: A graph mining approach for detecting identical design structures in object-oriented design models. Sci. Comput. Program. **95**, 406–425 (2014). https://doi.org/10.1016/j.scico.2013.09.015

18. Wang, D., Tan, A.H.: Creating autonomous adaptive agents in a real-time first-person shooter computer game. IEEE Trans. Comput. Intell. AI Games **7**(2), 123–138 (2015)

Exploring Frequent Itemsets
in Sweltering Climates

Ping Yu Hsu[1], Chen Wan Huang[1(✉)], Ming Shien Cheng[2],
Yen Huei Ko[1], Cheng-Han Tsai[1], and Ni Xu[1]

[1] Department of Business Administration, National Central University,
No. 300 Zhongda Road, Zhongli District, Taoyuan City 320, Taiwan (R.O.C.)
105481015@cc.ncu.edu.tw
[2] Department of Industrial Engineering and Management,
Ming Chi University of Technology, No. 84 Gongzhuan Road, Taishan District,
New Taipei City 24301, Taiwan (R.O.C.)
mscheng@mail.mcut.edu.tw

Abstract. With digital transformation and in the highly competitive retail market, it is important to understand customer needs and environmental changes. Moreover, obtain more profits through novel data mining technology is essential as well. Thus, the following questions should be addressed. Does climate influence the purchasing willingness of consumers? Do consumers buy different products based on the weather temperature? Few studies have used weather data and multilevel association rules to determine significant product combinations. In this study, real retail transaction records, temperature interval, and hierarchy class information were combined to develop a novel method and an improved association rule algorithm for exploring frequently purchased items under different weather temperatures. Twenty-six significant product combinations were discovered under particular temperatures. The results of this study can be used to enhance the purchasing willingness of consumers under a particular weather temperature and assist the retail industry to develop marketing strategies.

Keywords: Frequent itemsets · Multilevel association rules · Apriori algorithm

1 Introduction

In the new retail era, a priority for retail companies is to obtain higher profit through data analytics. In a highly competitive market, consumers have a wide product options, and it is very important for companies to know more about their products and customers in a highly competitive market.

In particular, the selection of suitable marketing promotion strategies in different situation and condition has been a perplexity for the retail industry. Because retail company need to undergoes various challenges of environmental and economic changes. For example, climate and temperature factors often affect our decision-making. Consumers will reduce to go out and shopping when the weather is bad. And when the temperature is hot, consumers will buy more cold drinks and so on. Thus, it is

© Springer Nature Singapore Pte Ltd. 2019
Y. Tan and Y. Shi (Eds.): DMBD 2019, CCIS 1071, pp. 240–247, 2019.
https://doi.org/10.1007/978-981-32-9563-6_25

must to know whether climate influences people's purchasing willingness and whether a high or low temperature influences their purchasing products.

Point out from past researches. The weather directly and indirectly affects production and consumption decisions [4]. A study provided strong evidence that fluctuations in the temperature can influence the sales of seasonal garments [2]. And another research has provided evidence that the weather affects a consumer's decision [6]. On the other hand. For the different product combination, whether the weather temperature influences the willingness of consumers to purchase. This is worth exploring and there are less studies in this field currently. Previous studies pertaining to the relation between sales and climate, and the influence of climate on sales was mostly analyzed using the association rule method have been conducted. However, less research has used temperature and hierarchy association rules to explore the commodity combinations purchased in retail stores.

This study used real transaction records in analysis. The temperature and product hierarchy class information were combined after data cleansing, and the improved algorithm was used to analyze the information of association rules. Customer transaction records were divided into high-temperature and non-high-temperature datasets based on the weather situation at which the purchase was made. By using a data mining algorithm and filtering method, the items frequently purchased under different temperature conditions were determined. The result of this search found 26 combinations of products purchased at a high temperature. For the retail industry, different product promotion and marketing strategies can be developed in different temperature conditions to increase customer willingness and sales volumes. Thus, when the temperature attains a certain high-temperature value in the future, a particular combination of products can be suggested to the consumers and increase company revenue.

This paper is organized as follows: (1) Introduction: This part explains the research motivation, research gap and objectives; (2) Related Work: This part contains a review on weather and consumer activity, weather and association rules; (3) Research Design: This part describes the content of the research framework design in this study; (4) Experiment Result: Describes the research data acquisition process. Numerical tables and values are also used to present the experimental results; and (5) Conclusion: The contribution of the novel methodology in this study is proposed, and future research suggestion.

2   Related Work

Climate change affect retail activity and cause uncertainty situation of sales.

2.1   Weather and Consumer Activity

Weather and temperature will affects people's moods. Such as that during sunny weather investors tend to socialize and communicate more which increases the amount of effective information. Overall, substantial research in psychology has confirmed that weather can influence an individual's mood. For instance, Persinger and Levesque [8] examined the effects of temperature, relative humidity, wind speed, sunshine hours and so on.

Weather can also affect economic activity and consumer decision-making. And directly and indirectly affects production and consumption decision making in every economic activity [4]. The previous researcher has illustrated evidence as well to suggest that weather affects consumer decision making. And provide empirical evidence to explain how the weather affects consumer spending [6]. And the changes in weather temperatures will influence sales, the fluctuations in temperature can impact sales of seasonal garments. During sales periods when drastic temperature changes occurred, more seasonal garments were sold [2]. Retail sales volume and weather temperatures will have the correlation and be an important research issue.

2.2   Weather and Association Rules

Academic research on weather-related issues, there are many algorithms that can be used and of course association rules algorithms are also quite useful in this topic. The previous research has presented the framework that assess and correlate weather conditions and their effects. An Apriori rules mining algorithm is employed subsequently, in order to obtain logical interconnections between the failure occurrences and the environmental data [9]. Alternatively, thru association rules to discover the insight in the weather information. And applied association rules to discover affiliation between several thresholds of synoptic stations [7].

Base on a Weather Research and Forecasting simulation, to apply association rules algorithm Apriori process, and other method such as search and fuzzy clustering. To provide a one-day ahead wind speed and power forecast [11].

2.3   Multilevel Association Rules

An association rule plays an important role in recent data mining techniques. Agrawal and Imielinski [1] have presented an efficient algorithm that generates all significant association rules between items in the database, to incorporate buffer management and novel estimation and pruning techniques and show the effectiveness of the algorithm. Srikant and Agrawal [10] have proposed the architecture of taxonomy, which is applied to association rules algorithms. To present a new interest-measure for rules which uses the information in the taxonomy, and given a user-specified "minimum interest-level", to prunes a large number of redundant rules.

From a multilevel point of view, researcher can study product inventory and shelf space. The method to make decisions about which products to stock and how much shelf space allocated to the stocked products be indicated, to provide a multilevel association rules. And find the relationships between products and product categories to allocate the products selected in the assortment stage [5]. In order to study the relationship between the product and the hierarchy. A high-level patterns that retain a significant degree of novelty with respect to their frequent descendants be provided, and propose a novel type of high-level itemset, namely "Expressive Generalized Itemset", and the proposed patterns are more expressive than traditional method [3].

3 Research Design

However, less research applied the use of temperature and hierarchy association rules to explore the commodity combination of retail stores. The method of this study is to experiment and explore the associated product portfolio by using customer transaction records and data mining algorithm.

3.1 Research Exploration and Definition

In order to find frequent itemsets, the simple method is to analyze all transaction data at once. The objective of this research is to explore important product combinations at a high temperatures. Therefore find out the frequent itemsets of all the data firstly. Then deduct those not significant items, and prune the repetition items of different temperature intervals. Afterward explore the significant product combination of the specific temperature.

3.2 Raw Data Filtering and Cleansing

The raw transaction records contains many fields, such as transaction period, cashier number, membership card number, serial number, product number and so on. This study filters fields first from the raw transaction records, to keep transaction date, product number and product class information.

Defining thirty-five degrees Celsius as a high temperature from the definition of Central Weather Bureau. And according to the different temperature information and merge into the analysis database.

3.3 Independent Sample T-Test

The purpose of this study is to find the momentous product of significant temperature. And adopt independent sample t-test to inspect whether there is significant difference between the high temperature sales average and the low temperature sales average. It is hypothesized that:

H1. There is significant difference under different temperature situations.

The product and classification of significant differences in sales volume and temperature are necessary information of this study. And as filtering criteria to compare with outcome of an improved association rules algorithm. Conversely, the product and classification with not significant difference will not be recorded in the database.

3.4 Hierarchical Association Rules

Due to the wide variety of retail products. This study apply association rules algorithm to find the frequent itemsets. According product classification to provide the corresponding hierarchy relation. Once data cleansing completed, by means of hierarchy association rules to explore frequent itemsets. However, there are some product combinations are no impact on sales and temperature. And they are not the objectives of this study. Via hypothesis testing as the filtering condition to prune those unfit product and find out the frequent product combinations of hot weather. This study proposed an improved Apriori algorithm, to combine with the independent sample t-test. And as the filtering criteria to reduce the scanned operation of database (Fig. 1).

```
Algorithm Improved-APRIORI pseudo-code
─────────────────────────────────────────────
APRIORI( I, minsup):
    Frequent Itemsets ← ∅
    C^(1) ← {∅}
    foreach i ∈ I do
        if i ∈ SignificantItem then
        // Let the items contain at least one significant difference item
            Add i as child of ∅ in C^(1) with sup(i) ← 0
    K ← 1
    While C^(k) ≠ ∅ do
        Compute Support(C^(k), D)
        foreach leaf X ∈ C^(k) do
            if sup(X) ≥ minsup then F ← F ∪ {(X, sup(X))}
            else remove X from C^(k)
        C^(k+1) ← Extend Prefixtree(C^(k))
        k ← k + 1
    return F^(k)
```

Fig. 1. Improved Apriori algorithm.

3.5   Duplication Criteria Filtering

The frequent itemsets of different temperature ranges will appear duplicate items in the previous step. The occurrence of the duplicate items represents that customers purchase the same products in different temperature conditions.

In order to screen out specific product combinations that occur in different temperature ranges. Consequently filter those unqualified temperature criteria to discover the significant products.

3.6   Research Framework

The temperature and product hierarchy class information were combined after raw data cleansing, and the improved algorithm was used to analyze the information of association rules. By using an improved algorithm, statistical hypothesis testing and filtering method to find the pattern. Here is the research framework of this paper (Fig. 2):

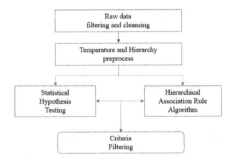

Fig. 2. Research framework.

4   Experiment Result

4.1   Raw Data Preprocessing

The research materials are selected from retail industry real transaction database. Those products include foodstuffs (fruits, vegetables, meat and beverages) and household items. To preprocess the error column and missing value data, and integrate items that appear in the same transaction serial number. And resolve the problem of transaction serial number containing multiple identical items.

The 1.6 million transaction records are retained for analysis after data cleansing.

4.2   Temperature Data Preprocessing

According to the purposed research period, the daily temperature data are combined with the basic transaction and hierarchy information after the above collation.

The maximum daily temperature of thirty-five degrees Celsius defined by the Central Weather Bureau and as a high temperature day. The consolidation transaction data are excerpted as shown in Table 1.

Table 1.   Transaction details.

Trans_D	SerNo	Price	QTY	Temp	Class_L	Class_M	Class_S
11230	45219	70	1	Low	14	14	14999
20816	26872	54	1	High	1	1	1999
20804	20868	88	1	Average	1	1	1999

4.3   Independent Sample T-Test

In order to find out the sales volume of products alter with the high or low temperature. The independent sample t-test was used to validate their significant.

4.4   Hierarchy Class Association Rules

The research apply hierarchy class association rules to find out the frequent itemsets. The item and data are excerpted as shown in Table 2.

Table 2.   Transaction table.

Tran_SerNo	Product
2061109988	10001035
2052231128	ClassM_513
2081626506	ClassS_30114

4.5   Improved Apriori Algorithm Experiment Results

Above steps obtain the transaction data of hierarchy node with classification of significant temperature. And through the improved Apriori algorithm and the setting of minimum threshold. To discover the frequent itemsets of high and low temperature. The results excerpted as shown in Table 3.

Table 3. Apriori result table.

Items	Support	Count
{3891, ClassM_301}	0.024091342	633
{ClassM_515, ClassS_5151}	0.020818268	547
{10003481, ClassM_409, ClassS_4091}	0.020970504	551

In order to filter out significant product combination that occur over different temperature group. Therefore sift those non-high temperature products and find out the objective of this paper. To preprocess the duplicate itemsets of high and low temperature frequent itemsets, and explore the final high temperature frequent significant product combination. Through the method of this study, the experimental result obtained twenty-six significant product combinations as shown in Table 4.

Table 4. Experimental result for signification product combination.

Items	Support	Items	Support
{CM_408, CS_4083}	0.0303	{10001454, CM_401, CS_4011}	0.0226
{CM_804}	0.0286	{CM_401, CM_519}	0.0222
{CM_804, CS_8042}	0.0285	{CM_403, CS_4013}	0.0221
{CM_408, CS_4081}	0.0261	{CM_401, CM_403, CS_4013}	0.0221
{CM_301, CM_408}	0.0259	{CM_505, CS_5056}	0.0221
{CM_401, CS_4031}	0.0240	{CM_401, CM_516}	0.0220
{CM_401, CM_403, CS_4031}	0.0240	{CM_402, CS_4021}	0.0218
{CM_408, CS_4085}	0.0230	{CM_401, CS_3013}	0.0212
{CM_408, CS_4011}	0.0228	{CM_301, CM_401, CS_3013}	0.0212
{CM_401, CM_408, CS_4011}	0.0228	{CM_301, CM_352, CM_401}	0.0204
{10001454}	0.0226	{CM_401, CS_5161}	0.0204
{10001454, CM_401}	0.0226	{CM_401, CM_502, CS_4011}	0.0204
{10001454, CS_4011}	0.0226	{CM_401, CM_516, CS_5161}	0.0204

This study dig out the product combinations that only appear at the specific temperature, the customers will frequently purchase those product combinations in hot weather. For retail stores, the finding assist to develop marketing strategies such as product portfolio promotions in advance. To enhance the willingness of customers buying and increase the sales revenue.

5 Conclusion

In this study, temperature information was combined with customer transaction records to divide customer transaction data into high-temperature and non-high-temperature datasets. Increase hierarchy class data and apply improved association rules algorithm. Moreover, twenty-six frequently purchased combinations of products in hot weather were found and identified. The improved algorithm proposed in this study reduces the time and effort required to repeatedly scan a database and thus lowers the runtime and operation cost. The study results can be used by retail stores to develop effective marketing strategies, such as product portfolio promotions, in advance for increasing the willingness of customers to purchase and enhancing the sales revenue of a company.

Subsequent research can apply the proposed algorithm to an e-commerce online platform. Through the way of web display to provide product marketing strategy, to avoid excessive changes in the shelf display. The purchasing rules between product and temperature can be considered to combine various recommendation system algorithms for providing a better marketing strategy.

References

1. Agrawal, R., Imieliński, T., Swami, A.: Mining association rules between sets of items in large databases. ACM SIGMOD Rec. **22**, 207–216 (1993)
2. Bahng, Y., Kincade, D.H.: The relationship between temperature and sales: sales data analysis of a retailer of branded women's business wear. Int. J. Retail Distrib. Manag. **40**, 410–426 (2012)
3. Baralis, E., Cagliero, L., Cerquitelli, T., D'Elia, V., Garza, P.: Expressive generalized itemsets. Inf. Sci. **278**, 327–343 (2014)
4. Bertrand, J.-L., Brusset, X., Fortin, M.: Assessing and hedging the cost of unseasonal weather: case of the apparel sector. Eur. J. Oper. Res. **244**, 261–276 (2015)
5. Chen, M.-C., Lin, C.-P.: A data mining approach to product assortment and shelf space allocation. Expert Syst. Appl. **32**, 976–986 (2007)
6. Murray, K.B., Di Muro, F., Finn, A., Popkowski Leszczyc, P.: The effect of weather on consumer spending. J. Retail. Consum. Serv. **17**, 512–520 (2010)
7. Nourani, V., Sattari, M.T., Molajou, A.: Threshold-based hybrid data mining method for long-term maximum precipitation forecasting. Water Resour. Manag. **31**, 2645–2658 (2017)
8. Persinger, M.A., Levesque, B.F.: Geophysical variables and behavior: XII: the weather matrix accommodates large portions of variance of measured daily mood. Percept. Mot. Skills **57**(February), 868–870 (1983)
9. Reder, M., Yürüşen, N.Y., Melero, J.J.: Data-driven learning framework for associating weather conditions and wind turbine failures. Reliab. Eng. Syst. Saf. **169**, 554–569 (2018)
10. Srikant, R., Agrawal, R.: Mining generalized association rules. Future Gener. Comput. Syst. **13**, 161–180 (1997)
11. Zhao, J., Guo, Y., Xiao, X., Wang, J., Chi, D., Guo, Z.: Multi-step wind speed and power forecasts based on a WRF simulation and an optimized association method. Appl. Energy **197**, 183–202 (2017)

Topic Mining of Chinese Scientific Literature Research About "The Belt and Road Initiative" Based on LDA Model from the Sub Disciplinary Perspective

Jie Wang⬡, Yan Peng(✉)⬡, Ziqi Wang⬡, Chengyan Yang(✉)⬡, and Jing Xu⬡

School of Management, Capital Normal University,
105 North Road of the Western 3rd-Ringroad, Beijing, China
{pengyan, 1162905152}@cnu. edu. cn

Abstract. At present, many domestic and overseas scholars have already done a large amount of studies about "The Belt and Road Initiative". Based on the LDA topic mining model, this paper classifies and analyzes scientific research literature of "the Belt and Road Initiative", from January 2014 to December 2018, in different disciplines areas. With the help of literature measurement method and LDA theme model. This paper analyses and digs in depth theme and structure of research literature of "the Belt and Road Initiative" in different disciplines. This work helps to promote the further comprehend the research status, research focus and development direction of "the Belt and Road Initiative" in different disciplines, and it can also provide a useful reference to merge studies and practices in different disciplines of "the Belt and Road Initiative".

Keywords: B&R · LDA · Topic mining · Sub disciplinary · Perplexity · Subject strength

1 Introduction

From September to October in 2013, when President Xi Jinping attended the meetings of Kazakhstan and Indonesia, he proposed major initiatives to jointly build "the Silk Road Economic Belt" and "the 21st-Century Maritime Silk Road" (call "the Belt and Road Initiative" for short.) It has received a lot of attention from the international community. The Belt and Road Initiative (B&R) is an important decision made by China's Communist Party Central Committee and the State Council for coordinating the situation at home and abroad to fit the profound change of global situation [1].

Since the introduction of B&R of the top-level design in China was come up, B&R has become a hotspot in academia. Many scholars have studied B&R in different perspectives. At present, the research of B&R presents a multi-knowledge structure. With many disciplines such as politics, economy, culture, education, etc., the content and structure of B&R are more complex and diverse. It is difficult to meticulous identify and comprehensive master the knowledge structure and internal structure of B&R.

Y. Tan and Y. Shi (Eds.): DMBD 2019, CCIS 1071, pp. 248–258, 2019.
https://doi.org/10.1007/978-981-32-9563-6_26

This paper classifies and analyzes scientific research literature of B&R, from January 2014 to December 2018, in different disciplines areas. With the help of literature measurement, LDA (Latin Dirichlet Allocation, LDA) theme model [2], this paper analyses and digs in depth theme and structure of research literature of "the Belt and Road Initiative" in different disciplines. This paper is helpful to further comprehend the research status, research focus and development direction of B&R in different disciplines. It can also provide a useful reference to merge studies and practices in different disciplines of B&R.

2 Current Research Status

Digging into the study topic, discipline structure and evolution of a certain field from the scientific literature have formed a relatively mature method system. Wang [3] summarizes the main research methods into two categories: one is keyword frequency analysis and co-word analysis, the other is data mining based on LDA model and HDP model, etc. Keyword frequency analysis and co-word analysis [4] are simple, easy to implement, and high reliable. However, there are still some problems such as strong subjectivity, lack of comprehensive analysis, and insufficient depth of research in selection of keywords. LDA [5] topic model is a useful technology in text data mining area, a remedy for the shortcomings of traditional methods, and is a typical generated probabilistic topic model. LDA model assumes that the potential topic is probability distribution of a series of words, and the document is probability distribution of a series of potential topics. At present, the LDA topic model has come into wide use in scientific intelligence analysis to discover subject topics, dig study's hotspots, and forecast study trends. In LDA model of scientific literature, two important probability distributions: topic and words polynomial distribution ϕ, and documents and topic polynomial distribution θ are obtained by selecting topics of the scientific literature corpus [6].

Many Scholars have done a lot of work about the optimization or improvement of LDA Model, meanwhile many researchers do a large amount of work about using LDA topic model in the literature topic discovery and the topic of sub-discipline mining. Yang [7] proposes a dynamic author topic (DAT) model, which integrates time and author characteristics to reveal the relationship among text content, theme and author on the basis of LDA model. Doing research on the scientific literature in a certain field is conducive to use cross-disciplinary for opening study's mind and promoting more cross-disciplinary to be formed and to be developed. Wang [3] compares and analyzes the differences between the global topic and the discipline topic based on the LDA topic model and topic extraction, taking the domestic knowledge flows study area as example, and shows that the global topic and the discipline topic have apparent differences in description, content correlation, content quality and discipline distribution, as the result. Liu [8] analyzes the differences and the similarities between China's study and international's study in the field of bio-diversity research from the perspective of subject classification. Literature concludes that China has a deficiency in the fields of biodiversity conservation and evolutionary biology, and is active in the fields of entomology and mycology.

Based on the above work, this paper uses the LDA topic model to mine the research topic of scientific literature of B&R from the perspective of discipline classification.

3   Sub Discipline Topic Selection in the Belt and Road Research Area

The purpose of this study is to show the knowledge structure and topic hotspots of B&R research field, through the LDA topic model and perspective of different subject categories.

The first step is getting bibliographic information of scientific literatures about B&R by using literature database. Second one is merging and removing duplication of research information etc., and classifying literature information, according to the classification number, to get experimental data from selection of topic. Third is normalizing research literature by word segmentation and model training, etc., and setting relevant parameters of LDA model. The last step is doing sub-discipline topic selection experiment about B&R field, screening and finding hotspots topics in different disciplines of B&R field.

3.1   Research Data Sources

This study takes the largest academic journal library in China, CNKI (China Knowledge Network) as the main data source. The search strategy is precisely B&R in the limited topic. The statistical time is from 2014 to 2018 (until December in 2018). In this research, we select nearly 20 thousand papers most relevant to the B&R topic, and select the 12 disciplines with the largest number of publications. The information of literature is exported through the information export function of CNKI, including the title, author, keywords, abstract, and the publication year. After filtering out the meeting notice, the summary of the manuscript, the repeated publication of the article and deleting the documents that lack of title, keywords and incomplete topic content, the distribution of the documents classified by topic is shown in Table 1.

Table 1. Distribution of documents by subject

Serial number	Subject classification name	Quantity
1	Economic reform	7603
2	Industrial economy	2192
3	Trading economy	1604
4	Financial	1190
5	Chinese politics and international politics	1048
6	Transportation economy	1016
7	Macroeconomic management and sustainable development	920
8	Business economy	769
9	Road and water transport	638
10	Agricultural economy	506
11	Culture	457
12	Building science and engineering	431

The number of scientific literature published in a certain field can indicate in some degree the development speed and research level of this research field. Between 2014 and 2018, nearly 50 thousand articles were published in this field in all. From 460 in 2014 to 15,144 in 2018, which shows a rapid growth trend. The annual distribution of the scientific literature about B&R strategy is shown in Fig. 1. It can be foreseen that with the continuous advancement of B&R construction and the in-depth development of cooperation between China and countries along this route, research in this field will maintain a rapid development trend in the future.

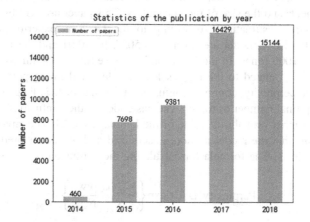

Fig. 1. Statistics of the publication by year

3.2 Literature Acquisition

(1) Select the keyword in the result document to calculate the word frequency and get the domain dictionary.
(2) Using Python's jieba [9] word segmentation package segments words of the original document's abstract, and take the domain dictionary got in the previous step as the user dictionary of the word segmentation to improve the effect of the word segmentation.

3.3 Theme Mining

This section introduces the method of setting parameters in the LDA model and the method of determining the number of topics in this experiment. The number of topics is determined by considering synthetically the confusion index and the topic intensity, and the two topics with higher intensity of the topic, selected in each category, are analyzed.

3.3.1   Corpus Selection and Model Parameter Settings

Guan [10] compares the topic selection effects based on LDA model under three different semantic situations of keywords, abstracts, keywords and abstracts. The conclusion is that the accuracy of topic selection is higher, the number of valid topics is large, semantic information of the topic is the clearest when using abstract as corpus to construct LDA text corpus. Therefore, this paper chooses to use the abstract text to build a corpus of LDA topic models. After pre-processing the abstract, the parameter training of the LDA topic model is implemented based on the open source JGibbLDA.

In the LDA model, the variable parameters to be set include the hyper-parameters α, β and the number of potential topics T. The values of α and β are related to the number of subjects and the size of the vocabulary. This paper uses Gibbs Sampling for parameter posterior estimation. According to the Griffiths experiment, the super-parameters of the LDA model are set as $\alpha = 50/T$, $\beta = 0.01$ and the number of iterations are set to 1000 times. In the LDA model, the setting of the number of potential topics T is directly related to the topic selection effect and fitting performance. This paper selects the perplexity, common confusion index in statistical language models, to calculate the optimal number of topics. The basic idea of the confusion index is giving the test's set a higher probability value of the language model. The smaller degree of confusion means that the model is more generalized and has better predictive performance for new texts. The formula for calculating the confusion is as follows:

$$\text{Perplexity}(D) = \exp\left\{ \frac{\sum_{d=1}^{M} \log p(w_d)}{\sum_{d=1}^{M} N_d} \right\} \tag{1}$$

In formula (1), D is the test set in the corpus, M is the total number of documents, N_d is the number of words in each document d, w_d is the word in document d, and p(w_d) is the probability of the word w_d in the document.

At the same time, the topic intensity describes the research heat of the topic in a certain period of time. If the number of documents on a topic is more, the intensity of the topic is higher. The topic intensity calculation formula is as follows:

$$P_k = \frac{\sum_{i}^{N} \theta_{ki}}{N} \tag{2}$$

In formula (2), P_k represents the strength of the kth topic, θ_{ki} represents the probability of the kth topic in the i-th document, and N is the number of documents. In the LDA results, the average intensity of the subject strength can be used to measure the overall subject research heat without the number of topics. The average value of the topic intensity is calculated as follows:

$$Average(P_k) = \frac{\sum_{k=1}^{T} P_k}{T} \tag{3}$$

In formula (3), P_k represents the intensity of the kth topic, and T represents the number of topics in the LDA result.

In this study, we implemented the calculation code of the average value of the topic intensity in python, and used the experimental data as input to obtain the average result of the topic intensity under different theme settings, as shown in the Fig. 2:

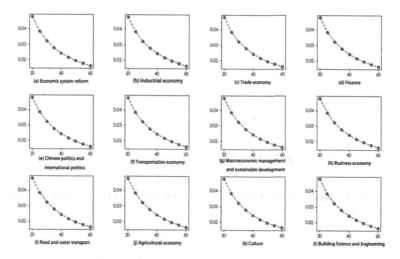

Fig. 2. The average intensity of the theme under different themes

The experiment found that in all subjects, with the number of topics T increased, the average value of the topic strength decreased. It can be concluded that with the number of topics T increases, the degree of confusion is generally towards down, but the average intensity of the topic is decreasing, which tends to cause too many topics, the selected topics are too similar, and the distinction between topics is not significant.

Therefore, in the practice, the total number of topics cannot be too large. In this paper, after selecting 10, 20, $\cdots$, 90, and 100 as the total number of topics, the initial experiment found that the total number of topics was at 20–60, so the final selection of the number of topics was determined as 20, 25, 30, 35, 40, 45, 50, 55, 60. The confusion index in different topic settings, results of the research, is shown below (Fig. 3):

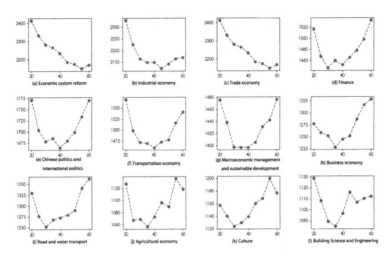

Fig. 3. The perplexity of different themes

According to the experimental results, the total number of topics in each subject is shown in Table 2.

Table 2. Total number of topics in each subject

Serial number	Subject classification name	Total number of topics
1	Economic reform	55
2	Industrial economy	45
3	Trading economy	50
4	Financial	30
5	Chinese politics and international politics	40
6	Transportation economy	40
7	Macroeconomic management and sustainable development	30
8	Business economy	35
9	Road and water transport	30
10	Agricultural economy	35
11	Culture	30
12	Building science and engineering	35

3.3.2   Topic Model Results and Analysis

According to the calculation results of the confusion degree, the number of topics T in the experiment is set, and the LDA topic experiment is practiced by using the parameters of model. The experimental results are obtained from the documents of 12 disciplines - topic distribution $\theta_{i,j}$ and topic - word distribution $\varphi_{i,j}$. The disciplines-topic distribution and topic-word distribution of 12 disciplines are the results of this experiment.

Calculating the intensity of all the topics selected by the LDA model, the intensity distribution of the topic is shown in the following figure (Fig. 4):

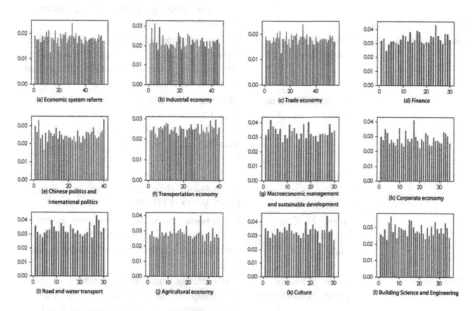

Fig. 4. Intensity distribution of discipline classification topics

In the classification of each subject, two topics with strong topic intensity and clear topic are selected as hot topics. The identification of two hot topics in each discipline and several top high-probability keywords (term probability is eliminated) are shown in Table 3. It is apparent that the terms in the discipline are highly correlated.

Table 3. The subject mining results of sub disciplines

Discipline classification	Discipline identifier	The most relevant terms related to the research topic (top 10)
Economic reform	Topic 29: Globalization	The global, world economy, Developing country, Economic Globalization, TPP, Developed country, trading, financial crisis, Global governance, Global value chain
	Topic 11: Open to the outside world	Open to the outside world, Open economy, New period, Reform and Opening, New pattern, layout, reform, Major strategy, participate actively, all-dimensional opening pattern, inland
Industrial economy	Topic 3: Energy cooperation	Energy cooperation, International, Protection, global cooperation, Renewable Energy, clean, climate change, Geopolitics, Energy resources, Global energy governance
	Topic 6: Made in China	Made in China 2025, Internet +, Global, Transformation and Upgrade, China Manufacturing, Industry 4.0, China Manufacturing, New Energy Vehicles, Innovation Drives
Trading Economic	Topic 20: FTA	TPP, Economy, Free trade zone, Asia Pacific, Trans-Pacific Partnership Agreement, Asia, GCC, WTO, TTIP, RCEP
	Topic 23: Cross-border e-commerce	Cross-border e-commerce, platform, Internet +, Development Opportunities, credit, Pay, enterprise, Logistics, customs clearance, big data
Financial	Topic 23: RMB internationalization	RMB internationalization, currency, opportunity, Chinese economy, SDR, US dollar, international status, Belt and Road strategy, progress, financial market
	Topic 15: Asian Investment Bank	Asian Investment Bank, Asia, Asia Infrastructure Investment Bank, Future, Multilateral, Dominant, Infrastructure, Developing Countries, Government, Member States
Chinese politics and international politics	Topic 39: Building a community	Community of destiny, cooperation and win-win, peace, mutual benefit, common development, co-construction, sharing, community of interests, new international relations, developing countries
	Topic 2: Realize the Chinese dream	Chinese dream, strategic conception, great rejuvenation of the Chinese nation, the 18th National Congress, the party Central Committee, strategic layout, four comprehensive, strategic decision-making, two overall situations, self-confidence

4   Conclusion

Since B&R strategy, the national top-level design in China, has come up, it has become a research hotspot in academia. The relevant cooperation of "The Belt and Road Initiative" strategy has been steadily advanced and it has been widely welcomed and actively participated at home and abroad. This thing is conducive to get the development direction and the research trend of the study in the field of different disciplines about B&R strategy.

At present, many scholars have studied B&R strategy from different disciplines' perspective. Based on the LDA topic mining model, this paper selects 24 hot topics from 12 hot disciplines in B&R strategic field from the perspective of disciplines extraction topic, and analyzes the research content of hot topics in different disciplines. In the practice, this research removes the duplication from the keywords dictionary as a new user dictionary of word segmentation to improve the effect of word segmentation. It is more effective to get research hotspots. At the same time, because the selection effect of LDA topic model is closely related to the setting of potential topic number, this paper determines the optimal number of topics by calculating the confusion index, comprehensively considering about the influence of topic intensity. Among the each of selected topics, two topics with strong topic intensity and clear topic were selected as hot topics, and were analyzed one by one. There are still some shortcomings in the research methods in this paper, such as the intersection of discipline classification and the selection of hot topics, which need to be improved in future research.

Acknowledgements. This work was supported by Social Science Project of Beijing Education Commission (SM201910028017) and Capacity Building for Sci-Tech Innovation - Fundamental Scientific Research Funds of Beijing Education Commission (Grant no. 19530050187 & 19530050142).

References

1. Huang, K., Zhao, F.: Quantitative research on the policy text of "one belt and one road initiative" – based on the perspective of policy tools. J. Intell. **37**(1), 53–57 (2018)
2. Blei, D.M., Ng, A., Jordan, M.I.: Latent Dirichlet allocation. J. Mach. Learn. Res. **3**(2003), 993–1022 (2003)
3. Wang, Y., Fu, Z., Chen, B.: Subject recognition of scientific documents based on LDA theme model: a comparative analysis of global and discipline perspectives. Inf. Stud.: Theory Appl. 39(7) (2016)
4. Tang, G., Zhang, W.: Research progress and analysis of subject subject evolution based on co-word analysis. Libr. Inf. Work **59**(5), 128–136 (2015)
5. Blei, D.M., Ng, A.Y., Jordan M.I.: Latent Dirichlet allocation. JMLR.org (2003)
6. Wang, S., Peng, Y., Wang, J.: Application of LDA-based text clustering in network public opinion analysis. J. Shandong Univ. (Sci. Ed.) **49**(09), 129–134 (2014)
7. Yang, R., Liu, D., Li, H.: An improved subject model integrating external features. Mod. Libr. Inf. Technol. **32**(1), 48–54 (2016)

8. Liu, A., Guo, Y., Li, S., et al.: A bibliometric analysis of biodiversity in China. J. Ecol. **32**(24), 7635–7643 (2012)
9. Jieba. https://pypi.org/project/jieba. Accessed 28 Aug 2017
10. Guan, P., Wang, Y., Fu, Z.: Analysis of the effect of subject extraction of scientific literature based on LDA subject model in different corpuses. Libr. Inf. Work **60**(2), 112–121 (2016)

Early Warning System Based on Data Mining to Identify Crime Patterns

Jesús Silva[1](✉), Stefany Palacio de la Cruz[2],
Jannys Hernández Ureche[2], Diana García Tamayo[2],
Harold Neira-Molina[2], Hugo Hernandez-P[3], Jairo Martínez Ventura[3],
and Ligia Romero[2]

[1] Universidad Peruana de Ciencias Aplicadas, Lima, Perú
jesussilvaUPC@gmail.com
[2] Universidad de la Costa, St. 58 #66, Barranquilla, Atlántico, Colombia
{spalacio1,jhernand4,dgarcia34,hneira,
lromeroll}@cuc.edu.co
[3] Corporación Universitaria Latinoamericana, Barranquilla, Colombia
hhernandez@ul.edu.co, jairoluis2007@hotmail.com

Abstract. The analysis of criminal information is critical for the purpose of preventing the occurrence of offenses, so the crime records committed in the past are analyzed including perpetrators. The main objective was to identify crime patterns in the city of Bogota, Colombia, supported using Early Warning System based on data mining (CRISP-DM method). The research results show the identification of 12 different criminal profiles demonstrating that the Early Warning System is applicable since it managed to significantly reduce the time devoted to the processes of registering complaints and searching for criminal profiles.

Keywords: Criminal patterns · Early warning · Data mining · Data grouping

1 Introduction

Most of the studies about violence in Colombia seek to understand the causes of the criminal acts, as well as their implications, to offer recommendations of public policies. Many of these studies consider homicide rates, or some of the individual variables related to crime, as violence indicators [1]. Murder is widely recognized as the most serious crime and the most homogeneous in time, and the one that enables more reliable comparisons. But it is important to note that there are other manifestations of violence and criminality that deeply affect the population and must be included within the same index in order to have a complete view about the behavior of security problems. In this sense, the measurement of crimes allows to identify the main characteristics and trends of several crimes in order to develop policies for their prevention and repression and mitigate the effects of these problems on society [2, 3].

The criminal statistics analysis can be divided into two broad areas. The first one considers the official figures obtained from the registry the police department and justice institutions in each country [4]. This information offers advantages such as its

© Springer Nature Singapore Pte Ltd. 2019
Y. Tan and Y. Shi (Eds.): DMBD 2019, CCIS 1071, pp. 259–268, 2019.
https://doi.org/10.1007/978-981-32-9563-6_27

national coverage, which normally follows a relatively constant accounting method and constitutes one of the main sources for the international comparative analysis. But it also presents disadvantages like the fact that people not always report crimes to authorities (mainly robberies, common injuries, and sexual offenses), and there may be errors or manipulations when entering the information to the system. In addition, regulatory changes can modify the definition of crimes, affecting the respective series.

The second area involves the victimization surveys which are intended to characterize issues related with crime, based on information collected directly from the population to give input to the authorities, improving the decision-making process. These surveys have an impact on subjects as territorial control, prevention, and follow-up of the crime, and the measurement of not denounced criminality [5].

Among the advantages of these surveys, the method applied enables to capture part of the cases that were not reported, assess the involved institutions and obtain observations that were not considered in official statistics. Additionally, these surveys allow to measure the perception of victims about crime, criminality issues, and the performance of responsible institutions. However, the victimization surveys also present disadvantages since they do not extend to the entire country and mainly focus on urban centers, they are not systematically and regularly carried out (usually take more than six months), and may have biases in so far as they are not effective for capturing, among others, crimes like murder and sexual crimes because of the survey nature [6, 7].

In this sense, the purpose of this research is the use of an early warning system for the exploitation of criminal information to obtain behavior patterns that facilitate the generation of prevention strategies. At the same time, the research seeks to study the added value of the use of data mining in the detection of criminal patterns in order to characterize them managing to extract conclusions in the prevention of crimes, that is, to apply data mining for explaining the past through historical information, understand the present, and predict future information.

2 Theoretical Review

Currently, there is no universally accepted definition of what criminal analysis is. In some police departments, it is considered as the study of police reports and the information extraction to enables the capture of criminals, in particular, serial killers. In other agencies, the criminal analysis consists of extracting statistical data from the bases of criminal acts that occur within an area and divide them into criminal families and times of the year. Whatever the definition of criminal analysis, its objective is to find relevant information within the data contained in each of the criminal acts and disseminate it among officers and investigators to assist in the capture of potential criminals, as well as stop criminal activity [8, 9].

The formal definition of criminal analysis employed in this work is the following: Criminal Analysis is a set of processes and analysis techniques aimed to provide timely and relevant information concerning the facts and correlations of criminal tendency to operational and administrative staff during the planning of actions to prevent and avoid criminal activities, and to clear up the cases [1].

2.1   Legislation and Policies Against Crime in Colombia

In the context of an internal armed conflict that continues bleeding the country with a lengthy trace of deaths, abductions, disappearances, displacement, etc., with the complex social reality that generates a large mass of excluded population and high rates of criminality caused by the drug-trafficking mafias and common delinquents, and with the low results presented by the criminal justice system, the State has chosen to apply legal reforms for a solution. In the last twenty years, Colombia has redacted four criminal procedure codes (Decree 050/1987, Decree 2700/1991, Law 600/2000, and law 906/2004), two penal codes (decree 100/1980 and law 599/2000), two codes of minors (Decree 2737/1989 and Law 1098/2006), without including the countless partial reforms. In general, just from the year 2000 to 2006 more than 50 criminal laws were issued, including the international conventions and protocols related to the subject [1, 10, 11].

2.2   CRISP-DM Method

The CRISP-DM Method (Cross-Industry Standard Process for Data Mining) was used for developing the research. It is a free distribution method used for data mining projects, developed in 1999 by the consortium of European companies named Pete Chapman and Randy Curber (NCR - Denmark), AG (Germany), Julian Clinton, Thomas Khabaza and Colin Shearer (SPSS - England), OHRA (Holland), and Thomas Reinartz and Rüdiger Wirth (DaimlerChrysler). This method consists of six phases [12, 13]:

PHASE 01: UNDERSTANDING THE BUSINESS Objectives and requirements from a non-technical view.

- Setting up the business objectives (objectives, needs, and success criteria)
- Assessment of the situation (requirements, assumptions, restrictions)
- Generation of the project plan (plan, tools, equipment, and techniques)

PHASE 02: UNDERSTANDING THE DATA Getting familiar with the data bearing in mind the business objectives

- Initial data collection
- Description of the data
- Exploring the data
- Verification of data quality

PHASE 03: ORGANIZING THE DATA: Obtaining the minable view or dataset

- Selection of data
- Data Cleaning
- Construction of Data
- Data Integration
- Data formatting

PHASE 04: MODELING Apply the data mining techniques to the dataset

- Selection of the modeling technique
- The evaluation design
- Construction of the model
- Evaluation of the model

PHASE 05: EVALUATION of the previous phase to determine if the dataset is useful for business needs

- Evaluation of Results
- Review of the Process
- Establishment of the next steps or actions

PHASE 06: IMPLEMENTATION Exploit usefulness of models, integrating them into the decision-making tasks in the organization.

- Implementation Plan
- Monitoring and Maintenance Planning
- Generation of final report
- Revision of the draft.

3 Materials and Methods

3.1 Database

From a total of 5,684 complaints a month, through sampling calculations, the result shows a total of 359 complaints a month in the year 2018, which will be used [14].

3.2 Methods

Regarding the development of the system, it was decided to opt for the OPEN UP Method because they are especially oriented to small projects, which is a tailor-made solution for that environment. For the development of the research, the CRISP-DM method was applied as a free distribution tool for data mining projects [12].

3.3 Indicators

According to a previous work [15], the indicators used in the early warning system are shown in Table 1.

Table 1. Early warning indicators of the system used for the study.

Variable	Dimension	Indicator	Description
Early Warning System based on data mining	Access to the Early Warning System	security Index of access to the system	Indicates the percentage of people in the Lambayeque Police Region who have access to the system according to the levels of security
	Characteristics of the Early Warning System	Index of satisfaction for the use of the Early Warning System	Indicates the percentage of people in the Lambayeque Police Region who feel satisfied with the use of the Early Warning System
		Usability of the Early Warning System	Indicates whether the Early Warning System is easy to be learnt and used by people
		Query Response Time	Delay time of the system to give a response to a user request
		Number of reports by persons denounced	Indicates the number of reports issued by the system about persons who have been denounced
		Time of dedication in the registration of the complaint	Time spent on the process for registering a complaint
		Time in information searches by criteria	Indicates the time it takes the system to determine a person with a criminal profile
Process of detection of criminal patterns	Denounce	Delay time in the issue of resolution of denounces	Delay time in response to a denounce
		Index of errors in data matches of the one involved by case file	Indicates the percentage of denounces registered with errors (duplication and inconsistency of data)
	Desktop Materials	Cost per desktop materials	Indicates the amount of material used in the process for registering denounces

4 Results

The scope of the Early Warning System includes the registration process and storage of complaints, determination of groups with similar characteristics that were involved in a complaint and determining criminal profiles. It also allows the maintenance of intervention unit, crime, type of crime, the aggrieved, involved, and attorney. It also presents the maintenance of user (police), consultations, the generation of reports, and finally allows to assess the possible suspects in a committed crime.

4.1 Selection of Data

The fields that were considered for the study were chosen considering the following factors:

- Variables that each field or attribute can take.
- Data quality.
- Importance of the attribute for the study according to the objectives of the project, delivery information and its significance.

The fields that are considered for the present study are the following [14]:

- Age: Numeric field that indicates the age of the suspect. This attribute takes values from 0 years up to the age of 100 years.
- Sex: Numeric field that indicates the sex of the detainee.
- Marital status: Describes the marital status of the involved at the time of committing a crime.
- Psychic condition: Psychic condition of the involved at the time of committing a crime.
- Level of instruction: A variable that describes the level of education of the involved.
- Recidivism: Represents the type of recurrence of the suspect in relation to a crime committed.
- Incidence Time Zone: This variable indicates the time of the offense, which was specified by ranges of 6 h each group.
- Type of weapon: Referred to the type of weapon used by the involved (delinquent) at the time of committing a crime.
- Id of the crime: Identification variable of the crime that was committed, which has already been established by the NPP.
- Id of the unit of intervention: Indicates the commissioner that registered the complaint.
- Aggression: Type of aggression caused to the victim at the time of a crime.
- Crime circumstances: Referred to the additional crimes committed according to the circumstance of the development of the main fact.
- Stolen amount: In the case of offenses against property or other type of crime incurred in the robbery of goods, a range of the value stolen was specified.

4.2 Selection of the Modeling Technique

At this stage, the techniques and algorithms used in the development of data mining were selected. Initially, non-supervised techniques were used, and it was necessary to make a data grouping or clustering to ease the discovery of hidden information. The algorithm used was K-Means, one of the most commonly used methods and the most popular of the "partition" grouping methods which is also the most widely used algorithm in the studies consulted for the research [13].

The next step was the validation of groups or clusters with specialists in the criminal field. Such validation enabled the classification technique. Finally, neural networks were applied as part of the supervised classification technique considering the previously defined cluster to achieve the identification of decision rules that help explain the composition of each group.

4.3   Clustering

The K-means algorithm was used in order to complete the data grouping, resulting in 12 criminal profile groups through their most significant features.

Firstly, before having analyzed the cluster, it was found that most of the clusters show a greater proportion of men who have committed a crime (sex = 2), which was the most frequent, according to records obtained from the regional police and with a smaller percentage for women, see Table 2.

Table 2. Distribution of recidivism of the involved.

Recidivism	
Description	%
Male	75.8
Feminine	23.8
No record	3.4
Total	100%

Therefore, males are expected to obtain more frequency in all the runs of clustering within formed groups since it was the most predominant sex in the records of complaints.

Figure 1 shows the scatter chart for the cluster analysis.

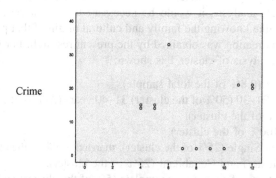

Fig. 1. Distribution of the cluster according to Crime.

The x-axis was represented by the 12 clusters, whereas the ordered ones were represented by the offenses. Since the 2 represented the offense of injury, it was observed that the clusters 1, 7, 8, 9 and 10 were the ones which formed this group of offenses, although cluster 9 obtained a greater proportion (55.6%) of the total number of crimes by injury. The x-axis was represented by the 12 clusters, whereas the ordered ones were represented by the offenses. Since the 2 represented the offense of injury, it was observed that the clusters 1, 7, 8, 9 and 10 were the ones which formed this group of offenses, although cluster 9 obtained a greater proportion (55.6%) of the total number of crimes by injury.

On the other hand, the 14 represented the offense of simple and aggravated robbery, and was contained in clusters 3, 4 and 5, cluster 4 obtained the highest proportion (43.8%) and cluster 3 obtained the lowest proportion with 3% of the total number of crimes by simple and aggravated theft.

The 15 represented the offense of simple and aggravated robbery, and was contained in clusters 3, 4 and 5, with the highest proportion in cluster 4 (60.2%) and the lowest proportion in cluster 3 with 6.1% of the total number of crimes by simple and aggravated robbery.

The 20 represented the offense of simple and aggravated damage. It was contained by cluster 12, which contained such offenses (100).

The 21 represented the crime of Usurpation/Extortion and was contained by clusters 11 and 12 with the highest proportion in cluster 12 (76.2%) and the lowest proportion in cluster 11 with 23.8% of the total number of crimes.

The 33 represented the crime of domestic violence, and was contained by the clusters 2 and 6, with the highest proportion in cluster 6 (94.3%) and the lowest proportion in cluster 2 with 5.7% of the total number of crimes by family violence.

4.4   Interpretation of the Clusters

Based on the information obtained from both the analysis of graphs prior research and collaboration of specialists, a first interpretation of each formed cluster was obtained. It is convenient to highlight, as discussed with the specialists, that for performing a deeper and accurate analysis, it would be necessary to have access to each suspect individual history like knowing the family and cultural origin of that person. However, a very good approximation was obtained by the present research. For reasons of space and privacy, just analysis of cluster 1 is shown:

- Cluster size = 4 (0.8% of the total sample).
- Average age = 21–30 (50% of the cluster) 31–40 years (25% of the cluster) and 41–100 years (25% of the cluster).
- Sex = Male (100% of the cluster).
- Marital Status = Single (25% of the cluster), married (25% of the cluster), Divorced (25% of the cluster) and Cohabitant (25% of the cluster).
- Level of education = Secondary incomplete (5% of the cluster) and technical (75% of the cluster).
- Recidivism = Recidivist (75% of the cluster) and multi-recidivist (25% of the cluster).

- Psychic Condition = Sick/altered (50% of the cluster), drugged (25% of the cluster) and Fair (25% of the cluster). ϖ Incidence Time Zone = 8:01 a.m.–2:00 p.m. (25% of the cluster) 2:01–8:00 p.m. (50% of the cluster) and 2:01 a.m.–7:59 a.m. (25% of the cluster).
- Intervention Unit = El Porvenir Police Station (75% of the cluster) and La Victoria Police Station (25% of the cluster).
- Type of Weapon = None (50% of the cluster) and knife (50% of the cluster).
- Aggression = Both types of aggression, namely physical and psychological integrity (100% of the cluster). Crime = Crimes against Life, Body and Health (100% of injuries).
- Crime circumstances = Crime Against Freedom (25% of the cluster of Violation to Personal Freedom offense, and 50% of the 95 cluster of Home Violation offense) and simple and aggravated robbery (25% of the cluster).
- Stolen amount = There was no robbery (75% of the cluster), robbery valued at more than 100 and less than 500 soles (25% of the cluster).

4.5 Classification

After obtaining the groups, the supervised classification was performed using Neural Networks with MATLAB R2010a [16–18]. The layers of hidden inputs and outputs were defined using 13 entries corresponding to the variables defined in the clustering stage. To know the optimum number of hidden layers [9], the number of inputs were contrasted with the outputs, resulting in a hidden network in order to optimize the work and preventing the overfitting within the training. The layers of output are 12 referring to groups that are already established. 150 iterations were used, avoiding to give a small number as the network cannot achieve the purpose of training. In addition to the process of neural network, three phases were considered: Training phase in which the 359 valid complaints were used to determine the parameters that define the neural network model. Later, the validation phase was applied to avoid overfitting, so this stage allowed to control the learning process.

5 Conclusions

The identification of criminal patterns was achieved with the support of the Early Warning System based on Data Mining developed for the Police Region of Bogota, whereby 12 criminal groups were defined, with different characteristics and behaviors, allowing the validation of pre-existing knowledge. In addition, it was possible to characterize those involved in a crime based on their most relevant attributes.

The use of the Early Warning System demonstrated its efficiency as it was contrasted with the information obtained prior the use in the Police Region, resulting in a considerable reduction of time were the system spent 4.9 min in the process of registering complaints and more of 1 h and a half for the processes of criminal profile searches. The system allowed maintaining the information ordered and updated in such a way that information can be accessed quickly, managing to reduce an average of 11 min in manual searches.

References

1. Green, W.J.: A History of Political Murder in Latin America: Killing the Messengers of Change. State University of New York Press, Albany (2015)
2. Jaramillo, J., Meisel, A., Ramírez, M.T.: More than 100 years of improvements in living standards: the case of Colombia. Cliometrica, October 2018. Online First
3. Karl, R.A.: Forgotten Peace: Reform, Violence, and the Making of Contemporary Colombia. University of California Press, Oakland (2017)
4. Roskin, M.G.: Crime and politics in Colombia: considerations for US Involvement. Parameters: US Army War Coll. Q. **34**(1), 126–134 (2001)
5. Santos Calderón, E.: El país que me tocó (Memorias). Penguin Random House Grupo Editorial (2018)
6. Calderón, M., Marconi, S.: Santuarios de la memoria: historias para la no repetición. Relatos de actos humanitarios en la vereda Beltrán y el municipio de Marsella, Risaralda (Tesis de Pregrado). Universidad Santo Tomás, Bogotá (2017)
7. Martínez, L.: Violencia y desplazamiento: Hacía una interpretación de carácter regional y local. El caso de Risaralda y su capital Pereira, en Revista Estudios Fronterizos, vol. 7, no. 14, julio–diciembre 2006
8. Martínez, S.: Núcleos urbanos y de frontera en el Centro Occidente Colombiano. Un proyecto de institucionalización del Estado Nación en el siglo XIX, en Americanía. Revista de Estudios Latinoamericanos. Nueva Época, no. 3, enero–junio 2016
9. Núñez, M.: Contexto de violencia y conflicto armado, Monografía Político Electoral del Departamento de Risaralda, 1997–2007. Misión de Observación Electoral, Corporación Nuevo Arcoiris, CERAC, Universidad de los Andes, Bogotá (2017)
10. Ovalle, L.: Memoria y codificación del dolor. Muertes violentas y desapariciones forzosas asociadas al narcotráfico en Baja California, en Revista de Estudios Fronterizos (2010)
11. Palacios, M.: Violencia púbica en Colombia 1958–2010. Fondo de Cultura Económica, Bogotá (2012)
12. Huber, S., Seiger, R., Kuhnert, A., Theodorou, V., Schlegel, T.: Goal-based semantic queries for dynamic processes in the Internet of Things. Int. J. Semant. Comput. **10**(2), 269 (2016)
13. Wirth, R., Hipp, J.: CRISP-DM: towards a standard process model for data mining. In: Proceedings of the 4th International Conference on the Practical Applications of Knowledge Discovery and Data Mining, pp. 29–39 (2000)
14. Merchan-Rojas, L.: Conducta Criminal: una Perspectiva Psicológica Basada en la Evidencia. Acta Colombiana De PsicologíA **22**(1), 296–299 (2019)
15. Azevedo, A.I.R.L., Santos, M.F.: KDD, SEMMA CRISP-DM: a parallel overview. IADS-DM (2008)
16. Vásquez, C., et al.: Cluster of the Latin American universities top100 according to webometrics 2017. In: Tan, Y., Shi, Y., Tang, Q. (eds.) DMBD 2018. LNCS, vol. 10943, pp. 276–283. Springer, Cham (2018). https://doi.org/10.1007/978-3-319-93803-5_26
17. Viloria, A., Mercedes, G.-A.: Statistical adjustment module advanced optimizer planner and sap generated the case of a food production company. Indian J. Sci. Technol. **9**(47), 1–5 (2016)
18. Varela, I.N., Cabrera, H.R., Lopez, C.G., Viloria, A., Gaitán, A.M., Henry, M.A.: Methodology for the reduction and integration of data in the performance measurement of industries cement plants. In: Tan, Y., Shi, Y., Tang, Q. (eds.) DMBD 2018. LNCS, vol. 10943, pp. 33–42. Springer, Cham (2018). https://doi.org/10.1007/978-3-319-93803-5_4

Fuzzy C-Means in Lower Dimensional Space for Topics Detection on Indonesian Online News

Praditya Nugraha⬡, Muhammad Rifky Yusdiansyah⬡,
and Hendri Murfi(✉)⬡

Department of Mathematics, Universitas Indonesia, Depok 16424, Indonesia
{praditya.nugraha,muhammad.rifky51,
hendri}@sci.ui.ac.id

Abstract. One of the automated methods for textual data analysis is topic detection. Fuzzy C-Means is a soft clustering-based method for topic detection. Textual data usually has a high dimensional data, which make Fuzzy C-Means fails for topic detection. An approach to overcome the problem is transforming the textual data into lower dimensional space to identify the memberships of the textual data in clusters and use these memberships to generate topics from the high dimensional textual data in the original space. In this paper, we apply the Fuzzy C-Means in lower dimensional space for topic detection on Indonesian online news. Our simulations show that the Fuzzy C-Means gives comparable accuracies than nonnegative matrix factorization and better accuracies than latent Dirichlet allocation regarding topic interpretation in the form of coherence values.

Keywords: Topic detection · Clustering · Fuzzy c-means ·
Low dimension space · Online news

1 Introduction

Nowadays Technologies of information and communication are well developed. It gives so much easiness, and everyone could use it without having any hard effort. Based on Internet World Stats Usage and Population Statistics, internet users in the world until June 2018 reached 4,2 billion users, or equivalent to 55,1% of the total world population. The rapid development cause easiness in getting any information on the internet. For example, we do not have to buy a newspaper anymore to know what is happening in the world; we can see online news instead. The information we get from online news comes in the form of articles that contain news' title and text. We can understand the articles by knowing what topic is discussed in the articles. Because we have a large number of articles, we are going to use a topic detection method. This method examines words in documents to get topics contained in the documents, shows how the topics relate to each other and how their trends change over time [1].

Fuzzy C-Means (FCM) is one of the clustering-based methods for topic detection. FCM is a soft clustering method that allows textual data to belong in more than one clusters. The means of all members in a cluster would be the center of the cluster called

Y. Tan and Y. Shi (Eds.): DMBD 2019, CCIS 1071, pp. 269–276, 2019.
https://doi.org/10.1007/978-981-32-9563-6_28

centroids. In a topic detection case, the centroids are interpreted as topics. Therefore, FCM implicitly assumes that textual data may have more than one topics. In general, FCM works well for low dimensional data and fails for high dimension data. FCM generates only one topic from high dimensional textual data for the topic detection problem. To overcome this problem, we can choose the small fuzzification constant approach to one. However, this setting forces the problem to be hard clustering where the textual data contain only one topic. Another approach is transforming the textual data into lower dimensional space, i.e., Eigenspace using singular value decomposition (SVD) and the method called Eigenspace based Fuzzy C-Means (EFCM) [2, 3].

In this paper, we apply EFCM for topic detection on Indonesian online news. We evaluate the accuracy of EFCM and compare it to other commonly used topic detection methods such as the standard nonnegative matrix factorization (NMF) and latent Dirichlet allocation (LDA). Our simulations show EFCM results the comparable accuracies to ones of NMF and the better accuracies than ones of LDA regarding topic interpretation in the form of coherence value.

The outline of this paper is as follows: In Sect. 2, we present the reviews of FCM, SVD, and EFCM. Section 3 describes the results of our simulations and discussions. Finally, a general conclusion about the results is presented in Sect. 4.

2 Methods

Besides another direct method for seeking topic detection, there is also some indirect method, such as Clustering based method. The common clustering method used for seeking topic detection in any textual data is fuzzy c-means algorithm (FCM) [4] (Sect. 2.1). However, FCM does not give a good result in high dimension data [5]. Since textual data usually constructed of high dimensional data. So that we transform the data into a lower dimension using two different way. Therefore, we use truncated singular value decomposition (SVD) (Sect. 2.2). We called this a fuzzy c-means in lower dimensional data (Sect. 3.1).

2.1 Fuzzy C-Means

Fuzzy C-Means (FCM) produce some clusters by divide dataset into membership values $M = [m_{ik}]$, which allows one data belong to two or more clusters. This method based on the minimization of the objective function as expressed:

$$J_{FCM}(M, V) = \sum_{i=1}^{c} \sum_{k=1}^{n} m_{ik}^{w} ||x_k - v_i||^2 \qquad (1)$$

where $||\cdot||$ is any norm represent similarities between the centroid and data, and w is a fuzzification constant to decide the level of fuzziness.

As the objective function is minimized, values in high membership area are given to data near the cluster centroid. The membership values and the centroids are computed iteratively using alternating optimization.

Algorithm 1. Fuzzy C-Means

Input : A, c, w, max number of iterations (T), threshold (ε)

Output : m_{ik}, v_i

1. set $t = 0$
2. initialize v_i
3. update $t = t + 1$
4. calculate $m_{ik} = \left[\sum_{j=1}^{c} \left(\frac{\|x_k - v_i\|_2}{\|x_k - v_j\|_2} \right)^{2/w-1} \right]^{-1}, \forall i, k$
5. calculate $v_i = \frac{\sum_{k=1}^{n} ((m_{ik})^w x_k)}{\sum_{k=1}^{n} (m_{ik})^w}, \forall i$
6. if a stopping, i.e., $t > T$ or $\|M^t - M^{t-1}\|_F < \varepsilon$, is fulfilled then stop, else go back to step 3

2.2 Singular Value Decomposition

Singular Value Decomposition (SVD) is a method of the decomposing matrix into the other three matrices. It is one of the matrix factorizations that commonly used in the data processing. The formal definition of SVD given in Definition 2.1.

Definition 2.1. Let **A** be an $m \times n$ matrix. SVD of **A** is

$$A = U\Sigma V^T \tag{2}$$

where U is an $m \times m$ orthogonal matrix, Σ is an $m \times n$ diagonal matrix and V is an $n \times n$ orthogonal matrix. The diagonal elements of matrix Σ are called singular values of **A** and sorted from largest to smallest [6].

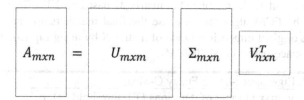

Fig. 1. An illustration of SVD

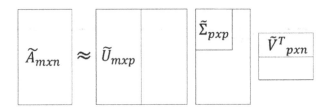

Fig. 2. An illustrations of truncated SVD

Matrix factorization using SVD illustrated in Fig. 1. Therefore, the useful property of the SVD is SVD allows us to get an approximation of any matrices into a smaller size. The singular values on matrix Σ are sorted in descending way from largest to smallest, then the best p-rank approximation matrix to the matrix $\mathbf{A}$ can be created by taking the first p singular values of matrix Σ where $p < <min(m, n)$. This type of SVD is also known as truncated SVD [6]. Let $\tilde{\mathbf{A}}$ be the p-rank approximation matrix to the matrix $\mathbf{A}$, then $\| \mathbf{A} - \tilde{\mathbf{A}} \|$ is the $p + 1$-th singular value of Σ. Truncated SVD illustrated in Fig. 2.

2.3   Eigenspace-Based Fuzzy C-Means

Since FCM giving bad result in high dimensional data. For example, in high dimensional space, FCM runs into the center of gravity of the entire data [7]. In high dimensional data, FCM produced only one centroid. Therefore, it gives one topic only. To solve this problem, we are trying to transform the high dimensional data into a lower dimension before FCM is used. Truncated SVD used to transform data into a lower dimension. SVD decomposes the matrix X_{mxn}, into $U\Sigma V^T$ as described in Definition 2.1. $\tilde{V}$ is a $p \times n$ matrix where $p < <min(m, n)$., then we interpret $\tilde{\Sigma}\tilde{V}^T$ as the lower dimensional data of $\mathbf{X}$. By using the truncated SVD, we transform the matrix $\mathbf{X}$ into the matrix $\tilde{\Sigma}\tilde{V}^T$.

Next, we use FCM on the lower-dimensional data of $\tilde{\Sigma}\tilde{V}^T$ (using Truncated SVD). In this step, we calculate the membership matrix $\tilde{M}$ based on $\tilde{\Sigma}\tilde{V}^T$. Therefore, after the convergent of the FCM algorithm, we use the final membership values to calculate centroids in the original dimensional data of matrix $\mathbf{X}$ by using Eq. (3). This algorithm is called Eigenspace-based FCM (EFCM).

Algorithm 2. Eigenspace-based Fuzzy C-Means

Input   : A, c, w, max number of iterations (T), threshold $(\varepsilon), p$

Output: $\mathbf{t}_i$

1.   transform A : $\tilde{R} = \text{TruncatedSVD}(X, p)$

2.   perform FCM : $m_{ik} = \text{FCM}(\tilde{M}, c, w, T, \varepsilon)$

3.   calculate the topics : $\mathbf{t}_i = \dfrac{\sum_{k=1}^{n}((m_{ik})^w \mathbf{x}_k)}{\sum_{k=1}^{n}(m_{ik})^w}$

---

3   Results and Discussion

In this section, we are going to analyze and compare the accuracies of the EFCM with the standard LDA and NMF method. The comparison uses a measurement unit called *topic interpretability*.

3.1   Topic Interpretability

Topic Interpretability is a quantitative method to seek the interpretability of a topic by calculating the coherence scores of words that construct the topic. One of the common formulas which used to estimate coherence scores is PMI. Suppose a topic t consists of an n-word that is $\{w_1, w_2, ...w_n\}$, the PMI of the topic t is

$$PMI(t) = \sum_{j=2}^{n} \sum_{i=1}^{j-1} \log \left(\frac{P(w_j, w_i)}{P(w_i)P(w_j)} \right) \tag{3}$$

where $P(w_j, w_i)$ is the probability of the word w_i appears together with the word w_j on the corpus, $P(w_i)$ is the probability of word w_i appears in the corpus, and $P(w_j)$ is the probability of word w_j appears in the corpus. Corpus is a database consisting of some text-based documents which used for a reference to calculate PMI. In this experiment, we use a corpus consisting of 3.2 million Indonesian Wikipedia documents.

In this simulation, we use the datasets from much credible news corporate's News feeds in January, February, and March in 2014. These datasets consist of Titles and the news' text. Dataset has 50304, 46834, and 31855 News, Respectively.

The datasets have been in the form of a list of sentences separated by Titles, and Text type. So, the process that needs to be done is only to form the word-document matrix A and do the weighting process. In this simulation, the weighting process is performed using the term frequency-inverse document frequency (TF-IDF) [8].

To convert text into vector representations, we use two processes, i.e., pre-processing and word-based tokenization. Firstly, we convert all words into lowercase; we replace two or more repeating letters with only two occurrences. There are 5370 Indonesian stop words that existed in the dataset, lower than two texts or more than 95% of text are filtered. For the final step, we used the term frequency-inverse document frequency for weighting the dataset. Besides using scikit-learn packages for preprocessing and tokenizing steps, we also used the natural language toolkit from nltk [9].

In this simulations, we also inspect two direct methods for topic detection such as LDA and NMF, in comparison with EFCM All of these methods uses a Python-based library called scikit-learn in using the methods [10]. We use the default values for the Dirichlet parameters, which provided by scikit-learn. For NMF, the words vectors dimensions are reduced into unit length as the topic increased in simulations, respectively. We used default parameters from scikit-learn for NMF. For EFCM parameters, we arrange the fuzzification constant $w = 1.5$, the maximum number of iteration $T = 1000$, the threshold $\varepsilon = 0.005$ and we are reducing its dimension into 5 using Truncated SVD.

These methods produce topics consisting of words, in each topic forming one vector. LDA method giving ten most probable words per topic as results. On the other hand, NMF and EFCM giving ten most frequent words as results. And after the results are collected, then the words in each topic are going to calculated by Normalized PMI to get coherence score.

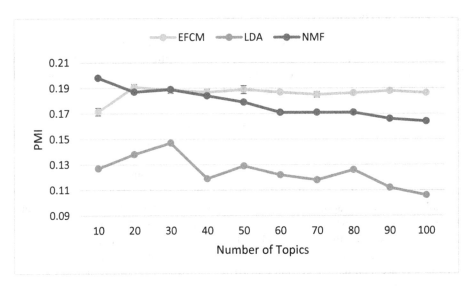

Fig. 3. Comparison in PMI score between EFCM in lower space, Latent Dirichlet Allocation, and Nonnegative Matrix Factorization in various numbers of topics for January Dataset.

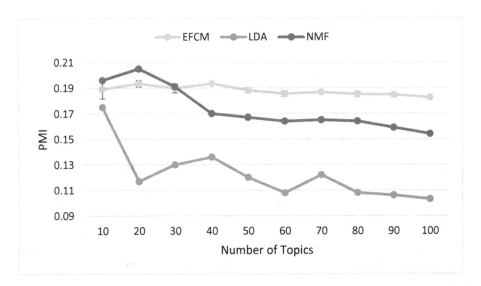

Fig. 4. Comparison in PMI score between EFCM in lower space, Latent Dirichlet Allocation, and Nonnegative Matrix Factorization in various numbers of topics for February Dataset.

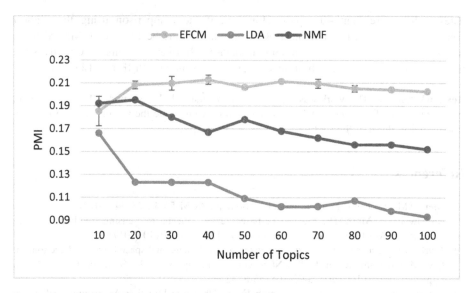

Fig. 5. Comparison in PMI score between EFCM in lower space, Latent Dirichlet Allocation, and Nonnegative Matrix Factorization in various numbers of topics for March Dataset.

Figures 3, 4 and 5 show the comparison of Coherence scores for the number of topic $c \in \{10, 20, ..., 90, 100\}$. From those figures, we see that the result gives a similar plot for each figures and the scores didn't strictly increase with the increasing number of topics. In some method, the scores are decreasing instead. We see that NMF tends to decrease with the increasing number of topics while EFCM with Truncated SVD dimension reduction tends to increase at first, then shows a static score. This method and NMF, as the topic increased but their coherence score tends to decrease. As the LDA method has the lowest score compared other methods, and also decreasing its score as the topics increased (Table 1).

Table 1. The comparison of the optimal topic interpretability in PMI scores

	January	February	March
LDA	0.1470	0.1750	0.1660
NMF	0.1980	0.2050	0.1950
EFCM	0.1908	0.1936	0.2130

4   Conclusions

In this paper, we examine the fuzzy c-means method in lower dimensional space, called EFCM, for topic detection on Indonesian online news. We use interpretability of the extracted topics in the form of coherence scores to compare EFCM with the standard NMF and LDA method. Our simulations show that EFCM gives best results in all datasets for more than 40 topics. However, if the comparison using the highest coherence score for each method, EFCM only attains the best result for March' news which is the lowest size dataset among all datasets. In overall results, EFCM gives the best accuracies between the best accuracies of the standard NMF and LDA methods.

Acknowledgment. This work was supported by Universitas Indonesia under PIT 9 2019 grant. Any opinions, findings, and conclusions or recommendations are the authors' and do not necessarily reflect those of the sponsor.

References

1. Blei, D.M.: Probabilistic topic models. Commun. ACM **55**(4), 77–84 (2012)
2. Muliawati, T., Murfi, H.: Eigenspace-based fuzzy c-means for sensing trending topics in Twitter. In: AIP Conference Proceedings, vol. 1862, p. 030140 (2017)
3. Murfi, H.: The accuracy of fuzzy c-means in lower-dimensional space for topic detection. In: Qiu, M. (ed.) SmartCom 2018. LNCS, vol. 11344, pp. 321–334. Springer, Cham (2018). https://doi.org/10.1007/978-3-030-05755-8_32
4. Bezdek, J.C.: Pattern Recognition with Fuzzy Objective Function Algorithms. Advanced Applications in Pattern Recognition. Springer, New York (1981). https://doi.org/10.1007/978-1-4757-0450-1
5. Winkler, R., Klawonn, F., Kruse, R.: Fuzzy c-means in high dimensional spaces. IJFSA **1**, 1–16 (2011)
6. Burden, R.L., Faires, J.D.: Numerical Analysis. Cole Cengage Learning, Boston (2011)
7. Hofmann, T., Schölkopf, B., Smola, A.J.: Kernel methods in machine learning. Ann. Stat. **36** (3), 1171–1220 (2008)
8. Manning, C.D., Schuetze, H., Raghavan, P.: Introduction to Information Retrieval. Cambridge University Press, Cambridge (2008)
9. Loper, E., Bird, S.: NLTK: the natural language toolkit. In: Proceedings of the COLING/ACL 2006 Interactive Presentation Sessions, pp. 69–72 (2006)
10. Pedregosa, F., et al.: Scikit-learn: machine learning in python. J. Mach. Learn. Res. **12**, 2825–2830 (2011)

Mining Tasks

A Novel Big Data Platform for City Power Grid Status Estimation

Wuhan Lin[1], Yuanjun Guo[2(✉)], Zhile Yang[2], Juncheng Zhu[3], Ying Wang[3],
Yuquan Liu[1], and Yong Wang[1]

[1] Guangzhou Power Supply Co. Ltd., Guangzhou 510620, China
[2] Shenzhen Institute of Advanced Technology Chinese Academy of Sciences,
Shenzhen 518055, China
yj.guo@siat.ac.cn
[3] Zhengzhou University, Zhengzhou 450001, China

Abstract. This paper proposed a Big Data platform solution for city power grid status estimation, with an in-memory distributed computing frame for power grid heterogeneous data analysis. A big data management system is proposed for managing high volume, heterogeneous and multi-mode power data. In addition, a high efficiency data-computing engine, a parallel computing model for power engineering and a 3D display system are consisted. Based on the platform, three real-system application cases are given in order to illustrate the practicability of the proposed platform, including high efficiency power grid load shedding computing and analysis, voltage sag association rule mining with 3D display, and real-time dynamic evaluation of operation status of wind farm units. As a conclusion, big data platform is a key technology to sufficiently realize the potential advantages of power data resources, and it can provide solid technological bases for the development of modern Smart Grid.

Keywords: Big data · Power grid · State estimation

1 Introduction

Along with the rapid development of modern power system, the size and span of the power grid has been significantly increased, as well as the types and quantities of the power equipment both in transmission and distribution networks [1–3]. As a result, the complexity and growth rate of power data is a major challenge for state assessment and monitoring of urban power grids. For a long time, the operation companies lacks accurate understanding of the equipment reliability and the risks caused by faults, therefore, there is not enough information support for decision-making, causing a large number of assets being retired in advance. According to statistics, the average working life of power transformers in China is only about 20 years, while the advanced international power supply enterprises can use their equipment nearly 40 years [4]. At the same time,

© Springer Nature Singapore Pte Ltd. 2019
Y. Tan and Y. Shi (Eds.): DMBD 2019, CCIS 1071, pp. 279–288, 2019.
https://doi.org/10.1007/978-981-32-9563-6_29

the various information of the grid operation cannot be highly shared with each other, making it inefficient to manage and monitoring the entire grid operation status [5,6].

Supervisory control and data acquisition (SCADA) devices are mainly used in traditional power industries to collect data and to secure grid operations [7]. Unlike traditional SCADA systems, the Phasor Measurement Units (PMU) are able to measure the voltage phasor of the installed bus and the current phasors of all the lines connected with that bus. In particular, PMUs are collecting data at a sampling rate of 100 samples per second or higher, therefore, huge amount of data need to be collected and managed. For example, the State Grid Corporation of China owns over 2.4 hundred million smart meters, making the total amount of collected data reaches 200 TB for a year, while the total number of data in information center can achieve up to 15 PB [8].

At present, Fujian Power Grid Corporation in China has established an online monitoring and fault diagnosis system for power transmission and transformation equipment [9]. North China Power Grid Corporation have established state maintenance assistant decision-making systems, which realized Integrated analysis of various existing data [10]. Guangdong Power Grid Corporation built a remote diagnosis center for high-voltage equipment, realizing remote monitoring of hundreds of substations, and connected thousands of monitoring equipments to achieve information integration [11]. Guangxi Power Grid Corporation established an equipment security early warning decision platform, realizing integration of online monitoring system, SCADA system and production system [12]. Yunnan Power Grid Corporation established a technical supervision data analysis center to achieve data automatic acquisition and centralized management [13].

For other companies all over the world, Alstom has developed a substation equipment condition monitoring system, which uses a variety of smart sensor technologies to monitor the operation status of the main equipment [14]. General Electric has developed an integrated substation monitoring and diagnosis system, including multiple gas transformer, switch gear monitoring and other solutions [15]. The existing big data acquisition and conversion devices for power grids at home and abroad mainly focus on some key equipments, thus a large number of weak correlations variables, spatio-temporal power data have been ignored. Under Big data environment, power operators must make timely and accurate decisions based on a large amount of integrated information.

Therefore, this paper proposes a big data platform for urban power grid state assessment. An overall framework has been designed based on big data fusion management and analysis technology, which is suitable for acquisition and processing massive volume of power data with multi-mode heterogeneous and low-value density characteristics. Moreover, an efficient data analysis engine is integrated as well as a three-dimensional panoramic display system. As a result, power system big data can be collaboratively analyzed to extract useful information, and the model calculation results are presented in an optimal visual form to help maintain and enhance the situational awareness of the grid operator and improve the decision-making.

2   Power Big Data Platform

In this paper, a Big Data platform was designed to meet the special condition of power grid in South China, such as large-scale, complex geographical and weather condition and AC/DC mixed operation over long distances. Big Data technologies are applied to this power network to assist with condition monitoring and state estimation of the transmission and distribution systems, collecting multi-platform power data and realizing high efficiency process and analysis of data from the power grid at different levels.

2.1   Big Data Platform Architecture

As shown in Fig. 1, the proposed big data platform consist of database, data interface, big data management, analytic engine, application and display layer.

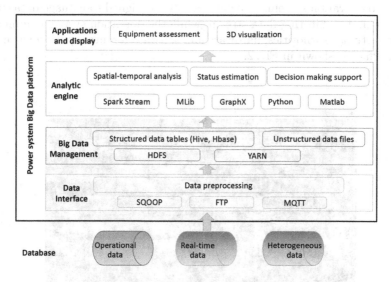

Fig. 1. Big Data processing and analysing platform for electric power system condition monitoring

The database for this platform covers the big data information of the city grid for power grid operation, management, and maintenance, as well as related heterogeneous data such as meteorological, geographic information, video images and others. Power big data not only presents the "4V" characteristics of massive data, high-speed, multi-class, low-value density, but also dynamic interactivity, sustainability and consumer empathy under the information era.

The power big data platform designed in this paper adopts structured or unstructured file management based on distributed file system HDFS and database management system based on Hive/HBase/Impala. Real-time data such as inspection GPS service data, space-time location data, equipment monitoring image and construction operational video live broadcast can be accessed through wearable devices. Data verification, cleaning, evaluation and data conversion are implemented in the data interface layer for subsequent images. As potential applications, deep learning methods can be used to analyze power big data (including electrical load, power generation, electricity consumption, electricity market, etc.) with human social activities based on local population, age distribution, occupational category, behavioral activity patterns and others. The multi-mode data management structure provides a sufficient data source for building a rich knowledge model of power grid and human society.

2.2   Power Big Data Management and Operation Interface

In order to carry out real-time operations and monitoring of the various functional tasks of the power system, this paper implemented a big data management portal and operation system. The data portal is designed to manage data, tasks, operation and monitoring data stream, while the operation system can monitor the user activity, authorities, hardware and software resources performance. The user interface is shown in Fig. 2.

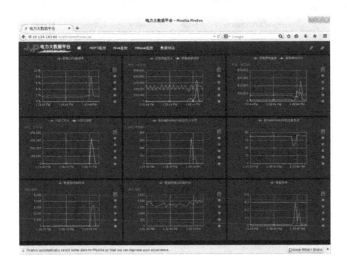

Fig. 2. User interface of operational and maintenance system for power gird Big Data platform

2.3   Key Technologies of Power Big Data Analysis Engine

The power grid big data analysis engine provides key techniques for the computation and analysis of the entire grid state assessment and analysis. Apache

Spark, due to its flexible cache distributed memory structure, is suitable for fast and efficient calculation of massive data, and has a standard interface of various popular programming languages (such as Java, Scala, Python, SQL, etc.). Therefore, the open source of Apache Spark is adopted in this platform, with the real-time streaming data computing Spark stream, machine learning algorithm library Spark MLlib, parallel graph computing tool Spark GraphX, Python, Matlab and deep learning framework Caffe and TensorFlow. Moreover, Machine learning algorithms and big data-based algorithms are integrated into the analysis engine, including collaborative filtering, classification, clustering, dimensionality reduction, and correlation analysis algorithms.

A special module of spatio-temporal correlation analysis is designed to reveal the strong or weak correlations between variables of power grid data, establishing a spatio-temporal network model and process-oriented data model. Therefore, customers electricity usage patterns in a certain neighborhood can be analyzed, and the overall impact of the space-time changes around on the status of transmission lines can be measured. A multi-factor related information database for the whole network can be established to support other required analysis and forecasting.

A large-scale graph parallel computing module constructs the grid topology map based on GraphX through the CIM files, with graph parallel computing scheme, which can realize specific applications such as island topology identification, impact range analysis, and spatio-temporal geographic process. An abstraction of power systems including power transmission lines, power equipment, substations, etc. can be formed in terms of graphs, which allows the fast graphic calculation methods to be applied to analyzed the potential range caused by any equipment failures, the risks and range of power outage, as well as the optimization of power grid structure.

Other key modules can be setup in order to meet different requirements, for example, integrated learning and fusion analysis is aimed at various complex professional models and different objects in power systems, including power system parameter design, stability analysis, risk assessment, transmission lines evaluations, power flow, active and reactive power monitoring and others. Fault diagnosis based on deep learning is mainly used for images or video data, which mainly come from infrared images and surveillance videos on power equipments. Spark parallel computing framework is suitable to process deep learning neural networks, which greatly improves computational efficiency and accuracy.

3   3D Panoramic Display System

With the help of three-dimensional panoramic display system, a large amount of information can be expressed effectively in a visible way, helping electric power operators analyze and utilize the power data, understand the real-time status of the power grid, and quickly find any potential problems. The 3D display system applies geographic information system technology, uses geographic scene as information carrier and geographic element as the basis of data organization.

The structure is shown in Fig. 3, consist of a data storage layer, an interface layer, a calculation layer, and a three-dimensional display layer.

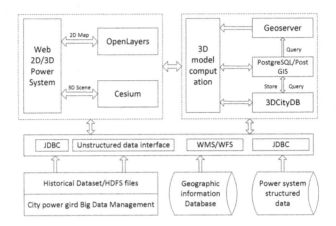

Fig. 3. Framework of 3D display system

The main data used include historical databases and big data files stored on distributed clusters. In addition, for the 3D display of geographic information, geospatial information database and related model-calculated grid-structured data is essential.

The computational layer and the three-dimensional display layer of the system communicate with each other to jointly display the three-dimensional environment of urban power grid. OpenLayers is adopted to support multiple map sources and the map background overlaying with other layers. Cesium supports the map engine of 3D scenes to realize the visualization of dynamic geospatial data. Through the calculation and display of the power grid 3D model (including telegraph towers, transmission lines, substations, electrical equipment, etc.), the 3D panoramic environment of the urban power grid is displayed.

Combined with the big data analysis engine and the calculation results of the professional model, the spatial information and power equipments visualization support modules are provided, with the entire grid system can be displayed on the geographic map. Therefore, the needs of real-time detection, analysis, decision making and statistical display of various geographic information-based power systems can be met.

4   Application Studies

4.1   High Efficiency Load-Shedding Calculation

The calculation of load-shedding in power system can quantify how much loss the real system is undergoing after equipment failure in a objective way, thus it can measure the operation risks and provide significant information for decision-making of equipment reconditioning or replacement. The actual reduction of load-shedding for different types on each electrical point is needed for the calculation, thus it is very time-consuming to calculate power grid risk evaluation with plenty of pre-defined fault scenarios. In the proposed platform, a calculation scheme based on Spark and Compute Unified Device Architecture (CUDA) is applied, as shown in Fig. 4.

Table 1. The comparisons of parallel computing with single machine results

Machines	Model	Memory	Cores	Executor	Time
3 machines	YARN	60G	30	30	11 min
Single	Local	200G	20	20	2.5 h

The complete load-shedding scheme contains off-line test stage and on-line parallel computing stage. The computation tasks are firstly divided into different working regions on Spark, then Matlab algorithms are packed and called, further processing of each computation task is transmitted to working threads on every division, where parallel computing is realized. After that, results at each step are collected progressively, thus the risk evaluation tasks for multiple scenarios can be finished. For real-system cases, total number of 6000 scenario files with 1.2 GB size are calculated according to the flowchart given in Fig. 4, and the comparison with calculation time on a single machine is given in Table 1.

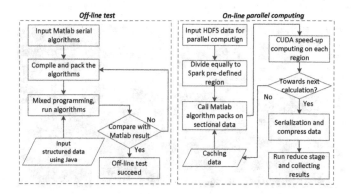

Fig. 4. Load-shedding calculation flowchart for parallel computing

It can be easily seen that parallel computing is able to solve the problem of low efficiency when risk evaluation in multi-scenarios is taken in power system. The load level of each electrical point can be monitored dynamically, and the topology change of power grid due to any system maintaining or multiple elements quit can also be calculated with high efficiency, the computation time is greatly shortened.

4.2   Voltage Sag Association Rule Mining and 3D Display

With the support of Big Data platform, transmission line trips records, power quality data, weather data and other related data can be collected, in order to monitor and analyze the transients. In addition, a three dimensional visualization system is developed to merge together all the analysis results with geographic, landforms and even weather conditions, then display in a very intuitive way. Therefore, situation awareness of system operators are greatly enhanced. In order to analyze the correlation between transmission line trips and voltage dips, multi-source data is needed, consisting of (1) transmission line tripping data, recording tripping time, fault description, fault type, and so on; and (2) voltage disturbances data, including monitoring location, disturbance type, happening time, lasting time, magnitude. The first step to analyze the transients is data fusion, combining two sets of data according to the unified time tags. It is necessary to combine substation coordinates, maps and other geographic information with these transients, thus the transmission lines status and the affected substation can be shown in terms of voltage dip magnitudes and durations. Therefore, the possible influence of transmission line trippings to the substations can be visualized to system operators. Big Data platform employes a 3D simulation display system, using data from the management layer as well as the model output directly from the computing engine, including 3D models of power line and electric equipment, 3D building models, geospatial data and power attribute data. Geospatial data as 3D virtual environment, can show geographic objects (e.g. roads, bridges, rivers) around electricity network. The generated 3D virtual environment with power transmission line situation is given in Fig. 5.

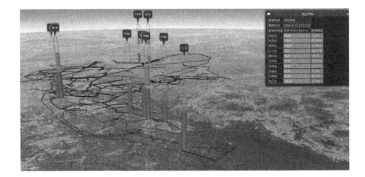

Fig. 5. 3D display of voltage dips and breakdown transmission lines with geographic information (Color figure online)

In Fig. 5, the green line represents the normal operational transmission line, while the red lines are with the appearance of the line trips. In order to show the voltage transients status, a cylinder with blue color shows the voltage dip magnitude, and the pink cylinder is the duration, and the name of the affected transmission lines are shown in the floating red tags above the cylinders. Therefore, the affected area can be directly visualized through the 3D virtual environment, and the dynamic change of the power grid operational status is easier to control for the system operators. If any transient happened, actions can be taken in time to prevent any enlarger of the accident.

4.3 Real-Time Dynamic Evaluation of Operation Status of Wind Farm Units

In recent years, China's research and investment in wind power has entered a new stage of development. However, the damages of wind turbines caused by faults and abnormal operation conditions have also increased year by year. Moreover, wind power plants are generally located in remote and harsh areas, which cause it is difficult to carry out maintenance on the unit in time, which increases the capital and labor costs. Therefore, the big data platform system proposed in this paper can process massive wind turbine data quickly and efficiently, helping detect and prevent potential faults and optimize the operation schemes in real time.

Most of the existing wind farm equipment management and operation optimization assessment methods mainly use power curve, temperature curve, etc., and comparing the operating state of the unit through observation. The method is simple, but the processing efficiency of massive data is low, and it is easy to ignore potential hidden dangers, resulting in inaccurate, slow and inefficient detection results. In this case, a dynamic principal component analysis method is run on Big Data platform. A time-lag data matrix of time series variables is constructed to reveal the dynamic relationship. The original data is simplified by extracting the principal components of the time-lag data matrix. The real-time operating state of the wind turbine is determined based on the extracted component information. Along with the high-efficiency data processing of the big data platform, the operation status of the wind turbine can be detected timely and potential faults and hazard can also be warned. Therefore, the operators can carry out maintenance and repairing actions as soon as possible.

5 Conclusions

This paper studies the key technologies of power big data fusion management and analysis, considering the operation and structural characteristics of urban power grids, and builds a big data platform with efficient analysis engine and 3D panorama display system. The whole system structure, key technologies and specific application scenarios of the power grid big data platform were introduced in this paper. With the help of big data platform, the internal relationships between the grid database and the knowledge base can be revealed, fast and effective decision-making assistance can be provided, thus the waste of power equipment resources can be reduced.

At present, the application research of big data technology in the field of power systems is still in the exploration stage. There are still many problems that need to be solved urgently. It is necessary to continue to carry out more in-depth and extensive research work to promote the healthy, rapid and harmonious development of smart grid big data technology in the future.

Acknowledgments. This research is financially supported by the National High Technology Research and Development Program (2015AA050201), China NSFC under grants 51607177, China Postdoctoral Science Foundation (2018M631005), Natural Science Foundation of Guangdong Province under grants 2018A030310671.

References

1. Zhou, K., Fu, C., Yang, S.: Big data driven smart energy management: from big data to big insights. Renew. Sustain. Energy Rev. **56**, 215–225 (2016)
2. Munshi, A.A., Yasser, A.R.M.: Big data framework for analytics in smart grids. Electr. Power Syst. Res. **151**, 369–380 (2017)
3. Chen, C.P., Zhang, C.Y.: Data-intensive applications, challenges, techniques and technologies: a survey on big data. Inf. Sci. **275**, 314–347 (2014)
4. Miao, X., Zhang, D.: The opportunity and challenge of big data's application in distribution grids. In: 2014 China International Conference on Electricity Distribution (CICED), pp. 962–964. IEEE (2014)
5. Peng, X., Deng, D., Cheng, S., Wen, J., Li, Z., Niu, L.: Key technologies of electric power big data and its application prospects in smart grid. Proc. CSEE **35**(3), 503–511 (2015)
6. Tu, C., He, X., Shuai, Z., Jiang, F.: Big data issues in smart grid-a review. Renew. Sustain. Energy Rev. **79**, 1099–1107 (2017)
7. Korres, G.N., Manousakis, N.M.: State estimation and bad data processing for systems including PMU and scada measurements. Electr. Power Syst. Res. **81**(7), 1514–1524 (2011)
8. Xue, Y., Lai, Y.: Integration of macro energy thinking and big data thinking part one big data and power big data. Autom. Electr. Power Syst. **40**, 1–8 (2016)
9. Wang Liwei, C.X., Ting, C.: Real-time monitoring development scheme for electric transmission and transformation equipment status estimation in Fujian power gird. Power Grid Environ. **6**, 56–58 (2016)
10. Hao, L., Jiang, C., Jia, L., Xun, G.: Research on wide-area damping control system of Mengxi power delivery passageway in North China power grid. Power Syst. Technol. (2017)
11. Xie Yishan, Y.Q., Chenghui, L.: Research of unified condition monitoring information model in data platform of power transmission equipment remote monitoring and diagnosis. Power Syst. Prot. Control **11**, 86–91 (2014)
12. Chen Hongyun, C.R., Yujun, G.: Development and realization of power grid security early warning and decision support system. Guangxi Electr. Power **34**, 10–12 (2011)
13. Cao Min, W.Y., Feng, Y.: Research and development of Yunnan power grid supervision and data analysis centre. Yunnan Electr. Power **41**, 194–196 (2013)
14. Christopher English: Alstom brings to market digital substation 2.0 featuring smart technologies (2014). https://www.alstom.com/press-releases-news/2014/8/
15. General Electric's: GE automation & protection (2018). http://www.gegridsolutions.com/MD.htm

Taxi Service Pricing Based on Online Machine Learning

Tatiana Avdeenko$^{(\boxtimes)}$ 🆔 and Oleg Khateev

Novosibirsk State Technical University, 20 K. Marksa Avenue,
630073 Novosibirsk, Russia
tavdeenko@mail.ru

Abstract. In present paper we propose an approach to the dynamic taxi service pricing based on data mining from the large database of cases accumulated in the company. We believe that historical data on decisions made by experienced managers in the past contain valuable information on the most subtle parameters that influence decision making, but cannot be taken into account in the direct optimization formulations proposed in the literature. The analysis of existing clustering methods is carried out and the Enhanced Self-Organizing Incremental Neural Network (ESOINN) is chosen which is the most suitable for online learning with a previously unknown number of clusters. A comparative analysis of the accuracy of the proposed method has been carried out. It is discovered that ESOINN shows the best accuracy of prediction in a realistic situation with acceptable waiting time. We also presented a web application implementing the proposed approach.

Keywords: Taxi service · Dynamic pricing · Cluster analysis · ESOINN · Decision making · Data mining

1 Introduction

Society continues to move toward automating many areas of daily life. Recently, taxi services have become one of the industries in which automation has taken the leading position (see, for example, Yandex Taxi [1]). One of the methods that successful taxi services must have is a dynamic tariff (dynamic pricing). By dynamic tariff we mean a special tool that is activated automatically when the demand for the company's services exceeds the supply. The application of dynamic pricing means that, despite the increased demand, the company will be able to provide the car within a few minutes. At times of high demand, the dynamic fare calculation algorithm raises the price (for example, when it is very cold), and thus encourages drivers to take the order. The price increase is carried out in such a way that it is possible to cut off the orders of customers who are not so important to leave quickly and who can wait for some time. Thus, with the help of dynamic pricing, supply and demand are automatically balanced, which leads to the formation of a "fair" price for the order.

Despite the urgency of this problem, there have been surprisingly a small number of works devoted to research in this area. The papers [2, 3] presents a review of the different models developed for the taxicab problem. The presented models are grouped

© Springer Nature Singapore Pte Ltd. 2019
Y. Tan and Y. Shi (Eds.): DMBD 2019, CCIS 1071, pp. 289–299, 2019.
https://doi.org/10.1007/978-981-32-9563-6_30

in two categories, aggregated and equilibrium models. A multiperiod dynamic model of taxi services was proposed and improved in [4, 5]. Derived from this multi-period model the taxi system optimization was proposed based on the bilevel program (ASM and FLORA algorithms proposed in [6, 7] correspondingly). The paper [8] examines the effects of nonlinear fare structures in taxi markets using an extended taxi model with an explicit consideration of perceived profitability.

The above works use mainly the optimization formulation of the problem to calculate the dynamic price, and therefore have limitations for use under real conditions. First, the optimization models cannot be applied in real situation due to the issues of computational complexity and the limited time for making decision. Second, such models do not take into account all existing parameters presented in life situations, for example, the behavior of drivers is not taken into account, psychological and other aspects cannot be described in terms of a deterministic formulation. To overcome the problem of computational complexity, it is proposed to use simulation models [9], but they are also not very suitable for an online solution. Markov Decision Process proposed in [10] makes it possible to introduce uncertainty into the formulation of the problem, but is still quite far from being applicable under real conditions.

Unlike the methods described above we propose an approach to the dynamic pricing based on data mining from a large database of cases accumulated in the company. It is assumed that rational decisions on the tariffs have been made in the past, so the use of historical data is a determining factor for decision-making. We believe that historical data on decisions made by competent managers in the past contain valuable information on the subtlest parameters that influence decision making, but cannot be taken into account in the direct optimization formulations.

In order to make a decision on the tariff value, it is necessary to divide the whole set of orders into clusters, i.e. to solve a problem, which in machine learning and data mining is called clustering problem, or unsupervised learning. After that, the task of determining the dynamic tariff can be built on the basis of case-based reasoning, when the closest class of orders is searched for the current order, and then a decision on the tariff value is made on the basis of the retrieved cases.

The paper is organized as follows. After this short introduction we consider problem statement for dynamic tariff determination for taxi service. In Sect. 3 we analyze clustering methods and choose the most appropriate ESOINN method for the data clusterization. In Sect. 4 we describe the features used for data mining, and also give comparison of ESOINN with Random Forest. In Sect. 5 we consider web application implementing the proposed method. Conclusion is drawn in Sect. 6.

2 Problem Statement

When the company chooses an order price, there is a subtle nuance associated with its optimal value. On the one hand, the lower is the price of the order, the more likely is that the customer will choose this option. On the other hand, the low price repels the drivers. The drivers can refuse the cheap order. They think it is more attractive idea to wait for another order with more favorable price than to go several kilometers to the order in which it is necessary to transfer the client through a half of a kilometer.

Suppose that at the moment there is a great demand for the taxi services, for example, the football championship match has just ended, and there is a need to take all the fans home. The number of drivers available is not able to take out all people. In this case, an increase in the price of the order can solve the following problems. First, it will help to attract new drivers for the shift. Second, it will help to weed out the orders that are not urgent, that is, the client is not ready to go at an inflated price, and prefers to wait until the demand falls.

The total cost of the order can be calculated according to the formula:

$$TotalCost = SupplyCost + Price * Mileage + Mark\text{-}Up,$$

where *SupplyCost* is the cost of supplying the car, *Price* is the cost per kilometer, *Mileage* is mileage between the delivery and the destination addresses, *Mark-Up* is extra charge determining dynamic tariff. To make a decision about the mark-ups, it is necessary to study the demand for the company's services at the moment. The demand can be expressed, for example, as the number of orders to which a driver has not been assigned (the more "unassigned" orders are accumulated, the higher the demand is). The mark-up list could contain the following parameters determining the increased demand: low temperature, traffic jam, public holiday, etc.

The process of determining the list and values of the mark-ups made by the experienced manager is very responsible, and poorly formalized. Besides, the process is quite long and tedious since it is necessary, having previously studied all the aspects that can affect the demand, to monitor the situation throughout the time, that is, to monitor the weather forecasts, the number of orders that have not been assigned to the driver, etc. Also, it is worth considering that sometimes the demand is difficult to predict, and if the top manager does not see the positive dynamics of the demand in time, the order confirmation time can increase several times. Or, on the contrary, when the demand fell (or should fall soon), it is important not to miss the moment, and time to turn off the mark-up. Otherwise, the customers will be offered an unreasonably high price, which may lead to an outflow of customers from the company.

Thus, the influence of the human factor on the decision-making to determine the mark-up is very high, which is a vulnerable point of the company. In this paper, we propose the use of artificial intelligence methods to develop a method for determining the dynamic tariff. Since the company has accumulated enough information about decision-making in the past, it is advisable to use machine learning methods for its analysis and development of similar solutions in the present. Thus, the development of a dynamic tariff is proposed to be based on *Case-Based Reasoning* (CBR).

Case-Based Reasoning (CBR) solves new problems by adapting previously successful solutions to similar problems. The works of Shank [11, 12] are widely held to be the origins of CBR. In these papers it has been proposed to generalize knowledge about previous situations and store them in the form of scripts that can be used to make decisions in similar situations. The inference process in case-based approach involves four basic steps, forming the so-called case-based reasoning cycle (CBR-cycle) composed of 4R (Retrieve, Reuse, Revise, Retain) [13].

Data Mining is the process of automatic extraction of formal knowledge from cases. Through Data Mining we can solve the first two stages of CBR – retrieving and reusing cases of previous decision-making about dynamic tariff. As a criterion for the effectiveness of the proposed method, the criterion of the number of failures due to the order confirmation time is proposed. It is important to emphasize the cause of failures, which we strive to reduce - the waiting time, not the cost of the transfer. The reduction of waiting time is a major goal of the development of a dynamic tariff.

Prediction of order confirmation time can be carried out by two methods of Data Mining: cluster analysis and regression analysis. In case of cluster analysis, we get a sample of similar orders and calculate the confirmation time based on this sample, for example, you can take the average confirmation time in the cluster. Regression analysis allows directly building a forecast model for the continuous confirmation time. For the dynamic tariff computation, we have chosen cluster analysis.

3 Choice of Clustering Method

The criteria for choosing appropriate clustering algorithm to calculate dynamic tariff are the following: possibility of online training; division into clusters without prior knowledge of the input data; division into clusters with low-density overlap, discovery of structure of clusters contaminated with noise.

The K-means method splits a set of vector space elements into a known number of clusters k. The algorithm seeks to minimize the standard deviation at the points of each cluster [14, 15]. The advantages of the method are the simplicity of implementation, the existence of a large number of ready-made solutions. The main disadvantage is that we need to know a priori the number of clusters. The method is also very sensitive to the initial selection of cluster centers.

Kohonen neural network is a method of projecting a multidimensional space into a space with a lower dimension with a predetermined structure [16]. It is a single-layer network built of neurons. Each neuron is connected to all components of the m-dimensional input vector. The advantages of the Kohonen network consists in the ease of implementation, as well as the possibility of obtaining a good solution if the initial selection of cluster centers fails. Disadvantages of the method are uncontrolled learning rate as well as the need to specify a priori the number of clusters.

Thus, the K-means method and the Kohonen neural network are not suitable for the problem of online training required in the calculation of the dynamic tariff. A fundamental problem for dynamic tariff computation is how the system can adapt to new information without damaging or destroying what is already known.

We propose two neural networks in order to overcome the problems that do not allow for online training. SOINN (Self-Organizing Incremental Neural Network) first proposed in [17] is a two-layer neural network. The first layer studies the density distribution of the input data and uses nodes and edges to represent that distribution. The second layer divides the clusters by defining low-density zones of the input data and uses fewer nodes than the first layer to represent the topological structure of the input data. When the training of the second layer is finished, SOINN reports the number of clusters and provides the node prototypes for each one. So it uses the same learning algorithm for both layers. When an input vector is passed to SOINN, it finds the nearest node (winner) and the second closest node (second winner) of the input vector. Subsequently it decides if the input vector belongs to the winner cluster or to the second winner using the similarity threshold criterion. The first layer of SOINN adaptively updates the similarity threshold of each node because the distribution of input data is unknown.

The benefits of SOINN allowing to use it to compute the dynamic tariff are as follows: SOINN does not need a priori information about the number of clusters; it learns and adapts in real-time (online clustering). The disadvantages of SOINN are the following. First, it is difficult to choose when you need to finish the training of the first layer and start training the second. Second, if the learning results of the first layer have changed, the results of the second layer will be destroyed, that is, the second layer needs to be retrained. So, the second layer is not suitable for online learning. Third, the network has a large number of parameters that need to be determined manually. Therefore, the intra-class insertion in the second layer may not be optimal.

The ESOINN (Enhanced Self-Organizing Incremental Neural Network) algorithm continues to develop the idea of SOINN that clusters form the regions of high probability density, and the learning algorithm should identify a graph that could accurately describe such zones [18]. Most of ESOINN functionality is identical to SOINN, however, it attempts to overcome the shortcomings of SOINN. In ESOINN, a second layer was removed and a set of techniques was developed that would allow one layer to obtain better results than two layers. Thus, ESOINN has the following main differences from SOINN. First, there is only one layer, so there are fewer network settings. ESOINN adds a condition to create a link between the nodes that have won, i.e. in the SOINN connection was added without condition (necessarily). Third, an additional parameters are introduced - the number of trainings, in which nodes are divided into different clusters (links that do not carry any information are removed and noise nodes are also removed).

Thus, to solve the problem of dynamic tariff calculation, the most suitable clustering method is the ESOINN network. In the next section, we will look at the main features that have been used to train the network, as well as the results of testing the method and comparing it with the Random Forest method.

4   Results

4.1   Selection of Features for Cluster Analysis with ESOINN

The idea of case-based reasoning is that a current decision is made based on the history of decision-making in the past. It is assumed that the top manager made thoughtful and informed decisions about the value of the mark-up, spending a lot of intellectual effort on their formulation. Therefore, we could hope that if current decision is made in accordance with the cases stored in the knowledge base, it will be at least as good as the decisions taken in the past.

The knowledge base contains order stories described by a large number of features. The more features you choose to train the neural network, the more complex its structure becomes, and the more time it takes to train it. In case of ESOINN, the number of nodes in the network increase with the growth of the number of features.

Therefore, it is important to identify the most significant features for neural network training. It is reasonable to determine the parameters of the order, which probably will most strongly affect the time of the order confirmation. Consider such most significant features.

Time During the Day in Which the Order is Made
In Fig. 1 it is shown the daily graph displaying the time of the order confirmation during the working day. As you can see, the demand is observed in the morning and evening rush hour, when people go to work and return from the work. Also, there is a great demand closer to 11 pm, which is reflected in a corresponding increase in the time of the order confirmation. On the weekends, the latter effect is even more noticeable, because people relax, visit restaurants, usually it is delayed until late at night, when public transport is no longer functioning, so they have to resort to taxi services.

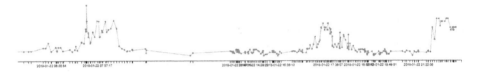

Fig. 1. Daily graph of the order confirmation time.

Price of the Order
Obviously, the higher is the cost of the order, the more attractive it is for the driver. On the other hand, the lower is the cost, the less attractive the order is for the driver. Therefore, the confirmation time could be increased significantly when the price is lower. In this case, it can be necessary to calculate the mark-up for each individual order, rather than for the entire neighborhood. Under the neighborhood we mean the area of the city, for which all outgoing orders have the same mark-up.

Weather Conditions
At low temperatures or in the rain, there is also a higher demand for the company's services, because it is uncomfortable to get to the public transport stop under these conditions. Since these circumstances occur from time to time, there is a reason to use taxi services.

The Number of Drivers in the Neighborhood
With an increasing number of orders (i.e. demand), the company providing taxi services must deliver a supply, that is, drivers who are ready to fulfill all orders. The fewer drivers are in a certain area, the higher the probability is that the order confirmation time will increase. Figure 2 shows the number of drivers in a certain neighborhood with purple dots, and the order confirmation time with pink dots. As you can see, the points with the maximum confirmation time correspond to the points with the minimum number of drivers.

Fig. 2. Dependence of order confirmation time on the number of drivers.

The Place from Which the Order Was Received

The place from which the order was received has a significant impact on the confirmation time. First, in the areas with a few orders there is small volume of data, which may lead to significant errors in predicting the dynamic tariff on the basis of case-based reasoning. Secondly, such areas are usually remote from the city and have a small number of drivers that also affect the order confirmation time.

4.2 Comparison of ESOINN and Random Forest

Random Forest was chosen for investigation and comparison of the ESOINN prediction accuracy. Strictly speaking, it is a method of classification and regression (not clustering), and it cannot be directly used for the dynamic pricing. However, using one of the features as an assessment of the quality of the division into classes (the time of order confirmation), we can compare the methods by prediction accuracy.

Comparison of ESOINN with Random Forest was made on the sample consisted of 2826 orders selected from the case base (Table 1). Half of the sample values were used for training, the other half for testing. For comparison, we take the model configurations that showed the best result. We compare them by the confirmation time. In the first experiment Random Forest showed better results for the training time while ESOINN showed slightly better classification accuracy. Note that all orders with both acceptable and excessively high confirmation time were included in the initial sample.

Table 1. Comparison of ESOINN and Random Forest

	Training time	Mean absolute error (MAE)	Mean square error (MSE)
ESOINN	0.25 s	2 min 00 s	4 min 19 s
Random Forest	0.13 s	2 min 45 s	4 min 19 s

In the second experiment (see Tables 2 and 3) we divided the sample by the time of confirmation of the order on the cases with acceptable confirmation time (less than 4 min), and critical (more than 4 min), after which customers tend to refuse the order. Note that it is 4 min that is an acceptable indicator of delay. In the first case of reasonable choice of the confirmation time ESOINN shows obviously better results (Table 2). In the second case of excessively long confirmation times Random Forest is better (Table 3).

Table 2. Comparison of ESOINN and Random Forest (confirmation time less than 4 min)

	MAE	MSE
ESOINN	51 s	1 min 18 s
Random Forest	2 min 12 s	3 min 02 s

Table 3. Comparison of ESOINN and Random Forest (confirmation time more than 4 min)

	MAE	MSE
ESOINN	7 min 11 s	9 min 27 s
Random Forest	5 min 8 s	7 min 47 s

To explain these results, consider the histograms of the value distribution in the two subsamples shown in Fig. 3. Remember that the algorithm based on the ESOINN assigns orders for forecasting to one of the clusters of previously executed orders, and on the basis of these orders finds the median confirmation time. It turns out that the denser are the confirmation times in the cluster, the better is the ESOINN forecast. The median time has been chosen rather than, for example, the average, to avoid emissions that would affect the end result.

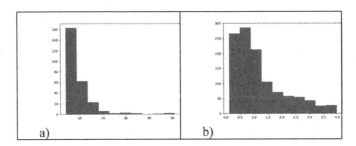

Fig. 3. Distribution of values in the sample of orders with waiting time: (a) more than 4 min; (b) less than 4 min.

As a result, we see that ESOINN is superior in cases where the order follows the behavior of the majority of previously implemented orders. To improve prediction, you should use a more reasonable metric instead of the median (for example, use a random forest inside the cluster). However, it may occur that there are few orders in the cluster, so it is impractical to build a random forest based on a small sample.

5   Web Application Implementing Taxi Service

To implement a service for the calculation of dynamic tariff based on ESOINN we developed a web application. The whole territory of the city (Novosibirsk, Russia) is divided into hexagons. Thus, in order not to include additional parameters (latitude and longitude) to ESOINN, we will train our neural network separately for each hexagon. Thanks to this solution, orders from the hexagon, which is located in a remote area (for example, in the Academic town), will not participate in the training of the neural network of the hexagon, which is located in the city center.

The mark-up is calculated not for each individual order, but for the entire hexagon, that is, each individual order will contribute to the computation of the total mark-up. This decision was made in order to take into account the speed of the appearance of orders in the hexagon. If in previous version of service each order was not connected with each other, now they are connected with the hexagon. The more orders for a certain time come to this hexagon, the higher is the demand in this hexagon. The confirmation time is directly involved in the training and is also necessary for further forecasting. Therefore, it is convenient to store this data together with the modified weights of the neural network.

Consider the demand of company's service on the 9th of May after the fireworks presented in Fig. 4. The map shows the hexagons, which have a mark-up. The brighter is the color of the hexagon, the higher is the mark-up in comparison with other hexagons. By clicking on the hexagon, on the right we can find the mark-up in it. In addition, we can refer to the more detailed information for a given hexagon: the number of orders that not yet confirmed; the number of available drivers; the computed hexagon mark-up; the predicted time of order confirmation; the median confirmation time.

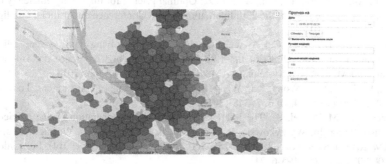

Fig. 4. Demand on the 9th of May after the fireworks

6   Conclusion

Thus, we have developed a dynamic pricing model and a web application based on the ESOINN clustering. At the initial stage (before the optimization of the program), the time of network training on 130 thousand orders was about 30 min, after the optimization of the program code – no more than two minutes. Finally two parameters were chosen as features for the neural network: the number of orders and the number of drivers in the neighborhood. Other parameters were excluded, because the number of orders depends on them, and the order confirmation time depends on the number of orders. With the introduction of the proposed algorithm, the volume of intellectual work of the company's top-manager has significantly decreased, the need for the demand control has disappeared. The percentage of failures due to confirmation time delay after the application of machine learning methods appeared to be not worse than manual computation of the mark-up. Further improvement is seen in the use of not only the past experience of the expert on the basis of case-based reasoning, but also, perhaps, in the integration of case-based reasoning with heuristic approaches to solving optimization problems.

Acknowledgement. The research is supported by Russian Ministry of Education and Science, according to the research project No. 2.2327.2017/4.6.

References

1. Yandex Taxi Homepage. https://taxi.yandex.ru/#index. Accessed 26 Feb 2019
2. Salanova, J.M., Estrada, M., Aifadopoulou, G., Mitsakis, E.: A review of the modeling of taxi services. Procedia Soc. Behav. Sci. **20**, 150–161 (2011)
3. Kim, Y.-J., Hwang, H.: Incremental discount policy for taxi fare with price-sensitive demand. Int. J. Prod. Econ. **112**(2), 895–902 (2008)
4. Yang, H., Wong, S.C., Wong, K.: Demand–supply equilibrium of taxi services in a network under competition and regulation. Transp. Res. Part B: Methodol. **36**(9), 799–819 (2002)
5. Yang, H., Ye, M., Tang, W.H.-C., Wong, S.C.: A multiperiod dynamic model of taxi services with endogenous service intensity. Oper. Res. **53**(3), 501–515 (2005)
6. Gan, J., An, B., Wang, H., Sun, X., Shi, Z.: Optimal pricing for improving efficiency of taxi systems. In: Proceedings of the 23th International Joint Conference on Artificial Intelligence, pp. 2811–2818 (2013)
7. Gan, J., An, B., Mia, C.: A scalable algorithm for solving taxi system efficiency optimization. In: Proceedings of the 13th International Joint Conference on Autonomous Agents and Multi-Agent Systems (AAMAS 2014), pp. 1465–1466 (2014)
8. Yang, H., Fung, C.S., Wong, K.I., Wang, S.: Nonlinear pricing of taxi services. Transp. Res. Part A Policy Pract. **44**(5), 337–348 (2010)
9. Maciejewski, M., Nagel, K.: The influence of multi-agent cooperation on the efficiency of taxi dispatching. In: Wyrzykowski, R., Dongarra, J., Karczewski, K., Waśniewski, J. (eds.) PPAM 2013. LNCS, vol. 8385, pp. 751–760. Springer, Heidelberg (2014). https://doi.org/10.1007/978-3-642-55195-6_71
10. Zeng, C., Oren, N.: Dynamic taxi pricing. Front. Artif. Intell. Appl. **263**, 1135–1136 (2014)
11. Schank, R.C., Abelson, R.P.: Scripts, Plans, Goals and Understanding. Erlbau (1977)
12. Schank, R.: Dynamic Memory: A Theory of Reminding and Learning in Computers and People. Cambridge University Press, Cambridge (1982)
13. Watson, I., Marir, F.: Case-based reasoning: a review. Knowl. Eng. Rev. **9**(4), 327–354 (1994)

14. MacQueen, J.: Some methods for classification and analysis of multivariate observations. In: Proceedings of the 5th Berkeley Symposium on Mathematical Statistics and Probability, pp. 281–297 (1967)
15. Lloyd, S.: Least square quantization in PCM's. IEEE Trans. Inf. Theory **28**, 129–137 (1982)
16. Gersho, A., Gray, R.M.: Vector Quantization and Signal Compression, 732 p. Springer, Heidelberg (1992)
17. Furao, S., Hasegawa, O.: An incremental network for on-line unsupervised classification and topology learning. Neural Netw. **19**, 90–106 (2006)
18. Furao, S., Ogurab, T., Hasegawab, O.: An enhanced self-organizing incremental neural network for online unsupervised learning. Neural Netw. **20**, 893–903 (2007)

PRESENT Cipher Implemented
on an ARM-Based System on Chip

Fernando Martínez Santa$^{(\boxtimes)}$ (ID), Edwar Jacinto (ID), and Holman Montiel (ID)

Universidad Distrital Francisco José de Caldas, Bogotá, Colombia
{fmartinezs,ejacintog,hmontiela}@udistrital.edu.co

Abstract. Ultra lightweight cryptography is a research area very active nowadays, mainly because it is widely used in application fields such as IoT (Internet of Things), intelligent sensors networks and smart grids, specially where embedded low-energy systems are involved. In this paper, the PRESENT ultra-lightweight cryptographic algorithm was implemented on a PSoC (Programmable System on Chip), it was totally performed in software by means of using C programming language, and using modular functions. The percentage of use of each implemented C function was analyzed as well as their execution time in order to define which of them are able to be implemented using the digital programmable blocks of the used SoC, and thus designing a new prototype optimized in space and in time.

Keywords: PRESENT · Ultra-lightweight cipher · Embedded system · System on Chip

1 Introduction

Recently, the ultra lightweight cryptography has been a fruitful researching area, mainly due to the accelerated development of application fields like IoT (Internet of Things) [11,14,16], smart grid [8], and the intelligent sensors networks [1,13], which require to ensure the security of their information as well as keep the energy consumption as low as possible [12,20]. Embedded systems are used for implementing applications in IoT and intelligent sensor networks, these systems use to have strict energy consumption requirements because most of times these devices use wireless communication and limited batteries. For that reason, implementing regular cryptographic algorithm is not reliable due to most of them spend a lot of execution time from the processor, which at the same time spends a lot of energy. Likewise, the kind of processors and the amount of memory used by the embedded system are usually simpler and smaller than the computers where generally the regular encryption algorithms are executed, what implies in most of cases that the implemented algorithm spends longer. Thinking of those kinds of implementations, some lightweight cryptographic algorithm have been developed, such as KLEIN [10], CLEFIA [15], SIMON [4], SPECK [4], among

© Springer Nature Singapore Pte Ltd. 2019
Y. Tan and Y. Shi (Eds.): DMBD 2019, CCIS 1071, pp. 300–306, 2019.
https://doi.org/10.1007/978-981-32-9563-6_31

others. Later, other named ultra-lightweight algorithms [13] were designed reducing even more the computing requirements of the system, for example Piccolo [18] and the one that is the aim of this paper PRESENT [5]. Lightweight ans ultra-lightweight cryptographic algorithms have been implemented in different kinds of embedded systems such as programmable digital devices (CPLDs and FPGAs) [2,9,15], microcontrollers such as ARM [6,7,19] and mixed systems (processor plus combinational logic circuits) using mainly SoC (System on Chip) [3,17]. The use of programmable logic devices and mixed devices that use them, is due to some of the parts of these cryptographic algorithms can be performed better by digital combinational circuits than a processor because they do not add any delay time.

PRESENT is an ultra lightweight block cipher, that has been implemented in both microcontrollers and programmable logic devices, so the purpose of this work is implementing in a PSoC® (Programmable System on Chip), due to it is a low cost integrated circuit and has not only a powerful processor but programmable logic too. This paper describes the implementation of the PRESENT algorithm totally in software (using only the core processor) and the time a features analysis in order to define which blocks are reliable to be implemented in hardware (using the programmable logic), for reducing the total execution time and therefore the power consumption.

2   Methodology

The implementation of the PRESENT algorithm is performed using a Cypress PSoC®. In this section, first the used PSoC® is described, then the algorithm itself and finally how it was implemented on the chip.

2.1   Programmable System on Chip

This implementation uses the PSoC® microcontroller of the 5LP family, specifically the CY8C5888LTI-LP097 which is composed by an 32-bit processor ARM® Cortex®-M3 and some different programmable digital and analog blocks. This chip has 256 KB of FLASH program memory and 64 KB of SRAM as data memory. For this implementation the CY8CKIT-059 board that includes itself the needed programmer.

Usually PSoC® microcontrollers are programmed in C language in spite of the compiler support C++, this is done with the purpose of optimizing the amount of memory occupied by the written algorithms. That is the reason this implementation of PRESENT was written and optimized in C language.

2.2   PRESENT Cipher

PRESENT is an ultra-lightweight cryptographic algorithm, this is also consider as a Block Cipher. PRESENT uses 64 bits for data and 80 bits for the key, and combine them in 32 steps named rounds until obtaining the encrypted data [5].

In each round the 64 most significant bits of the key are operated with the data by means of using a bit-by-bit XOR operation, the resultant data is passed through two operation blocks: sBoxLayer and pLayer, which update the data obtained. Also, there is a key-update block that changes the key in each round. This process is shown in the diagram of the Fig. 1.

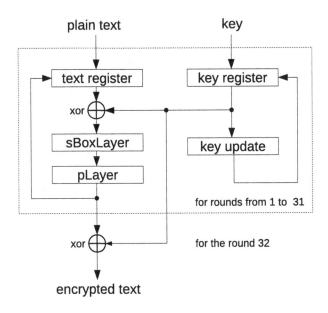

Fig. 1. General block working diagram of the PRESENT algorithm.

The sBoxLayer changes the input data in groups of 4 bits (nibbles), replacing each of the 16 possible combinations with the content shown in the Table 1, where x is each nibble and $S[x]$ is the replacement value. On the other hand, the pLayer exchanges or permutes the input bits one by one in the order shown in Table 2, where i is the bit position of the input data and $P(i)$ the output position.

In order to update the key, its register is rotated to the left by 61 bit positions, the left-most four bits are passed through the sBoxLayer, and the bits from the number 19 to the 15 are operated by means of a XOR with the counter of the current round (5 bits) [5].

After each round, both the processed data and the key are updated and the process is repeated for 30 more times, and in the last round (the number 32), the data is only exclusive-ored with the key without passing through the sBoxLayer and the pLayer. Finally, the data obtained after the round 32, is the encrypted data.

The decryption process starts from the encrypted data and the key obtained after the round 32 of the encryption process. This works similar than the encryption process but it needs the inverse functional blocks. Due to, as the cipher key

Table 1. Replacement nibble content performed by the sBoxLayer [5].

x	0	1	2	3	4	5	6	7	8	9	A	B	C	D	E	F
$S[x]$	C	5	6	B	9	0	A	D	3	E	F	8	4	7	1	2

Table 2. Permutation bit order performed by the pLayer [5].

i	0	1	2	3	4	5	6	7	8	9	10	11	12	13	14	15
$P(i)$	0	16	32	48	1	17	33	49	2	18	34	50	3	19	35	51
i	16	17	18	19	20	21	22	23	24	25	26	27	28	29	30	31
$P(i)$	4	20	36	52	5	21	37	53	6	22	38	54	7	23	39	55
i	32	33	34	35	36	37	38	39	40	41	42	43	44	45	46	47
$P(i)$	8	24	40	56	9	25	41	57	10	26	42	58	11	27	43	59
i	48	49	50	51	52	53	54	55	56	57	58	59	60	61	62	63
$P(i)$	12	28	44	60	13	29	45	61	14	30	46	62	15	31	47	63

as the decipher key is the same, the decipher has to run the 32 rounds of the key updating of the cipher before to start the decryption process.

2.3 Implementation

The PRESENT algorithm was implemented totally in software by means of using C language, using bit-by-bit operations and binary mask in order to maintain the code as close to the low level programming as possible and thus keep low the memory occupied space as well as the execution time.

The implementation includes a USB-UART interface to communicate with a computer and try the well working of it (See Fig. 2), sending plain text form the computer and receiving from the embedded system the encrypted text.

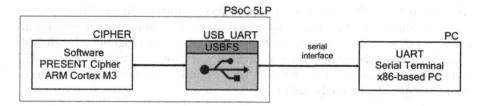

Fig. 2. Block diagram of the implementation of the PRESENT algorithm in the PSoC®

The PRESENT code was described as a modular system composed for the C functions: *data_xor_key()*, *sBoxLayer()*, *pLayer()*, *key_update()*, other some sub-functions and the main routine in charge of doing the management of the input and output data and the round repetitions. The data and the key registers were implemented by means of 32-bit integer arrays, and the *sBox* table by means of a constant fixed-size byte array. The four previously named functions implement the main functional blocks shown in the working diagram of the Fig. 1. The *data_xor_key()* function operates the data and the most significant bits of the key through and bit-wise exclusive OR, *sBoxLayer()* separates the data in nibbles and apply the replacement table *sBox*, *key_update()* updates the key in each round as described in the Sect. 2.2, and finally *pLayer()* permutes each of the 64 bits of the data, this function is a large code in C language due to the ARM® processor has no instructions to operate individual bits, so it was necessary to do a masking operation over the data by each bit to modify.

3 Conclusion

The final implemented PRESENT algorithm works perfectly ciphering and deciphering data over some different test done sending plain text from a computer and receiving encrypted text from the embedded system. The algorithm occupied approximately 2.13 kB of program memory of the total 256 kB, which represents only the 0.83% of the FLASH memory. It also uses only 64 bytes of RAM of the 64 kB available, in this sense the obtained values show that this algorithm is really an ultra-lightweight cipher. On the other hand, the total execution time after applying the encrypting process to a data of 64 bits is approximately 7 ms, when the processor works to 24 Mhz. This time, depending on the application could be very high, for instance, for a intelligent sensor connected to a wireless network with strict dropout requirements or any other real time application. For that reason, it is very important to look for reducing its execution time, then a time analysis was done for each created function for the PRESENT algorithm. Those results and as well as the program FLASH memory occupation for each function is shown in Fig. 3.

It is evident that *sBoxLayer()* and *pLayer()* functions are the ones that not only occupy the higher amount of memory (10.5 and 70.4% respectively) but last longer (13.8 and 80.5% respectively). In addition, as *sBoxLayer()* as *pLayer()* are able to be easily implemented on the programmable digital blocks of the used SoC, due to they are replacement and permuting simple operations. Then, if it is possible to implement these two functions on the programmable digital blocks instead of the processor the code occupancy could be reduced up to 80% and the most important, the execution time could be reduced up to 94%, which would be equivalent to only 420 μs.

4 Future Work

Develop a hybrid software-hardware implementation of the PRESENT cryptographic algorithm on the same PSoC® device, where the functions that use

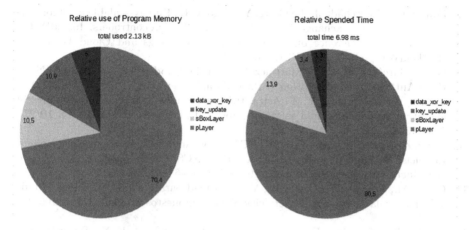

Fig. 3. Percentage of memory use of each function (left) and the execution time percentage (right)

more amount of memory and the ones that last longer are implemented using the programmable logic blocks and the rest as C language functions using the main processor.

Acknowledgments. This work was supported by Universidad Distrital Francisco José de Caldas, specifically by the Technology Faculty. The views expressed in this paper are not necessarily endorsed by Universidad Distrital. The authors thank ARMOS research group for the simulations and tests done.

References

1. Alizadeh, M., Salleh, M., Zamani, M., Shayan, J., Karamizadeh, S.: Security and performance evaluation of lightweight cryptographic algorithms in RFID. Kos Island, Greece (2012)
2. Arora, N., Gigras, Y.: FPGA implementation of low power and high speed hummingbird cryptographic algorithm. Int. J. Comput. Appl. **92**(16) (2014)
3. Bai, X., Jiang, L., Dai, Q., Yang, J., Tan, J.: Acceleration of RSA processes based on hybrid ARM-FPGA cluster. In: 2017 IEEE Symposium on Computers and Communications (ISCC), pp. 682–688. IEEE (2017)
4. Beaulieu, R., Treatman-Clark, S., Shors, D., Weeks, B., Smith, J., Wingers, L.: The SIMON and SPECK lightweight block ciphers. In: 2015 52nd ACM/EDAC/IEEE Design Automation Conference (DAC), pp. 1–6. IEEE (2015)
5. Bogdanov, A., et al.: PRESENT: an ultra-lightweight block cipher. In: Paillier, P., Verbauwhede, I. (eds.) CHES 2007. LNCS, vol. 4727, pp. 450–466. Springer, Heidelberg (2007). https://doi.org/10.1007/978-3-540-74735-2_31
6. De Clercq, R., Uhsadel, L., Van Herrewege, A., Verbauwhede, I.: Ultra low-power implementation of ECC on the ARM cortex-M0+. In: 2014 51st ACM/EDAC/IEEE Design Automation Conference (DAC), pp. 1–6. IEEE (2014)

7. Franck, C., Großschädl, J., Le Corre, Y., Tago, C.L.: Energy-scalable montgomery-curve ECDH key exchange for ARM cortex-M3 microcontrollers. In: 2018 6th International Conference on Future Internet of Things and Cloud Workshops (FiCloudW), pp. 231–236. IEEE (2018)

8. Gaona García, E.E., Rojas Martínez, S.L., Trujillo Rodríguez, C.L., Mojica Nava, E.A.: Authenticated encryption of PMU data. Tecnura 18(SPE), 70–79 (2014)

9. Gómez, E.J., Sarmiento, F.M., Ariza, H.M.: Functional comparison of the present algorithm on hardware and software embedded platforms. HIKARI CES 10(27), 1297–1307 (2017)

10. Gong, Z., Nikova, S., Law, Y.W.: KLEIN: a new family of lightweight block ciphers. In: Juels, A., Paar, C. (eds.) RFIDSec 2011. LNCS, vol. 7055, pp. 1–18. Springer, Heidelberg (2012). https://doi.org/10.1007/978-3-642-25286-0_1

11. Grillo, M., Pereira, W., Cardinale, Y.: Seguridad para la Autentificación, Cifrado y Firma en la Ejecución de Servicios Web Compuestos. Tekhné 1(18), 106–118 (2017)

12. Kerckhof, S., Durvaux, F., Hocquet, C., Bol, D., Standaert, F.-X.: Towards green cryptography: a comparison of lightweight ciphers from the energy viewpoint. In: Prouff, E., Schaumont, P. (eds.) CHES 2012. LNCS, vol. 7428, pp. 390–407. Springer, Heidelberg (2012). https://doi.org/10.1007/978-3-642-33027-8_23

13. Lee, Y.C.: Two ultralightweight authentication protocols for low-cost RFID tags. Appl. Math. Inf. Sci. 6(2S), 425–431 (2012)

14. Luhach, A.K.: Analysis of lightweight cryptographic solutions for Internet of Things. Indian J. Sci. Technology 9(28) (2016)

15. Mozaffari-Kermani, M., Azarderakhsh, R.: Efficient fault diagnosis schemes for reliable lightweight cryptographic ISO/IEC standard CLEFIA benchmarked on ASIC and FPGA. IEEE Trans. Ind. Electron. 60(12), 5925–5932 (2013)

16. Ning, H., Liu, H., Yang, L.: Cyber-entity security in the Internet of things. Computer 46(4), 1 (2013)

17. Raju, B.K.B., Krishna, A., Mishra, G.: Implementation of an efficient dynamic AES algorithm using ARM based SoC. In: 2017 4th IEEE Uttar Pradesh Section International Conference on Electrical, Computer and Electronics (UPCON), pp. 39–43. IEEE (2017)

18. Shibutani, K., Isobe, T., Hiwatari, H., Mitsuda, A., Akishita, T., Shirai, T.: *Piccolo*: an ultra-lightweight blockcipher. In: Preneel, B., Takagi, T. (eds.) CHES 2011. LNCS, vol. 6917, pp. 342–357. Springer, Heidelberg (2011). https://doi.org/10.1007/978-3-642-23951-9_23

19. Tschofenig, H., Pegourie-Gonnard, M.: Performance of state-of-the-art cryptography on ARM-based microprocessors. In: Proceedings of the NIST Lightweight Cryptography Workshop (2015)

20. Zhou, B., Egele, M., Joshi, A.: High-performance low-energy implementation of cryptographic algorithms on a programmable SoC for IoT devices. In: 2017 IEEE High Performance Extreme Computing Conference (HPEC), pp. 1–6. IEEE (2017)

Enhancing Quality of Movies Recommendation Through Contextual Ontological User Profiling

Mohammad Wahiduzzaman Khan, Gaik-Yee Chan[✉],
and Fang-Fang Chua

Faculty of Computing and Informatics, Multimedia University,
Cyberjaya, Malaysia
avy164@gmail.com, {gychan, ffchua}@mmu.edu.my

Abstract. Nowadays, the Internet and Cloud Computing technologies have enabled movies watching to be enjoyed anytime and anywhere in richer varieties and choices. As a result, one will be overwhelmed and tend to make poor choices. The increasing on-demand movies services motivates more choices of the programs to be offered and users need recommendation system to provide them with contextual and personalized suggestions. This paper proposes to use contextual ontological user profiling for movies recommendation which considers personalization in enhancing the effectiveness of the recommendations. We evaluated the performance of our proposed solution with few scenarios representing problems aroused from traditional collaborative and content-based recommendations. These problems are the cold start, data sparsity, over specialization, gray sheep and inefficiency issue. Evaluation results and analysis show that proposed work not only is capable of resolving these problems but is also competent in mitigating the scalability and inefficiency issue.

Keywords: Collaborative filtering · Content-based · Contextual · Ontology · User profiling

1 Introduction

Nowadays, the Internet and Cloud Computing technologies have enabled movies watching to be enjoyed anytime and anywhere in richer varieties and choices. As a result, one will be overwhelmed and then select poor choices. As such, recent research in providing intelligent recommender systems based on user preferences is vigorous as in [1–4].

There are two main traditional approaches for prediction and recommendation, namely, collaborative filtering and content-based. However, they have many drawbacks such as the cold-start, over specialization, new user and new item problems [5–8]. Generally, these approaches also suffer from scalability issue due to the time consuming and storage intensive matrix computations for user-to-user similarity and item-to-item rating comparisons. Consequently, the hybrid approach that combines content-based, collaborative filtering, and/or ontological approach evolve with the aim to improve the quality of recommendations. In particular, ontology helps in defining

© Springer Nature Singapore Pte Ltd. 2019
Y. Tan and Y. Shi (Eds.): DMBD 2019, CCIS 1071, pp. 307–319, 2019.
https://doi.org/10.1007/978-981-32-9563-6_32

relevant concepts and relationships among these concepts within specific domains in a semantically richer way, not only interpretable by human, but also by machines [9].

In this paper, we extend the work in [10] and propose a hybrid (combining content-based with collaborative filtering techniques) recommender system incorporated with contextual elements integrated with ontological user profile to enhance the quality of recommendation. As observed, by including contextual elements such as age, gender, location and occupation of the users on the movies dataset, the experimental evaluation demonstrated satisfactory results in terms of effectiveness and efficiency.

The organization of the paper is as follows, Sect. 2 discusses related works, Sect. 3 describes the proposed hybrid recommendation system, OCHRS, Sect. 4 evaluates and analyzes performance of OCHRS and Sect. 5 presents conclusion and future works.

2   Related Works

This section summarizes and discusses related research works on collaborative filtering, content-based and hybrid which is using both collaborative and content-based filtering as well as the ontology-based approach of recommendation.

The **collaborative filtering** approach is also known as user-based approach. In this user-based approach, items are filtered based on users' opinions. For example, research by [5] makes use of collaborative filtering to obtain similar items among users based on users' rating patterns. This approach helps to solve the over specialization problem, but the new user, data sparsity and gray sheep problems remained unresolved. In another research [11], collaborative filtering approach is also used to find similar items based on users' rating patterns. This approach finds the ratings of unrated items based on a user-item ranking matrix, thus is able to mitigate data sparsity problem. A user-based clustering algorithm is used in research by [7] to improve the computation for user-item rating matrix. Nevertheless, this user-based approach still does not resolve the new user and gray sheep problems. Research by [12] uses collaborative filtering approach with k-means clustering method to analyze users' web transaction history patterns. This approach is able to find subsequent relevant pages according to sequence of pages being visited earlier, hence solving the over specialization problem.

The problem with collaborative filtering approach is that when there is a new user, his/her preferences are unknown and this leads to the new user problem. Another problem with collaborative filtering approach is the data sparsity problem. This problem arises when available data in a recommender system is lesser than predicted data. This problem, perhaps, could be resolved by adding more personalized information like age and gender to the computation of recommendations. Additionally, collaborative filtering approach represents users and items as vectors with length m and size n, the number of users and the number of items in the system, respectively. Thus, the process to find users with similar preferences requires vector computations which lead to scaling and performance issues [4].

The **content-based** or cognitive filtering approach recommends item based on item and user profiles. Generally, user profile contains user ratings for an item. Hence, an item that matches the description of the item and the user's interests shall be recommended [13]. The item profile that is being created shall automatically be updated in response to the acceptance of items by the user. As a result, this approach depends mostly on item description and therefore, it is not able to solve the over specialization problem. The content-based approach is used in research by [14] to find relevant items. In this approach, a weighted value is assigned to every user for each feature of an item. However, this approach suffers from new user or cold start problem. Research by [8] proposes a content-based recommendation system. This system makes use of Term Frequency Inverse Document Frequency (TF-IDF) and cosine similarity to find relevant items through voluminous research papers in a digital library system. It is a vector space's model. Users' queries are represented as vectors of weights based on keywords, thus performance issue could not be avoided. Moreover, this approach could not resolve the over specialization and new user problems. The research by [6] has found that quality of recommendation could be improved over time through learning of users' profile based on implicit and explicit feedback. However, it could not resolve the cold start problem. These researches have shown that content-based approach faces limited content analysis issue. Moreover, these features being extracted have to be linked to the user profile meaningfully. Additionally, an algorithm, mainly machine learning-based, would help in learning users' profiles from the identified items. Only with learned user's profile, content-based recommender system is able to recommend items.

Generally, a **hybrid approach** towards recommendation is by combining collaborative filtering and content-based techniques with the goal to improve performance and eliminate shortcoming of these two traditional approaches. Research in [15] proposes to combine content-based and collaborative filtering approach that includes context aware features to mitigate the problem with cold start. However, this hybrid approach depends only on user's rating and knows little about user's behavior for generating quality recommendation. Besides, this hybrid approach could not avoid the scalability issue because of complex calculation involved in obtaining the closest neighbors when the number of users in the system increases. Similarly, research in [16] uses Boltzmann machines or probabilistic models that combine collaborative and content information coherently. These models could resolve the cold start problem but they still suffer from scalability issue. The hybrid approach used in research [4] is a combination of item-based collaborative filtering and content-based approach. This hybrid approach is able to operate with high load of users' requests and recommend items which are not so popular. However, it suffers from scalability issue as it uses three different matrixes for computation of recommendation which leads to extra consumption of processing space and time. Research in [17] proposes a recommendation model based on Bayesian ranking optimization techniques. This model uses meta-path-based latent features extraction method to extract implicit users' rating and other semantic information of items. Although this model could work under a heterogeneous information network environment with rich attributes, but it has scalability issue. It has to maintain the matrix for item-item similarity computation.

It can be seen from these researches that hybrid (content-based with collaborative filtering) recommendation systems which consider user behavior and users' rating in recommending items, still has scalability issue and perform poorly with a small amount of particular data leading to data sparsity problem.

The **ontology** approach is then used for finding new way to exploit user preferences and user behavior. User profile and items which are interrelated and belonging to a specific domain could be represented ontologically. Terms in ontology are semantically and hierarchically linked with each other. This hierarchical structure of ontology provides a convenient way to analyze preferences at various abstraction levels. Consequently, ontology with its great reasoning and modeling power could help in storing and exploiting users' preferences. Subsequently, the semantically richer level of knowledge could be made known and be reused.

The research in [18] introduces an ontology recommender system which recommends academic research papers on-line. In this method, user profiles for monitoring behavior are created and then profile visualization approach is used to obtain relevant feedback. Research papers are categorized into ontological classes and collaborative filtering is used to identify papers seen by similar users on topics of interest. This ontological collaborative filtering recommender system has shown to improve user profiling accuracy that could mitigate the problem with cold start but not over specialization. Another ontology-based collaborative filtering recommender system proposed by [3] uses the technique of spreading activation in creating ontological user profile. The research in [9] uses spreading activation for creating an ontological recommender system integrated with content-based approach. This recommender system is able to provide deep understanding of user's behavior, but it still could not resolve the over specialization problem.

The research in improving the quality of recommendation is on-going and we believe that by including more contextual elements such as age, gender, location and occupation of the users in our proposed ontological hybrid (content-based with collaborative filtering) recommender system, we shall be able to resolve problems of the traditional content-based or collaborative filtering approach and at the same time mitigating the scalability or inefficiency issue.

3   Our Proposed OCHRS

An overall architecture of the proposed ontological with contextual elements hybrid recommendation system, OCHRS is depicted Fig. 1 (read from left to right, top to bottom in the sequence of number 1, 2, and so on).

Referring to Fig. 1, the first step in our methodology is to gather users' and items' data. For experimental purpose, we use Movielens 100k dataset [19]. The Movielens main dataset contains three sub-data sets. The first sub-data set being movie data containing 9K records of movie id, titles and genre. The second sub-data set consist of 671 users' information that includes user id, age, gender, occupation and location. The third sub-data set have 100K records of 671 users' ratings on various movies. All these three data sets are combined to form into a single dataset namely the 'user-item' dataset as our raw dataset (see step 2 of Fig. 1). This 'user-item' raw dataset consists of about 100K records of 671 users having age, gender, location and occupation of each user as contextual elements. As per user rating, it consists of average of 80 different movies genre in which 20, the least in number, is rated by user id 1. These 100k records are stored in row and column formats before transforming to ontological user profile. This 'user-item' raw dataset has eight data columns, namely, Id of user, Id of movie, Rating for movie, Genre, Age, Gender, Occupation and Location. Each element in this raw dataset is linked hierarchically and semantically into an ontological structure as shown in Fig. 2. This linked structure serves as our ontological user profile (see step 3 of Fig. 1).

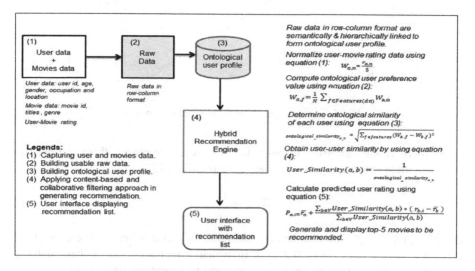

Fig. 1. Overview architecture of OCHRS

Referring to Fig. 2, there are seven classes, namely, user, movie, genre, age, gender, location and occupation which represent different abstract concepts in the movie and user domains. Each class contains property which could be semantically linked, for instance, the user class is linked to the movie class with 'watch' relationship. Translated semantically, it means user watches a movie. Basically, every class has specific attributes, take for instance the class for movie has id for movie and rating for movie as attributes and the values of rating attribute range from 1–5. The lowest rating for a particular movie is represented by attribute value 1 whereas the highest rating is represented by the attribute value 5.

Ontological user profile is made up of instances which are semantically and hierarchically linked data structures. Each instance represents a unique individual that belongs to a class having specific values for its attributes. One instance could be representing user id 6 who has given a rating of 4.5 for movie id 1250 whose genre incudes drama, adventure and war. Also, this user with id 6 has four contextual information; she is a female lawyer aged 50 years old staying at location with zip code 55117. These ontological structures serve as inputs to our hybrid recommendation engine for computation as in step 4 of Fig. 1.

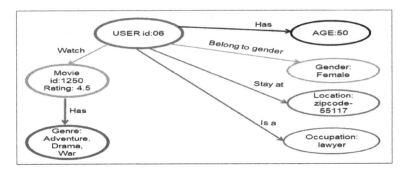

Fig. 2. An example of an ontological user profile

As users' ratings are given by users with different preferences, there is a need to normalize these subjectively rated values by using Eq. 1 [20]:

$$W_{a,n} = \frac{r_{a,n}}{5} \tag{1}$$

where $W_{a,n}$ represents a weighted preference value for user a of item n. These weighted preference values range from 0.2 to 1 in which 1 represents the highest and 0.2 represents the lowest user preference, as mapped to Movielens datasets ratings from highest to lowest preferences of 5 to 1. Users' ratings are normalized over a range from 0.2 to 1 so that these normalized weighted preference values will not exceed integer limit when used in Eq. 3. The term $r_{a,n}$ represents user a's rating of an item n on a scale of 0 to 5. Subsequently, ontology-based user profile can then be formulated using Eq. 2 [20]:

$$W_{a,f} = \frac{1}{N} \sum_{f \in Features(d_n)} W_{a,n} \tag{2}$$

in which N represents number of movies user a has watched, $W_{a,f}$ represents ontological user preference value for user a with feature f. The feature f represents movie genre, and d_n is a list of movies with genre f. For example, the ontology-based user profile having weighted preference or normalized rating value for users with id 1, 2, and 10 are 0.165, 0.103 and 0.165 respectively for the feature Adventure. These preference values range from 0 to 1 map to the lowest to highest preferences respectively. For example, user id 1 has rated 20 movies in which 8 belong to Adventure type movies.

Subsequent to normalized user rating computation is the calculation of ontological concept values for user id 1 based on Adventure concept using Eq. 2. For user id 1, $N = 20$ and d_n represents a list of movies with Adventure feature. The normalized rating values are then applied onto Eq. 2. So $W_{1,adventure} = 0.165$ is obtained. The highest preference value of 0.165 means user id 1 most probably likes to watch Adventure type of movies over other type of movies and least likely will watch Documentary type of movies with preference value of 0.

The proposed hybrid (content-based and collaborative filtering) recommendation engine makes use of data from ontology-based user profile for the generation of similar peers or neighbors. The similarities among users can be determined by the ontological similarity of each user profile based on Eq. 3 [3]:

$$ontological\_similarity_{a,b} = \sqrt{\sum_{f \in features} \left(W_{a,f} - W_{b,f} \right)^2}$$ (3)

where $ontological\_similarity_{a,b}$ is ontological similarity between two users, a and b. After all similarities are computed, the user-user similarity value can then be obtained by inversing this value using Eq. 4 [3]:

$$User\_Similarity(a, b) = \frac{1}{ontological\_similarity_{a,b}}$$ (4)

The rationale for inversing these values are to obtain meaningful values for recommendation computation. For instance, the higher the value, the closer is the similarity between the two users and vice versa. For example, 4.918 is obtained as the highest among 10 values, it represents similarity value for users with id 1 and id 6. This means ontologically, both of these users rated similar types of movies in a similar manner.

After the calculation for user-user similarity values, the next step is to get the k-most similar users, where $k = 5$ in our experiments. The 5-most similar users for a target user could be obtained by sorting in descending order, the user-user similarity values and then select the top five users from the results. The five similar users for user id 1 are user id 6, 10, 7, 2, and 8.

For generating top-5 movies to be recommended to a user, predicted user rating is to be calculated by using Eq. 5 [3]:

$$P_{a,i} = \bar{r}_a + \frac{\sum_{b \in V} User\_Similarity(a, b) * (r_{b,i} - \bar{r}_b)}{\sum_{b \in V} User\_Similarity(a, b)}$$ (5)

where $P_{a,i}$ represents predicted rating for movie i of user a. The similarity value of user a and b is represented by $User\_Similarity(a, b)$. It is found that recommendation not seen by target user before could be generated through this evaluation.

For example, the five movies with {movie ids, predicted rating, genre} are: {82, 4.928, Comedy | Drama}; {4873, 4.90, Animation | Drama | Fantasy}; {26587, 4.58, Crime | Romance | Drama}; {4979, 4.58, Drama | Comedy} and {923, 4.34, Drama | Mystery} which have not been watched by user id 1. However, the peers or neighbors of user id 1 (user id 6, 10, 7, 2 and 8), have watched these movies. However, a point to note is that none of these movies have genre Adventure which is the most preferred for user is 1.

Now for context-aware feature, we will consider user's age, occupation, gender and location. On the basis of these properties, we find similar users for target user id 1 whose age is tagged with numeric value 1, gender is 'F', occupation has numeric value 10 and location is 48067. From our experiment, it was found that user id 119 has the highest contextual similarity value with user id 1 (refer to Table 1, Row 3). This means user id 1 and user id 119 share highest contextual similarity compare to other four users. Both of them are in same age, gender and occupation group except for location. They share three common contextual similarities therefore similarity value of users with id 1 and id 119 is 3. On the other hand, user id 126 (Table 1, Row 4) does not share any contextual property with user id 1, the similarity value is therefore zero.

Table 1. Sample user-user similarity value based on contextual properties

User id	Age	Gender	Occupation	Location	Similarity value
619	18	F	1	94706	1
119	**1**	**F**	**10**	**77715**	**3**
126	18	M	9	97117	0
349	1	M	10	8035	2

Using this contextual similarity value of user id 1, we calculate 5-most similar users. The five most similar users for user id 1 are: user ids 99, 119, 484, 606 and 634. Subsequently, we compute target user predicted rating for the items which have been watched by these peers but not watched by user id 1 using Eq. 5. Results are tabulated as shown in Table 2 for top-5 items recommended based on contextual similarity.

Table 2. Top-5 recommended movies based on contextual similarity

Movie id	1233	**1085**	912	**1136**	1077
Target user predicted rating	4.09	4.09	4.05	3.98	3.98
Genre	Action | Drama | War	**Adventure | Drama**	Drama | Romance	**Adventure | Comedy | Fantasy**	Comedy | Sci-Fi

Results as tabulated in Table 2 are displayed through the user interface (see step 5 of Fig. 1). Table 2 shows the top-5 recommended movies for user id 1 based on contextual similarity with the peers. It seems that based on contextual similarity, the results are close to target user id 1's preference with two Adventure types of movies being recommended (Table 2 Columns 3 and 5, Row 3) while none of the adventure type of movie is recommended based on transaction history.

4   Performance Evaluation of OCHRS

This section evaluates the effectiveness and robustness of OCHRS based on a few scenarios. Its efficiency is evaluated through comparison with clustering techniques.

4.1   Scenario-Based Performance Evaluation of OCHRS

The effectiveness of OCHRS is evaluated based on several scenarios, where every scenario will be treated as a problem which occurred in traditional collaborative and content-based recommendations. These problems are the cold start, data sparsity, over specialization, gray sheep and inefficiency issue.

Now, let us use the original Movielens rating data sets to find similar users using Pearson co-relation formula and then query for top-5 item for user id 1 using traditional collaborative filtering approach. It is found that the recommendation system based on collaborative filtering approach fails to recommend any item. This data sparsity problem occurs because user id 1 provided only a few numbers of ratings and it is also a gray sheep problem because user id 1 rated movies which are not rated by any user in the system at all.

OCHRS, on the other hand, is able to recommend items to user as shown in Table 2. This is because OCHRS is not just considering user rating for finding similar users, but it is using deep ontological knowledge such as user preferences for each concept in the system and user contextual information like age, gender, location and occupation to find the similar users. OCHRS, thus, has overcome the data sparsity and gray sheep problems.

When a new user comes into a system, there is very little information about his preferences. This is the reason why traditional recommender systems fail to recommend item to the target users. However, OCHRS by considering user contextual information of age, gender, location and occupation, is able to find the similar users with ease. Therefore, OCHRS is able to recommend items to the users readily.

To further confirm OCHRS is able to resolve the new item problem, we manually added a movie into the system, for example, movie id: 5213565, genre: adventure and put 5.0 as rating for 5-most similar users for user id 1. Subsequently, our algorithm is run again to compute user id 1's predicted user rating. Predicted user rating of user id 1 for movie id: 5213565, is 4.92. Thus, it can be seen that OCHRS has successfully tackled the new item problem.

Over specialization problem is another problem faced by traditional recommendation systems. When a user enters into a system and rated similar genre movie, traditional recommender system will only recommend those movies having the same genre as preferred mostly by the user. By so doing, traditional recommendation systems are actually confining users' taste. However, OCHRS will not confine users' taste to a particular type of movie but rather, it broadens users' taste by recommending different genre items (genre such as Action, Animation, Children, Comedy and Crime in which user has lower weighted preferences) as shown in both Table 2.

Traditional recommender systems also face the inefficient performance issue as they require extensive memory storage and computation time. For instance, if the recommender system has m number of items, then it is required to maintain m * m matrix to find similar items. Additionally, for n number of users in the system, it requires n * n matrix to find similar users in the system. Therefore, in a hybrid recommendation system using content-based and collaborative based approaches, it is required to maintain two matrices of size m * m and n * n for computation of items' and users' similarity purpose. For OCHRS which uses ontological user profile whereby the representation of data are classes which could be stored as instances in the system, provides a fast and easy means of access as compared to the traditional columns and rows representation. OCHRS does not require to maintain m * m or n * n matrix for generation of recommendations for the target user. Hence, OCHRS helps to mitigate the scalability issue if not entirely eliminating it.

4.2   Efficiency of OCHRS as Compared with Clustering Techniques

To further confirm the performance of OCHRS in terms of efficiency, experiments based on the same 'user-item' ontological data sets are conducted by using different clustering technique such as k-means, Expectation Maximization (EM), FarthestFirst, and Mark Density for finding similar neighborhood users.

Our experiments test the various clustering algorithms on 671 user likelihood values for different concepts and contextual properties in our OCHRS using Weka 3.8 [21] tool on four datasets where each data set has one of the contextual element excluded. The results show that the four clustering algorithms have one thing in common, i.e. they require several iterations to generate clusters for the four data sets with combination of three different contextual properties. However, for OCHRS, finding similar users for any users with three similar contextual properties requires a single iteration only. This is because ontological user profile is designed in such a way that data represented as classes are stored as instances in OCHRS. This leads to only a one-time access for clusters computation.

4.3   Effectiveness and Robustness of OCHRS

More experiments are conducted to test the effectiveness of OCHRS in providing recommendation and its robustness in performance. For example, top 5 movies recommended by OCHRS based on three different user groups, i.e. 5, 10 and 15-most similar users are obtained. These user groups have commonality in terms of age, gender, occupation and location.

The robustness of OCHRS in recommending the most suitable item for the target user based on preferences can be seen when the most preferred type of movie, i.e. Adventure type of movie (Movie id 1085) is being recommended by all the three similar users groups for target user id 1. Additionally, OCHRS is able to recommend other types of movies such as Action, Animation, Comedy, Children and Crime that target user id 1 might have lower preferences. Thus, OCHRS is able to resolve the data sparsity, over specialization and gray sheep problems by using deep ontological knowledge based on user preferences for each concept in the system and user contextual information like age, gender, location and occupation to find the similar users.

Experiments are carried out with the 10 similar users group to see if different combinations of any two contextual information between age, gender and occupation has effect on the recommendation or not. Refer to Table 3 for the tabulated results. It is seen that different set of top-5 recommendation is presented when either occupation, or age, or gender is excluded (Table 3: Column 1, 4 and 7), although there may be some overlaps. This shows OCHRS is able to solve the traditional recommender system's problems such as data sparsity and cold start problem.

Table 3. Top-5 recommended movies (any 2 between Age, Gender and Occupation)

Include: Age, Gender Exclude: Occupation			Include: Occupation, Gender Exclude: Age			Include: Age, Occupation Exclude: Gender		
Movie id	Predicted user rating	Genre	Movie id	Predicted user rating	Genre	Movie id	Predicted user rating	Genre
1223	4.05	Adventure | Drama	1085	4.05	Adventure | Drama	969	4.06	Adventure | Comedy | Romance | War
745	4.05	Animation | Children | Comedy	345	4.0	Comedy | Drama	1235	4.05	Comedy | Drama | Romance
345	3.98	Comedy | Drama	969	3.98	Adventure | Comedy | Romance | War	745	4.05	Animation | Children | Comedy
903	3.92	Drama | Mystery | Romance | Thriller	903	3.92	Drama | Mystery | Romance | Thriller	919	4.05	Adventure | Children | Fantasy | Musical
923	3.88	Drama | Mystery	745	3.83	Animation | Children | Comedy	111	4.03	Crime | Drama | Thriller

To summarize, OCHRS is able to solve the new user and gray sheep problems not resolved by collaborative approach of [5, 7, 11, 12]; the new user or cold start problems not resolved by the content-based approach of [8, 14]; the scalability issue found in [4, 15–17] even though they are also using hybrid approach due to they have complex usage on matrix computation; and the over specialization problem still existed in [9] and [18] although they are using ontology with content-based or collaborative filtering approach respectively.

5   Conclusion and Future Work

This paper introduces an ontology-based context aware hybrid recommender system (OCHRS) which integrates user profiling and contextual elements for improving the quality of recommendation. Experimental results and performance evaluation analysis have showed that OCHRS is capable of resolving the cold start, data sparsity, over specialization, gray sheep problems and inefficiency issue. More experiments using large scale real-life datasets to provide statistics on MAE (mean absolute error) and RMSE (root mean squared error) that would determine recommendation accuracy will be presented as future works due to space limitation of this paper.

References

1. Bambini, R., Cremonesi, P., Turrin, R.: A recommender system for an IPTV service provider: a real large-scale production environment. In: Ricci, F., Rokach, L., Shapira, B., Kantor, Paul B. (eds.) Recommender Systems Handbook, pp. 299–331. Springer, Boston, MA (2011). https://doi.org/10.1007/978-0-387-85820-3_9
2. Carrer-Neto, W., Hernandez-Alcaraz, M.L., Valencia-Garcia, R., Garcia-Sanchez, F.: Social knowledge-based recommender system. Application to the movies domain. Expert Syst. Appl. **39**, 10990–11000 (2012)
3. Zhang, Z., Gong, L., Xie, J.: Ontology-based collaborative filtering recommendation algorithm. In: Liu, D., Alippi, C., Zhao, D., Hussain, A. (eds.) BICS 2013. LNCS (LNAI), vol. 7888, pp. 172–181. Springer, Heidelberg (2013). https://doi.org/10.1007/978-3-642-38786-9_20
4. Pripužić, K., et al.: Building an IPTV VoD recommender system: an experience report. In: Proceeding of 12th International Conference on Telecommunications, pp. 155–162 (2013)
5. Schafer, J.B., Frankowski, D., Herlocker, J., Sen, S.: Collaborative filtering recommender systems. In: Brusilovsky, P., Kobsa, A., Nejdl, W. (eds.) The Adaptive Web. LNCS, vol. 4321, pp. 291–324. Springer, Heidelberg (2007). https://doi.org/10.1007/978-3-540-72079-9_9
6. Lops, P., de Gemmis, M., Semeraro, G.: Content-based recommender systems: state of the art and trends. In: Ricci, F., Rokach, L., Shapira, B., Kantor, Paul B. (eds.) Recommender Systems Handbook, pp. 73–105. Springer, Boston, MA (2011). https://doi.org/10.1007/978-0-387-85820-3_3
7. Zhang, L., Qin, T., Teng, P.: An improved collaborative filtering algorithm based on user interest. J. Softw. **9**(4), 999–1006 (2014)
8. Philip, S., Shola, P.B., John, A.O.: Application of content-based approach in research paper recommendation system for a digital library. Int. J. Adv. Comput. Sci. Appl. **5**(10), 37–40 (2014)
9. Bahramian, Z., Ali, A.R.: An ontology-based tourism recommender system based on spreading activation model. In: The International Archives of the Photogrammetry, Remote Sensing and Spatial Information Sciences, XL-1-W5, pp. 83–90 (2015)
10. Khan, M.W., Chan, G.Y., Chua, F.F., Haw, S.C.: An ontology-based hybrid recommender system for internet protocol television. In: Badioze Zaman, H., et al. (eds.) IVIC 2017. LNCS, vol. 10645, pp. 131–142. Springer, Cham (2017). https://doi.org/10.1007/978-3-319-70010-6_13
11. Badaro, G., Hajj, H., El-Hajj, W., Nachman, L.: A hybrid approach with collaborative filtering for recommender systems. In: Proceedings of 9th International Wireless Communications and Mobile Computing Conference, pp. 349–354 (2013)
12. Kulkarni, K., Wagh, K., Badgujar, S., Patil, J.: A study of recommender systems with hybrid collaborative filtering. Int. Res. J. Eng. Technol. **3**(4), 2216–2219 (2016)
13. Pazzani, M.J., Billsus, D.: Content-based recommendation systems. In: Brusilovsky, P., Kobsa, A., Nejdl, W. (eds.) The Adaptive Web. LNCS, vol. 4321, pp. 325–341. Springer, Heidelberg (2007). https://doi.org/10.1007/978-3-540-72079-9_10
14. Uluyagmur, M., Cataltepe, Z., Tayfur, E.: Content-based movie recommendation using different feature sets. In: Proceedings of the World Congress on Engineering and Computer Science, vol. I, pp. 1–5 (2012)
15. Woerndl, W., Schueller, C., Wojtech, R.: A hybrid recommender system for context-aware recommendations of mobile applications. In: Proceedings of IEEE 23rd International Conference on Data Engineering Workshop, pp. 871–878 (2007)

16. Gunawardana, A., Meek, C.: A unified approach to building hybrid recommender systems. In: Proceedings of the third ACM Conference on Recommender Systems, RecSys 2009, New York, USA, pp. 117–124 (2009)
17. Yu, X., et al.: Recommendation in heterogeneous information networks with implicit user feedback. In: RecSys 2013, Hong Kong, China, pp. 347–350 (2013)
18. Middleton, S.E., De Roure, D., Shadbolt, N.R.: Ontology-based recommender systems. In: Staab, S., Studer, R. (eds.) Handbook on Ontologies. INFOSYS, pp. 477–498. Springer, Heidelberg (2004). https://doi.org/10.1007/978-3-540-24750-0_24
19. Grouplens, Movielens 100K Dataset. http://grouplens.org/datasets/movielens/100k/. Accessed Dec 2016
20. Cantador, I., Bellogín, A., Castells, P.: A multilayer ontology-based hybrid recommendation model. AI Commun. **21**(2), 203–210 (2008)
21. Weka (2017). Weka 3.8, Datamining Software. https://www.cs.waikato.ac.nz/ml/weka/. Accessed July 2017

Determinants of ResearchGate (RG) Score for the Top100 of Latin American Universities at Webometrics

Carolina Henao-Rodríguez[1]([⊠]), Jenny-Paola Lis-Gutiérrez[2],
Mercedes Gaitán-Angulo[2], Carmen Vásquez[3], Maritza Torres[4],
and Amelec Viloria[5]

[1] Corporación Universitaria Minuto de Dios, Bogotá, Colombia
linda.henao@uniminuto.edu
[2] Fundación Universitaria Konrad Lorenz, Bogotá, Colombia
{jenny.lis,mercedes.gaitana}@konradlorenz.edu.co
[3] Universidad Nacional Experimental Politécnica "Antonio José de Sucre",
Barquisimeto, Venezuela
cvasquez@unexpo.edu.ve
[4] Universidad Centroccidental "Lisandro Alvarado", Barquisimeto, Venezuela
mtorres@ucla.edu.ve
[5] Universidad de la Costa, Barranquilla, Colombia
aviloria7@cuc.edu.co

Abstract. This paper has the purpose of establishing the variables that explain the behavior of ResearchGate for the Top100 Latin American universities positioned in Webometrics database for January 2017. For this purpose, a search was carried out to get information about postgraduate courses and professors at the institutional websites and social networks, obtaining documents registered in Google Scholar. For the data analysis, the econometric technique of ordinary least squares was applied, a cross-sectional study for the year 2017 was conducted, and the individuals studied were the first 100 Latin American universities, obtaining a coefficient of determination of 73.82%. The results show that the most significant variables are the number of programs, the number of teacher's profiles registered in Google Scholar, the number of subscribers to the institutional YouTube channel, and the GDP per capita of the university origin country. Variables such as (i) number of undergraduate programs, (ii) number of scientific journals; (iii) number of documents found under the university domain; (iv) H-index of the 1st profile of researcher at the university; (vi) number of members of the institution; (v) SIR Scimago ranking of Higher Education Institutions; (vi) number of tweets published in the institutional account; (vii) number of followers in the Twitter institutional account; (vii) number of "likes" given to the institutional count, were not significant.

Keywords: ResearchGate · Universities · Google scholar ·
Academic assessment · Webometrics

© Springer Nature Singapore Pte Ltd. 2019
Y. Tan and Y. Shi (Eds.): DMBD 2019, CCIS 1071, pp. 320–327, 2019.
https://doi.org/10.1007/978-981-32-9563-6_33

1 Introduction

When discussing the impact of the publications, it is important to review the issue of bibliometric indicators, such as the ResearchGate metrics (RG). ResearchGate is a high-impact academic network that was founded in 2008 by Ijad Madisch, Sören Hofmayer, and Horst Fickenscher. During the first half of 2018, ResearchGate has more than 15 million members [1], and contains important academic networking tools, with a wide catalog of bibliometric indicators, among which, ResearchGate Score stands out [2]. The ResearchGate Score is the flagship indicator calculated using an undisclosed algorithm to measure the scientific reputation [3].

This paper aims to establish the variables that explain the behavior of ResearchGate for the Top100 Latin American universities positioned in Webometrics database in 2017. An ordinary least square model was applied with ResearchGate Score as the dependent variable and with the following explanatory variables: (i) number of post-graduate programs, (ii) number of teacher's profiles registered in Google Scholar, (iii) number of subscribers to the institutional YouTube channel, and (iv) GDP in each country where the institution is located.

The topic of ResearchGate has been treated by different authors, like [4] who indicated that RG is a research-oriented academic social network that reflects the level of research activity in the universities. The study suggests that academic social networks act as indicators in the assessment of research activities and may be useful and credible for acquiring scholar resources, staying informed about research results, and promoting the academic influence. In the same way, [2] conducted a study in which the RG Score was analyzed and revealed the main advantages and disadvantages of this indicator, concluding that it does not measure the prestige of the researchers but their level of participation in the platform.

From another perspective, [5] conducted a research that allowed to assess if the data of use and publication of ResearchGate reflected the existing academic hierarchies. This study concludes that the classifications based on the ResearchGate statistics correlate moderately well with other classifications of academic institutions, suggesting that the use of ResearchGate broadly reflects the traditional distribution of academic capital. At the same time, [6] presents a method to capture the structure of a full scientific community (the community of Bibliometrics, Scientometrics, Informetrics, Webometrics, and Altmetrics) and the main agents that are part of it (scientists, documents, and sources) through the lens of Google Scholar Citations (GSC). The method was applied to a sample of 814 researchers in Bibliometrics with a public profile created in Google Scholar Citations, and later used in the other platforms, collecting the main indicators calculated by each of them. The results obtained from this study were: (i) ResearchGate indicators, as well as the readers of Mendeley, present a high correlation with all indicators of GSC; and (ii) there is a moderate correlation with the indicators in ResearcherID.

Regarding the use of the Webometrics data, there are several researches that have already used them. For example, [7, 8] compared different rankings like the Ranking of Shanghai, QS World University Ranking, SCimago Institutions Rankings SIR, and the

Web Ranking of Universities-Webometrics, finding that the indicators associated with the research and institutional capacity stand out as common criteria in the reviewed evaluation methodologies. This research firstly states that Brazil occupies the first positions in the four rankings; and secondly, that there is a greater number of Latin American universities in QS (40%) and Webometrics (31%), while in the other two rankings it does not exceed 8%. [9] carried out a cluster analysis of the Top100 Latin American universities positioned in the Webometrics database in January 2017. The research included information about postgraduate programs and social networks on the web sites of these institutions and teachers. The variable with the highest correlation with the ranking is the number of postgraduate programs.

In the same way, the study of [10], analyzed a group of manageable visibility factors corresponding to the universities present in the Top100 of the Webometrics database of Latin America published in January 2017 for the identification of profiles. For this purpose, data was collected about the academic offer and scientific journals published on each university website; figures on documents and profiles found in Google Scholar; activity on social networks; and the institutional score reported by ResearchGate as a scientific network. Clusters were formed by quartiles to characterize the visibility profiles of Latin American universities considering the variables studied. The higher offer of postgraduate programs and the presence in scientific networks and Google Scholar characterize the best positioned universities.

For its part, [8] and [10], used the Data Envelopment Analysis (DEA) for processing the academic data published on the website of each university, the content and profiles shown in Google Scholar (GS), the data published by the university in ResearchGate as scientific network, and finally, the data of social networks such as the Twitter and Facebook accounts of the corresponding institutions. The authors found that the postgraduate offer, visibility in GS, and the use of scientific and social networks contribute favorably to the web positioning of Latin American universities.

From another approach, [11], also discussed the importance of webmetrics techniques for measuring the visibility, specifically in the case of university libraries in Sri Lanka. Similarly, [12] uses two quantitative techniques in Multi-criteria Decision Analysis (MCDA) which are the Entropy method and the Technique for Order of Preference by Similarity to Ideal Solution (TOPSIS), applied to the Webometrics ranking for universities in the world. These models help evaluators to apply a strategic vision for future developments by the use of the multi-criteria decision analysis method. The author concludes that Webometrics classification systems are perceived differently by different stakeholders and, therefore, can be approached in different ways.

2 Method

For the development of the document, an econometric exercise was performed to analyze the determinants of the ResearchGate (RG) score, since it is an indicator that considers the popularity and commitment to the RG community [13]. Therefore, it measures the number of publications, followers, and interactions within this scientific social researcher's network.

The econometric technique used was the ordinary least squares. A cross-section was made for the year 2017, where the individuals studied were the first 100 Latin American universities of the Webometrics Ranking for that year.

2.1   Data

The data used to build the model were obtained from the Web Ranking of Universities [14] and the statistics of the International Monetary Fund, published in April 2018. The analysis was made on the first 100 Latin American universities of the Webometrics Ranking.

2.2   Variables

As mentioned above, the dependent variable was the ResearchGate Score and the explanatory variables were: (i) number of postgraduate programs offered by the educational institution (po), (ii) number of teacher's profiles registered in Google Scholar (sp), (iii) number of subscribers to the institutional YouTube channel (YouTube), and (iv) GDP per capita at constant prices adjusted to the purchasing power parity in base dollars 2011 as a control variable of the level of economic development of the origin country of the university, since it is expected that RG score has a positive relationship with quality and impact of the research in its environment (pibppp).

2.3   Model

The log-log model is specified as follows:

$$\text{Lrgj} = \beta 0 + \beta 1 \, \text{lpoj} + \beta 2 \, \text{lgspj} + \beta 3 \, \text{lgyoutubei} + \beta 4 \text{lpibpppi} + \varepsilon_{jt}. \quad (1)$$

j corresponds to the university; i is the country of origin of the university; lrg is the logarithm of the ResearchGate score; lpo is the logarithm of the number of postgraduate programs offered by the university; lgsp is the logarithm of the number of teacher's profiles registered in Google Scholar; lgyoutube is the logarithm of the number of subscribers to the institutional YouTube channel; lpibppp is the logarithm of GDP per capita at constant prices adjusted to the purchasing power parity in base dollars 2011 of the country where the university is located and ε_{jt} is a random disturbance that is supposed $\varepsilon_{jt} \sim N(0, \sigma^2)$.

3   Results

The results of the model are available below. The proposed model considered all the variables in natural logarithm (Fig. 1).

```
   Source |       SS      df       MS              Number of obs =        85
----------+----------------------------------     F(  4,     80) =     60.21
    Model | 54.4667077      4  13.6166769          Prob > F       =    0.0000
 Residual | 18.0911795     80   .226139744         R-squared      =    0.7507
----------+----------------------------------     Adj R-squared  =    0.7382
    Total | 72.5578872     84   .863784371         Root MSE       =   .47554
```

```
       lrg |      Coef.   Std. Err.       t    P>|t|     [95% Conf. Interval]
-----------+----------------------------------------------------------------
       lpo |   .2328636   .0884934     2.63    0.010     .0567561    .4089711
      lgsp |   .7892523   .0635073    12.43    0.000     .6628688    .9156359
 lgyoutube | -.0914424   .0319105    -2.87    0.005    -.1549463   -.0279386
   lpibppp |   1.005943   .2797976     3.60    0.001     .4491277    1.562758
     _cons |  -5.339164   2.858211    -1.87    0.065    -11.02719    .3488574
```

Fig. 1. Results of the model

In this case, the significant variables are shown. Other variables initially considered as (i) number of undergraduate programs, (ii) number of scientific journals; (iii) number of documents found under the domain of the university; (iv) H-index of the 1st profile of university researcher; (vi) number of members of the institution; (v) Scimago SIR ranking of Higher Education Institutions; (vi) number of tweets published in the institutional account; (vii) number of followers in the Twitter institutional account; (vii) number of "likes" given to the institutional account, were not significant.

To validate the model, the relevant tests were performed for the MCO cases. When performing the Ramsey test, it can be concluded that there was no omission of relevant variables since the null hypothesis can not be rejected. Ramsey RESET test using powers of the fitted values of lrg.

Ho: model has no omitted variables

F (3, 77) = 1.85

Prob > F = 0.1454

The White's test shows that the variance of random perturbations, conditional on the values of the regressors, are constant. Since the null hypothesis of homoskedasticity can not be rejected, as shown below (Fig. 2):

White's test for Ho: homoscedasticity against

Ha: unrestricted heteroskedasticity

chi2(14) = 10.59

Prob > chi2 = 0.7180

```
-------------------------------------------------------
Source | chi2 df p
--------------------+----------------------------
Heteroskedasticity | 10.59 14 0.7180
Skewness | 5.45 4 0.2441
Kurtosis | 3.18 1 0.0745
--------------------+----------------------------
Total | 19.22 19 0.4428
-------------------------------------------------------
```

Fig. 2. Cameron & Trivedi's decomposition of IM-test

To test for normality of the errors, the tests of normality of kurtosis and pointing, and Shapiro-Wilk were made, concluding that the null hypothesis of normality to a significance level of 5% can not be rejected (Fig. 3).

```
------- joint ------
Variable | Obs Pr (Skewness) Pr (Kurtosis) adj chi2(2) Prob>chi2
-------------+---------------------------------------------------------
Residual | 85 0.1073 0.0939 5.27 0.0717
swilk residual
Shapiro-Wilk W test for normal data
Variable | Obs W V z Prob>z
-------------+---------------------------------------------------------
Residual | 85 0.97859 1,545 0,956 0.16950
```

Fig. 3. Skewness/Kurtosis tests for Normality

Finally, the variance inflation factor (VIF) was calculated to prove that there was no multicollinearity in the independent variables, concluding that there was no multicollinearity since the value was less than 3 (Fig. 4).

Variable	VIF	1/VIF
lgsp	1.51	0.660182
lpo	1.27	0.785697
lpibppp	1.18	0.849824
lgyoutube	1.05	0.950065
Mean VIF	1.25	

Fig. 4. Variance inflation factor (VIF)

Since the model does not violate any of the MCO assumptions, no transformation is necessary. The coefficient of determination is 73.82%. The significant variables in the model were the number of postgraduate programs offered by the university; the number of teacher's profiles registered in Google Scholar; the number of subscribers to the institutional YouTube channel and the GDP per capita of the country where the university is located.

Therefore, the equation of the model corresponds to:

$$Lrgj = -5.339164 + 0.2328636 * lpoj + 0.7892523 * lgspj + -0.914424 * lgyoutubei$$
$$+ 1.005943 * pibpppi + \varepsilon_{jt}.$$

(2)

The effects of each of the variables are explained below.

- An increase of 1% in the number of postgraduate programs offered by the university increases the ResearchGate score of the institution by 0.23%.
- An increase of 1% in the number of teacher's profiles registered in Google Scholar increases the ResearchGate score of the institution by 0.79%.
- An increase of 1% in the number of subscribers to the institutional YouTube channel reduces the ResearchGate score of the institution by 0.9144%.
- An increase of 1% in GDP per capita at constant prices adjusted to the purchasing power parity in base dollar 2011 of the country where the university is located increases the ResearchGate score of the institution by 1.006%.

4   Conclusions

The model of standard errors corrected for the model has a determination coefficient of 73.82%. The only variable that presented a negative relationship with the dependent variable was the number of subscribers to the institutional YouTube channel, that is, the popularity on YouTube of the institution has an inverse effect on the RG score, what can be explained since YouTube corresponds to a social network and not to an academic one.

Variables such as (i) number of undergraduate programs, (ii) number of scientific journals; (iii) number of documents found under the university domain; (iv) H-index of the 1st profile of researcher at the university; (vi) number of members of the institution; (v) SIR Scimago ranking of Higher Education Institutions; (vi) number of tweets published in the institutional account; (vii) number of followers in the Twitter institutional account; (vii) number of "likes" given to the institutional count, were not significant.

Considering the above, some of the recommendations for the institutions to increase their position in the RG score are: (i) increase the number of teachers with active profiles in academic social networks, and (ii) increase the number of postgraduate programs and postgraduate students.

For future researches, this analysis could be carried out on other rankings such as Shanghai, QS World University Ranking, and SCimago Institutions Rankings SIR. Those results could allow to provide recommendations to the academic authorities of higher education institutions to increase their visibility.

References

1. ReseachGate the network for science and research (2018). https://solutions.researchgate.net/recruiting/?utm_source=researchgate&utm_medium=community-loggedout&utm_campaign=indextop

2. Orduña-Malea, E., Martín-Martín, A., López-Cózar, E.D.: ResearchGate como fuente de evaluación científica: desvelando sus aplicaciones bibliométricas. El profesional de la información (EPI) **25**(2), 303–310 (2016)

3. Yu, M.C., Wu, Y.C.J., Alhalabi, W., Kao, H.Y., Wu, W.H.: ResearchGate: an effective altmetric indicator for active researchers? Comput. Hum. Behav. **55**, 1001–1006 (2016)

4. Yan, W., Zhang, Y.: Research universities on the ResearchGate social networking site: an examination of institutional differences, research activity level, and social networks formed. J. Inform. **12**(1), 385–400 (2018)

5. Thelwall, M., Kousha, K.: ResearchGate: disseminating, communicating, and measuring scholarship? J. Assoc. Inf. Sci. Technol. **66**(5), 876–889 (2015)

6. Martín-Martín, A., Orduña-Malea, E., Ayllón, J.M., López-Cózar, E.D.: The counting house: Measuring those who count. Presence of bibliometrics, scientometrics, informetrics, webometrics and altmetrics in the Google Scholar citations, Researcherid, ResearchGate, Mendeley & Twitter (2016). arXiv preprint arXiv:1602.02412

7. Torres-Samuel, M., Vásquez, C.L., Viloria, A., Varela, N., Hernández-Fernandez, L., Portillo-Medina, R.: Analysis of patterns in the university world rankings webometrics, Shanghai, QS and SIR-SCimago: case Latin America. In: Tan, Y., Shi, Y., Tang, Q. (eds.) DMBD 2018. LNCS, vol. 10943, pp. 188–199. Springer, Cham (2018). https://doi.org/10.1007/978-3-319-93803-5_18

8. Torres-Samuel, M., et al.: Efficiency analysis of the visibility of Latin American universities and their impact on the ranking web. In: Tan, Y., Shi, Y., Tang, Q. (eds.) DMBD 2018. LNCS, vol. 10943, pp. 235–243. Springer, Cham (2018). https://doi.org/10.1007/978-3-319-93803-5_22

9. Vásquez, C., et al.: Cluster of the Latin American universities Top100 according to webometrics 2017. In: Tan, Y., Shi, Y., Tang, Q. (eds.) DMBD 2018. LNCS, vol. 10943, pp. 276–283. Springer, Cham (2018). https://doi.org/10.1007/978-3-319-93803-5_26

10. Torres-Samuel, M., Vásquez, C., Viloria, A., Lis-Gutiérrez, J.P., Borrero, T.C., Varela, N.: Web visibility profiles of Top100 Latin American universities. In: Tan, Y., Shi, Y., Tang, Q. (eds.) DMBD 2018. LNCS, vol. 10943, pp. 254–262. Springer, Cham (2018). https://doi.org/10.1007/978-3-319-93803-5_24

11. Ramanayaka, K.H., Chen, X., Shi, B.: Application of webometrics techniques for measuring and evaluating visibility of university library websites in Sri Lanka. J. Univ. Libr. Assoc. Sri Lanka **21**(1), 1–17 (2018). https://doi.org/10.4038/jula.v21i1.7908

12. Jati, H.: University webometrics ranking using multicriteria decision analysis: entropy and TOPSIS method. Soc. Sci. **13**(3), 763–765 (2018). http://docsdrive.com/pdfs/medwelljournals/sscience/2018/763-765.pdf

13. Manoj, M.: Webometrics as a tool for measuring popularity of websites: an analysis of websites of IISERs in India. Int. J. Sci. Res. Sci. Technol. **4**(2), 1472–1476 (2018)

14. Cybermetrics Lab Webometrics Ranking of World Universities (2018). http://www.webometrics.info/en/About_Us

BIOG: An Effective and Efficient Algorithm for Influence Blocking Maximization in Social Networks

Kuei-Sheng Lin and Bi-Ru Dai[(✉)]

Department of Computer Science and Information Engineering,
National Taiwan University of Science and Technology, Taipei, Taiwan, R.O.C.
m10615082@mail.ntust.edu.tw, brdai@csie.ntust.edu.tw

Abstract. With the growing of social networks, information can be propagated and shared faster in social networks. Unfortunately, rumors and misinformations are spread faster simultaneously. Influence blocking maximization (IBM) problem is studied to find positive seeds to block the spread of negative information as much as possible. Some existing solutions can find effective positive seeds, but are too slow to handle large social networks. Some other solutions are efficient, but cannot effectively block negative information. In this work, we propose BIOG algorithm and devise a new graph structure $MIOG$ to solve IBM problem. As verified by experiments on real world datasets, BIOG algorithm is able to block negative information with high efficiency and effectiveness.

Keywords: Influence blocking maximization · Social network · Competitive influence diffusion

1 Introduction

In recent years, the popularity of social networks, such as Facebook, Twitter, is increasing and allows more information, such as idea, news, and merchandising, to be propagated faster. As a result, viral marketing becomes one of the most important marketing strategies in social networks. A critical issue for the success of viral marketing is how to find proper initial customers who can spread product information as far as possible. Some researches [4,5,7,8] regard it as Influence Maximization (IM) problem which aims to find nodes with the highest influence. CELF algorithm uses the submodularity property to improve the efficiency of the greedy algorithm [5], PMIA algorithm utilizes MIA structure to approximate influence spread of each node [8], and TIM algorithm converts this problem to the maximum coverage problem [7].

Although social networks bring a lot of benefits to people, they also provide a place for rumors and misinformations to be diffused. Therefore, some works [1,3,6,9–11] aim to block the spread of unwanted information in social networks. However the traditional greedy algorithm for Influence Blocking Maximization

© Springer Nature Singapore Pte Ltd. 2019
Y. Tan and Y. Shi (Eds.): DMBD 2019, CCIS 1071, pp. 328–337, 2019.
https://doi.org/10.1007/978-981-32-9563-6_34

(IBM) problem is slow because it needs to estimate influence spread of each node by a large amount of Monte-Carlo simulations [1]. CMIA-O algorithm can effectively solve IBM problem using MIA structure [10]. However, the time cost is usually not satisfactory when the negative influence spreads from high density regions. CMIA-O needs to construct a lot of subgraphs and estimate many nodes. CB-IBM algorithm requires the community information as input [6], which is usually not available in general datasets. [9,11] aim to prevent the target region from being influenced by bad information.

In this work, algorithm BIOG is proposed to block the negative influence efficiently. A new graph structure $MIOG$ is designed to approximate the influence region. In addition, candidate nodes will be selected to reduce the number of nodes to be estimated. Finally, the efficiency and effectiveness of BIOG is verified by experiments with real world datasets.

The rest of paper is organized as follows. The problem definition is introduced in Sect. 2. Our BIOG algorithm is presented in Sect. 3. Experiments are shown in Sect. 4. Finally, conclusion and future work are in Sect. 5.

2   Problem Definition

For a given competitive diffusion model, such as COICM [1], The goal of IBM problem is finding a positive seed set $\mathbf{S}_{\mathcal{P}}$ to block negative influence as much as possible. The blocked negative influence is defined in [1] and is briefly stated as follows.

Definition 1 (Blocked negative influence). Given a graph $G = (V, E)$, the diffusion model with propagation probability $pp_e(u, v)$ of each edge (u, v), a negative seed set $\mathbf{S}_{\mathcal{N}}$ and a positive seed set $\mathbf{S}_{\mathcal{P}}$, the blocked set $B(\mathbf{S}_{\mathcal{P}})$ is a set of nodes which are activated by $\mathbf{S}_{\mathcal{N}}$ when there is no positive seed, but are not activated by $\mathbf{S}_{\mathcal{N}}$ when $\mathbf{S}_{\mathcal{P}}$ joins to propagate positive influence. The blocked negative influence $\sigma(\mathbf{S}_{\mathcal{P}})$ is defined as the expected size of $B(\mathbf{S}_{\mathcal{P}})$.

According to Definition 1, the IBM problem is defined formally as follows.

Definition 2 (IBM problem). Given a graph $G = (V, E)$, the diffusion model with propagation probability $pp_e(u, v)$ of each edge (u, v), a negative seed set $\mathbf{S}_{\mathcal{N}}$ and a positive integer k, IBM problem aims to find the optimal positive seed set $\mathbf{S}_{\mathcal{P}}^*$ with the maximum size of k such that $\sigma(\mathbf{S}_{\mathcal{P}}^*)$ is maximized.

As mentioned in Sect. 1, existing solutions of IBM problem are generally slow on large networks because of requiring a huge number of times of simulations, estimations, or construction of subgraphs. In this work, motivated by CMIA-O [10] and PMIA [8], which are effective methods to solve IBM problem and IM problem, respectively. We propose BIOG to efficiently solve IBM problem under the competitive diffusion model COICM.

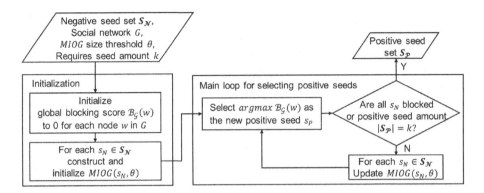

Fig. 1. The main flow chart of BIOG

3   Proposed Algorithm

We propose the algorithm BIOG, stands for Block Influence Out Graph, to approximate the blocked negative influence and identify nodes with higher potential of blocking negative influence. As shown in Fig. 1, BIOG consists of two phases. In the initialization phase, the global blocking score $\mathcal{B}_\mathcal{G}(w)$, which is designed to replace the tradition blocked negative influence, and a new graph structure $MIOG$ are initialized. In the second phase, positive seeds are selected and $MIOGs$ are updated iteratively.

3.1   Construct $MIOG$

To approximate the influence propagation, a graph structure $MIOA(u, \theta)$, which contains all maximum influence paths from u with propagation probability being greater or equal to θ, is introduced in [8]. However, $MIOA$ cannot work well in IBM problem. When there are multiple maximum influence paths between two nodes, $MIOA$ will randomly select one of them and some potential routes will be ignored. Therefore, we propose a new graph structure Maximum Influence Out-Graph, abbreviated as $MIOG$, to solve this problem.

The propagation probability of a path P, denoted as $pp_p(P)$, is defined as the multiplication of propagation probability of all edges in P. The influence path set $IPS(u, v, G)$ contains paths with the maximum propagation probability among all paths from node u to v. $MIOG(u, \theta)$ can be regarded as a subgraph of G, which is influenced by node u and is defined as follows,

$$MIOG(u, \theta) = \{IPS(u, v, G') \mid pp_{max}(u, v, G') \geq \theta \text{ and } v \in (V \setminus S_N)\}, \quad (1)$$

where $\mathbf{S}_\mathcal{N}$ is the negative seed set, $pp_{max}(u, v, G')$ is the maximum pp_p from u to v in G', and G' is the abbreviation of $G \setminus (\mathbf{S}_\mathcal{N} \setminus u)$. $MIOG$ can be generated by the variant of Dijkstra's algorithm where the weight of each edge (x, y) is $-log(pp_e(x, y))$ and all shortest paths are selected [8]. By maintaining multiple shortest paths in $MIOG$, the approximation of a node influence is more precise and robust than $MIOA$.

3.2   Compute Blocked Negative Influence

A good positive seed should block more negative influence. However, the calculation of blocked negative influence is usually time-consuming [1,10]. Therefore, we design a new blocking score to approximate blocked negative influence efficiently.

The activation probability $ap(v, \mathbf{S}_{\mathcal{N}}, G)$ is the probability of node v being influenced by $\mathbf{S}_{\mathcal{N}}$ in a graph G [10]. An idea to find positive seeds is finding nodes which block the most ap in $MIOG$ of negative seeds. Therefore, to estimate the possible amount of ap that is blocked by a node v, the local blocking score of a node v in $MIOG(s_N, \theta)$ is denoted as $\mathcal{B}_{\mathcal{L}}(v, MIOG(s_N, \theta))$, and is defined as

$$\mathcal{B}_{\mathcal{L}}(v, MIOG(s_N, \theta)) = \sum_{r \in N_r(v, MIOG(s_N, \theta))} ap(r, MIOG(s_N, \theta)) +$$

$$\sum_{cn \in N_{cn}(v, MIOG(s_N, \theta))} \left(\sum_{cr \in N_r(cn, MIOG(s_N, \theta) \backslash v)} ap(cr, MIOG(s_N, \theta)) * pp_e(v, cn) \right). \quad (2)$$

Note that $ap(r, MIOG(s_N, \theta))$ is the abbreviation of $ap(r, s_N, MIOG(s_N, \theta))$ and node s_N is a negative seed. The reachable node set $N_r(v, MIOG(s_N, \theta))$ contains node v and nodes which v can reach in $MIOG(s_N, \theta)$. $N_{cn}(v, MIOG(s_N, \theta))$ contains common neighbors of nodes v and s_N in the original graph G which are also in $MIOG(s_N, \theta)$. The first summation can be regarded as the blocking ability of v and the double summation can be regarded as the blocking ability of common neighbors of v and s_N. Common neighbors of v and s_N are included in local blocking score because s_N propagates the negative influence to other nodes through its neighbors. Thus, if v can block more neighbors of s_N, more negative influence will be blocked.

For the example in Fig. 2, the circle represents $MIOG(s_N, \theta)$, and a dotted line is an edge in the original graph but not in $MIOG(s_N, \theta)$. In $\mathcal{B}_{\mathcal{L}}(c, MIOG(s_N, \theta))$, the first summation sums up ap of reachable nodes of c, which are $\{c, f, m\}$, and the double summation sums up ap of $\{d, g, h\}$ multiplying $pp_e(c, d)$, where $\{d, g, h\}$ are reachable nodes of d, which is the common neighbor of c and s_N.

The global blocking score of node v, denoted as $\mathcal{B}_{\mathcal{G}}(v)$, is defined as the summation of $\mathcal{B}_{\mathcal{L}}(v, MIOG(s_N, \theta))$ for each negative seed s_N in $\mathbf{S}_{\mathcal{N}}$. The node with the maximum $\mathcal{B}_{\mathcal{G}}$ will be considered as the best choice of a new positive seed. The blocking score approximates the blocked negative influence by the summation of ap, which prevents from running time-consuming simulations, and reduces the number of nodes to be calculated by considering $MIOG$ of each negative seed independently. Therefore, the blocked negative influence can be estimated efficiently.

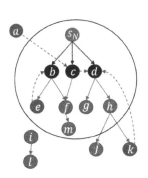

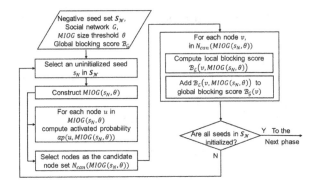

Fig. 2. An example of BIOG

Fig. 3. The flow chart of initializing $MIOG$

3.3   Initialize $MIOG$

Figure 3 shows steps to initialize $MIOG$. For each negative seed s_N, $MIOG(s_N, \theta)$ is constructed, candidate node set $N_{can}(MIOG(s_N, \theta))$ will be selected, and ap of each node will be computed. After that, local blocking scores are calculated and added into global blocking scores.

Candidate Node Set. To reduce the time for calculating blocking scores, nodes with higher capability of blocking negative influence, which are usually closer to the negative seed, will be selected as candidate nodes. Given the negative seed s_N, nodes are classified into 3 classes. Class 1 contains neighbors of s_N in $MIOG(s_N, \theta)$, and Class 2 contains nodes which can influence neighbors of s_N directly, i.e. in-neighbors of Class 1 nodes in the original graph G. Except s_N, remaining nodes are all in Class 3. As shown in Fig. 2, nodes b, c, d belong to Class 1, nodes a, e, k belong to Class 2, and other nodes f, g, h, i, j, l, m belong to Class 3. The blocking score of nodes i, j, l are equal to 0 because they do not have any reachable node or common neighbors with s_N. Nodes f, g, h, m do not have any common neighbor with s_N (if they have common neighbors, they belong to Class 2), so their blocking scores are less than Class 1 nodes which can reach them. Therefore, nodes in Classes 1 and 2 will be regarded as candidates.

After the activation probability of each node is obtained, local blocking scores and the global blocking score of each candidate node will be computed. By only calculating block scores of candidate nodes, the execution time will be reduced significantly.

3.4 Update *MIOG*

When a new positive seed s_P is selected, each $MIOG(s_N, \theta)$ will be updated, as shown in Fig. 4. First, ap of nodes which are influenced by s_P will be updated. Second, s_P will be removed from $N_{cn}(y, MIOG(s_N, \theta))$ of each candidate node y in $MIOG(s_N, \theta)$. Finally, blocking scores will be updated.

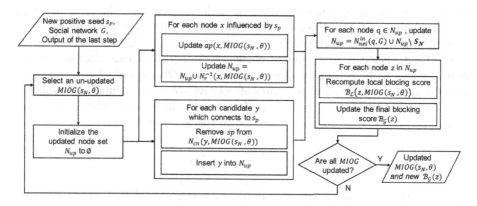

Fig. 4. The flow chart of updateing *MIOG*

Update ap and N_{cn}. When a positive seed s_P is selected, ap of s_P is set to 0 because s_P will not be influenced by s_N. Common neighbors of s_P and s_N are influenced by s_P directly, so their ap are multiplied by $(1 - pp_e(s_P, cn))$, where cn is a common neighbor, for each node x in reachable nodes of s_P and common neighbors, ap of x is recomputed and multiplied by $(1 - \sum_{i \in IN_P} pp_e(i, x))$ to reflect the influence of $\mathbf{S_P}$, where IN_P are positive in-neighbors of x. The reverse reachable node set $N_r^{-1}(x, MIOG(s_N, \theta))$, which is the set of candidate nodes that can reach x, will be put into the updating node set N_{up}. In addition to recompute ap, s_P will be removed from N_{cn} of candidate nodes to avoid recomputing the blocking score of s_P, which implies ap already been blocked by s_P, for remaining candidates. These changed candidate nodes will also be inserted into N_{up}. For the example in Fig. 2, assumes c is the new positive seed, and ap of c is set to 0. ap of its common neighbor d will be multiplied by $(1 - pp_e(c, d))$. After that, ap of reachable nodes of c and d, which are f, m, g, h will be recomputed and multiply positive influence. Finally c is removed from $N_{cn}(a, MIOG(s_N, \theta))$.

Extend N_{up} and Update Blocking Score. In addition to N_{up} which is found above, blocking scores of in-neighbors of N_{up} are changed by s_P indirectly. As shown in Fig. 2, assumes c is the new positive seed, $N_{up} = \{a, b, c, d\}$ which contains reverse reachable nodes of f, m, and d, i.e. $\{b, c, d\}$ and candidate node a which connects to c. Although k, which is an in-neighbor of d, is not changed by c, the blocking score of k will be changed by d. Therefore, k will also be put into N_{up}. In other words, in-neighbors of each node in N_{up} will also be put into N_{up} for updating blocking scores. Finally, for each node in N_{up}, the local blocking score is recomputed and the global blocking score is updated accordingly.

With the design of $MIOG$ and blocking scores, the approximation of negative blocked influence can be computed efficiently. In addition, candidate nodes can reduce nodes to be estimated, and updates of $MIOG$ and blocking scores can avoid recomputing the blocking scores of positive seeds which have been selected previously. In this way, subsequent positive seeds will be more effective because overlapped blocking regions are excluded in the following iterations.

4   Experiment

In this section, we will describe experimental settings, and experimental results will be illustrated.

4.1   Experimental Settings

To verify the efficiency and effectiveness of BIOG for IBM problem, we conduct experiments on three real-world networks. Email dataset is an email network of University Rovira i Virgili [2], with 1.1K nodes and 5.5K edges. NetHEPT dataset is a collaboration network extracted from High Energy Physics Theory of the e-print arXiv (http://www.arXiv.org), with 15.2K nodes and 37.2K edges. NetPHY dataset also comes from e-print arXiv, and is the collaboration networks extracted from Physics, with 37.2K nodes and 232K edges.

Our BIOG algorithm is compared with five algorithms, namely, CMIA-O [10], Greedy [6], HD, Proximity [3] and Random. In BIOG and CMIA-O, θ is set to 0.01 which is same as [10]. Greedy is mentioned in [6], and we combine it with CELF algorithm [5] with 1000 times of Monte-Carlo simulations to estimate the influence. HD selects top k high degree nodes as positive seeds. Proximity selects neighbors of negative seeds which have more influence by negative seeds [3]. Random randomly selects k nodes as positive seeds. Similar to [10], we use the competitive diffusion model COICM [1]. Propagation probability is generated by two popular diffusion models, WC model and IC model with $pp_e = 0.1$ [4]. All algorithms are written in Python. We run Monte-Carlo simulations 1000 times and take their average as the negative influence spread.

4.2   Results of Experiments

Since the Greedy algorithm is only feasible on small networks, the experiments of Greedy only run on Email dataset. We randomly select 50 negative seeds and aim to select 200 positive seeds. For larger datasets, NetHEPT and NetPHY, we randomly select 200 negative seeds and aim to select 200 positive seeds. Under IC model, as shown in Fig. 5, BIOG matches the blocked negative influence of Greedy and CMIA-O, and is better than other methods. Figure 5a presents that Greedy takes more than one day and CMIA-O takes 16 minutes; however, BIOG only takes 25 seconds to handle this. BIOG is much faster than CMIA-O, especially in large datasets, as shown in Fig. 5b and c. Results of WC model are similar, as shown in Fig. 6. As verified by these experiments, BIOG algorithm is able to block negative influence effectively and efficiently.

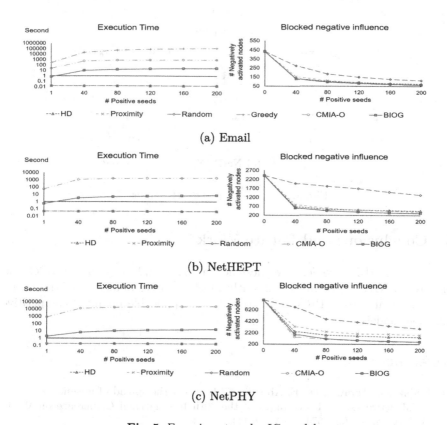

Fig. 5. Experiment under IC model

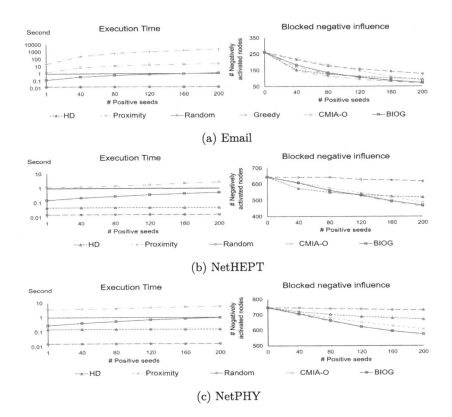

Fig. 6. Experiment under WC model

5   Conclusions and Future Work

To solve the IBM problem, we designed a new graph structure $MIOG$ and proposed BIOG algorithm to efficiently block negative influence. In the future work, we will extend BIOG algorithm to be applicable on more competitive diffusion models.

References

1. Budak, C., Agrawal, D., El Abbadi, A.: Limiting the spread of misinformation in social networks. In: Proceedings of the 20th International Conference on World Wide Web, pp. 665–674. ACM (2011)
2. Guimera, R., Danon, L., Diaz-Guilera, A., Giralt, F., Arenas, A.: Self-similar community structure in a network of human interactions. Phys. Rev. E **68**(6), 065103 (2003)
3. He, X., Song, G., Chen, W., Jiang, Q.: Influence blocking maximization in social networks under the competitive linear threshold model. In: Proceedings of the 2012 SIAM International Conference on Data Mining, pp. 463–474. SIAM (2012)

4. Kempe, D., Kleinberg, J., Tardos, É.: Maximizing the spread of influence through a social network. In: Proceedings of the Ninth ACM SIGKDD International Conference on Knowledge Discovery and Data Mining, pp. 137–146. ACM (2003)
5. Leskovec, J., Krause, A., Guestrin, C., Faloutsos, C., VanBriesen, J., Glance, N.: Cost-effective outbreak detection in networks. In: Proceedings of the 13th ACM SIGKDD International Conference on Knowledge Discovery and Data Mining, pp. 420–429. ACM (2007)
6. Lv, J., Yang, B., Yang, Z., Zhang, W.: A community-based algorithm for influence blocking maximization in social networks. Clust. Comput. 1–16 (2017)
7. Tang, Y., Xiao, X., Shi, Y.: Influence maximization: Near-optimal time complexity meets practical efficiency. In: Proceedings of the 2014 ACM SIGMOD International Conference on Management of Data, pp. 75–86. ACM (2014)
8. Wang, C., Chen, W., Wang, Y.: Scalable influence maximization for independent cascade model in large-scale social networks. Data Min. Knowl. Discov. **25**(3), 545–576 (2012)
9. Wang, X., Deng, K., Li, J., Yu, J.X., Jensen, C.S., Yang, X.: Targeted influence minimization in social networks. In: Phung, D., Tseng, V.S., Webb, G.I., Ho, B., Ganji, M., Rashidi, L. (eds.) PAKDD 2018. LNCS (LNAI), vol. 10939, pp. 689–700. Springer, Cham (2018). https://doi.org/10.1007/978-3-319-93040-4_54
10. Wu, P., Pan, L.: Scalable influence blocking maximization in social networks under competitive independent cascade models. Comput. Netw. **123**, 38–50 (2017)
11. Zhu, W., Yang, W., Xuan, S., Man, D., Wang, W., Du, X.: Location-aware influence blocking maximization in social networks. IEEE Access **6**, 61462–61477 (2018)

Retraction Note to: Chapters

Ying Tan and Yuhui Shi

Retraction Note to:
Chapter "Data Mining to Identify Risk Factors Associated with University Students Dropout" in: Jesús Silva, Alex Castro Sarmiento, Nicolás María Santodomingo, Norka Márquez Blanco, Wilmer Cadavid Basto, Hugo Hernández P, Jorge Navarro Beltrán, Juan de la Hoz Hernández, and Ligia Romero, Data Mining and Big Data, CCIS 1071, https://doi.org/10.1007/978-981-32-9563-6_5

Retraction Note to:
Chapter "Differential Evolution Clustering and Data Mining for Determining Learning Routes in Moodle" in: Amelec Viloria, Tito Crissien Borrero, Jesús Vargas Villa, Maritza Torres, Jesús García Guiliany, Carlos Vargas Mercado, Nataly Orellano Llinas, and Karina Batista Zea, Data Mining and Big Data, CCIS 1071, https://doi.org/10.1007/978-981-32-9563-6_18

1. Chapter 5.
The Editors have retracted this conference paper [1] because it contains material that substantially overlaps with content translated from another article by different authors [2]. The authors Jesús Silva, Alex Castro Sarmiento, Hugo Hernández P., and Ligia Romero agree to this retraction, the authors Nicolás María Santodomingo, Norka Márquez Blanco, Wilmer Cadavid Basto, Jorge Navarro Beltrán, and Juan de la Hoz Hernández have not responded to any correspondence from the editor/publisher about this retraction.

[1] Silva, Jesús, et al. "Data Mining to Identify Risk Factors Associated with University Students Dropout." International Conference on Data Mining and Big Data. Springer, Singapore, 2019. https://doi.org/10.1007/978-981-32-9563-6_5
[2] Reyes-Nava, A., et al. "Minería de datos aplicada para la identificación de factores de riesgo en alumnos." Res. Comput. Sci. 139 (2017): 177–189. http://dx.doi.org/10.13053/rcs-139-1-14

The retracted version of these chapters can be found at
https://doi.org/10.1007/978-981-32-9563-6_5
https://doi.org/10.1007/978-981-32-9563-6_18

Y. Tan and Y. Shi (Eds.): DMBD 2019, CCIS 1071, pp. C1–C2, 2021.
https://doi.org/10.1007/978-981-32-9563-6_35

2. Chapter 18.

The Editors have retracted this conference paper [1] because it contains material that substantially overlaps with content translated from another article by different authors [2].The authors Amelec Viloria, Tito Crissien Borrero, Jesús Vargas Villa, Maritza Torres, Nataly Orellano Llinas, and Karina Batista Zea agree to this retraction, the authors Jesús García Guiliany and Carlos Vargas Mercado have not responded to any correspondence from the editor/publisher about this retraction. Tito Crissien stated that he was not aware of this submission.

[1] Viloria, Amelec, et al. "Differential evolution clustering and data mining for determining learning routes in moodle." International Conference on Data Mining and Big Data. Springer, Singapore, 2019. https://doi.org/10.1007/978-981-32-9563-6_18

[2] Vega, Alejandro Bogarín, Cristóbal Romero Morales, and Rebeca Cerezo Menéndez. "Aplicando minería de datos para descubrir rutas de aprendizaje frecuentes en Moodle." Edmetic 5.1 (2016): 73–92. https://doi.org/10.21071/edmetic.v5i1.4017

Author Index

Acosta Ortega, Felipe   34
Akimenko, Tatiana A.   80
Anand Hareendran, S.   93
Ariza-Colpas, Paola   162
Au, Thien Wan   231
Avdeenko, Tatiana   289

Batista Zea, Karina   170
Bouza, Carlos   53

Cadavid Basto, Wilmer   44
Castro Sarmiento, Alex   44
Chan, Gaik-Yee   307
Chen, Luo   61
Chen, Wenshi   201
Cheng, Ming Shien   104, 240
Chua, Fang-Fang   307
Crissien Borrero, Tito   170

Dai, Bi-Ru   328
Dang, Anhong   211
de la Hoz Hernández, Juan   44
De-la-hoz-Franco, Emiro   162
Djailani, Irfandi   13

Gaitán-Angulo, Mercedes   53, 320
García Guiliany, Jesús   170
García Tamayo, Diana   259
Gong, Dunwei   135
Guiliany, Jesús García   179
Guo, Dongyu   201
Guo, Yuanjun   279

Haji Mohd Sani, Nurulhidayati   231
Henao-Rodríguez, Carolina   53, 320
Hernández P, Hugo   34, 44
Hernández Ureche, Jannys   259
Hernández-Fernández, Lissette   34
Hernandez-P, Hugo   259
Hsu, Ping Yu   104, 240
Hu, Weiwei   211
Huang, Chen Wan   240

Jacinto, Edwar   146, 220, 300
Jheng, Yang Ruei   104
Jiménez Delgado, Genett   34

Khan, Mohammad Wahiduzzaman   307
Khateev, Oleg   289
Ko, Yen Huei   104, 240

Larkin, Eugene V.   80
Lee, Soojung   154
Lin, Kuei-Sheng   328
Lin, Wuhan   279
Lis-Gutiérrez, Jenny-Paola   53, 320
Liu, Xinhui   201
Liu, Xiping   23
Liu, Yuquan   279
Llinas, Nataly Orellano   179
Lu, Mingyu   201
Luo, Zhi Chao   104
Lynden, Steven   70

Ma, Mengyu   61
Ma, Zhixian   191
María Santodomingo, Nicolás   44
Marínez, Fernando   146
Márquez Blanco, Norka   44
Martínez Santa, Fernando   300
Martínez Ventura, Jairo   259
Martínez, Fredy   146, 220
Mercado, Carlos Vargas   179
Mercado-Caruso, Nohora   34
Montero, Edgardo Sánchez   179
Montiel, Holman   220, 300
Murfi, Hendri   114, 124, 269

Navarro Beltrán, Jorge   44
Neira-Molina, Harold   259
Nugraha, Praditya   269
Nurrohmah, Siti   114

Orellano Llinas, Nataly   170
Oviedo-Carrascal, Ana Isabel   162

Palacio de la Cruz, Stefany   259
Palencia, Pablo   179
Peng, Yan   248
Phon-Amnuaisuk, Somnuk   3, 231
Prasad, Sreedevi   93

Rengifo Espinosa, Carlos   34
Rifky Yusdiansyah, Muhammad   269
Romero, Ligia   44, 259

Salma, Dilla Fadlillah   124
Saputro, Aditya Rizki   114
Sarwinda, Devvi   124
Shen, Chien-wen   13
Silva, Jesús   34, 44, 259

Tan, Ying   211
Tang, Cheng-Wei   13
Taveekarn, Waran   70
Torres Cuadrado, Esperanza   34
Torres, Maritza   170, 320
Tsai, Cheng-Han   240

Varela, Noel   179
Vargas Mercado, Carlos   170
Vargas Villa, Jesús   170

Vásquez, Carmen   179, 320
Viloria, Amelec   53, 170, 320
Vinod Chandra, S. S.   93

Wan, Changxuan   23
Wang, Jie   248
Wang, Ying   279
Wang, Yong   279
Wang, Ziqi   248
Wu, Ye   61

Xu, Haiguang   191
Xu, Jing   248
Xu, Ni   240

Yang, Chengyan   248
Yang, Yan   135
Yang, Zhile   279
Yao, Xiangjuan   135

Zea, Karina Batista   179
Zhong, Zhinong   61
Zhu, Jie   191
Zhu, Juncheng   279
Zhu, Yongkai   191

Printed in the United States
by Baker & Taylor Publisher Services